Acclaim for Richard Ellis's

TUNA

A *St. Louis Post-Dispatch* Best Book of the Year

"A researcher and painter, [Ellis] knows his stuff. And this is only the latest in his collection of information-rich, gee-whiz books on the marine world."

—*Los Angeles Times Book Review*

"Enlightening." —*Seattle Post-Intelligencer*

"Ellis's rapt description of the physiology that makes tunas one of the fastest things in the ocean—they're warm-blooded fish, odd as that seems—lends emotional urgency to his account of the collapsing tuna fishery. . . . Rich in description, *Tuna* is packed with information."

—*Orion* magazine

"Marvelous." —*The Plain Dealer*

"Ellis writes eloquently, even passionately, of the tuna's beauty and power." —*New York Post*

"With all the authority and grace for which his writings are renowned, Richard Ellis offers up an impassioned plea to protect and save one of the deep ocean's loveliest creatures." —Simon Winchester, author of *The Map That Changed the World*

"A thorough account of the history of recreational and commercial tuna fishing."
—*The Denver Post*

"Our foremost contemporary ocean chronicler, Richard Ellis, offers here an impassioned plea to consider the grandeur—and the tragic demise—of the swiftest, strongest fish that migrate across our water planet. This is nature writing at its best, from the heart."
—Dick Russell,
author of *Eye of the Whale*

RICHARD ELLIS

TUNA

Richard Ellis is the author of more than a dozen books. He is also a celebrated marine artist whose paintings have been exhibited in museums and galleries around the world. He has written and illustrated articles for numerous magazines, including *Audubon*, *National Geographic*, *Discover*, *Smithsonian*, and *Scientific American*. He lives in New York City.

www.richardellis.info/home.html

ALSO BY RICHARD ELLIS

Big Fish
Tiger Bone & Rhino Horn
Singing Whales and Flying Squid
The Empty Ocean
No Turning Back
Sea Dragons
Aquagenesis
Encyclopedia of the Sea
The Search for the Giant Squid
Imagining Atlantis
Deep Atlantic
Monsters of the Sea
Physty
Great White Shark (with John McCosker)
Men & Whales
Dolphins and Porpoises
The Book of Whales
The Book of Sharks

TUNA

TUNA

Love, Death, and Mercury

RICHARD ELLIS

VINTAGE BOOKS
A DIVISION OF RANDOM HOUSE, INC.
NEW YORK

FIRST VINTAGE BOOKS EDITION, JULY 2009

Originally published in hardcover in the United States by Alfred A. Knopf, a division of Random House, Inc., New York, in 2008.

The Library of Congress has cataloged the Knopf edition as follows:
Ellis, Richard.
Tuna : a love story / by Richard Ellis.—1st ed.
p. cm.
Includes bibliographical references.
1. Bluefin tuna. 2. Bluefin tuna fisheries. 3. Endangered species. I. Title.
QL638.S35E45 2008
333.95'6783—dc22 2007052253

Vintage ISBN: 978-0-307-38710-3

Author photograph © Rick Edwards, American Museum of Natural History
Book design by Christopher M. Zucker

www.vintagebooks.com

Printed in the United States of America
10 9 8 7 6 5 4 3 2

CONTENTS

INTRODUCTION

Besides high prices, the bluefin commands an awed, almost mystical respect and devotion among those who know the animal most intimately. One says, "If you talk to enough fishermen, you may sense how much we really love the bluefin and how much they mean to us. I think it's the way the Indians felt about the buffalo." But people love the bluefin in different ways. Fishers, conservationists, governments, and international treaty organizations continually embroil themselves in bitter international struggles over control and salvation of the fishery.

CARL SAFINA, *Song for the Blue Ocean*

FOR A LONG TIME, I wrestled with titles for this book. I tried *Horse Mackerel, The Quest for Tuna, The Perfect Fish, Maguro, Toro* (sounded like a bullfight), and many other unsuitable candidates. Nothing conveyed what was the intended spirit of the book: the world's best-loved fish (another rejected title) was in serious trouble. As Oscar Wilde said (although he was probably not referring to tuna), "each man kills the thing he loves," and even as tuna are being killed in prodigious numbers, there are five public aquariums in Japan and one in California that exhibit large tuna swimming in large tanks. The spectators might thrill at the sight of these magnificent creatures, but they also might consider having a piece for supper. Everybody eats tuna and everybody loves it: white meat, light meat, red meat, in sandwiches and salads, raw in sashimi and sushi. The Japanese regard a slice of uncooked, bright red tuna *(toro)* as the most delectable food item it is possible to eat. Like many other fishes, tuna contains omega-3 fatty acids, which are particularly healthy. (Does it contain mercury? Well, yes, but the scientists and the tuna industry are still arguing about how much is bad for you.) The fraternity of big-game fishermen and the writers who describe their exploits feel that tuna are among the world's most magnificent animals; they speak of them with unrestrained reverence. To biologists, the tuna is the epitome of hydrodynamic excellence; it is fast, powerful, streamlined, and equipped with specializations that enable it to perform its duties better than any other fish in the ocean.

For the most part, the common terms for a group of animals are uninspired: a pack of wolves, a herd of cattle, a flock of birds, a pod of whales, and so on. But when James Lipton entitled his 1968 book *An Exaltation of Larks,* he managed to elevate group nomenclature from dull to dazzling. The term actually appears in a 1486 work of uncertain authorship called *The Book of St. Albans,* but it is likely that nobody before Lipton ever referred to a bunch of larks as an "exaltation." (In my copy of *The Book of St. Albans,* a reprint annotated by Joseph Haslewood in 1810 and reprinted again in 1966 by Abercrombie & Fitch, the term is *an Exaltynge of Larkys,* which would appear to be "exalting" rather than "exaltation," but never mind.) In fact, larks are not flocking birds, so the question of what to call a collection of them probably never came up. The European skylark, however, is renowned for its burbling, melodic flight song, considered one of the most beautiful sounds in nature, and therefore "exaltation" seems to fit nicely. In Shelley's "To a Skylark," the "blithe spirit" pours forth "profuse strains of unpremeditated art"; Wordsworth's skylark's song is "a flood of harmony, with instinct more divine"; and Frederick Tennyson (Alfred's brother) wrote (in "The Skylark"), "How the blithe lark runs up the golden stair . . . And all alone in the empyreal air, fills it with jubilant sweet songs of mirth . . . Is it a bird or a star that shines and sings?"

Most people think of a *school of fish,* and indeed, most of the tuna literature refers to aggregations of these fish as "schools." My recommendations will not change the language, but I do feel that the mighty tuna deserves something better than the common collective that we use for minnows or sardines. Even if the term appears nowhere else but in a chapter heading of this book, I believe "a celebration of tuna" does justice to the character and accomplishments of these great fish, and for what it's worth, I considered it as another possible title.

In the process of writing most of my earlier books on various marine creatures, I found that detailed information was often more than a little difficult to find. For *The Book of Whales,* originally published in 1980, I had to search for studies in obscure scientific journals, often written in a foreign language, to get information on, say, Bryde's whale or Burmeister's porpoise. At that time, there was no comprehensive book about the whales and dolphins of the world; the closest thing was Beddard's *A Book of Whales,* published in 1900. There were even some rare whales, known as beaked whales, about which hardly anything at all was known, and a beaked whale previously unknown to science was discovered in 1994.

There is altogether too much information about tuna. Commercial and recreational fishermen, ichthyologists, ecologists, fisheries biologists, nutritionists, conservationists, and aquaculturists all have something to say about tuna—usually bluefins, but there is also a lot of published information about yellowfins, skipjacks, and albacore. The International Commission for the Conservation of Atlantic Tunas (known as ICCAT, pronounced "eye-cat") has published an almost endless series of reports, any and all of which are available to researchers, professional or amateur. You don't need special permission to enter the ICCAT website; all you have to do is open it (www.iccat.es), click on "main menu" and then "publications," and you will find thousands upon thousands of discussions of everything about tuna (and swordfish, marlins, and sharks too). Many of the papers are specialized ("Estimates of the Abundance and Mortality of West Atlantic Bluefin Tuna Using the Stock Synthesis Model"), but there are also many user-friendly items, such as Jim Joseph's "Brief History of Tuna Research." If I had tried to use all the available material in the ICCAT reports, I would be reading into the next century, and if I ever finished reading, I would have to produce a five-thousand-page book—and then it would be too technical anyway. So I read a good many of the ICCAT papers (and much other stuff as well), and tried to condense the information into a book that could be read and understood by somebody who didn't happen to be an ichthyologist, a nutritionist, a chemist, or a fisheries biologist.

Many people in those professions—and lots of others—helped me in the preparation of this book: Al Anderson, Pete Barrett, Ted Bestor, Barbara Block, Roberto Mielgo Bregazzi, Jack Brink, Alex Buttigieg, Gary and Karen Cannell, Sienen Chow, Al Craig, Alessandro de Maddalena, Sylvia Earle, Jess Farley, Chuck Farwell, Harry Fierstine, Becky Goldburg, Jim Joseph, Molly Lutcavage, Terry Maas, Brian MacKenzie, Josué Martínez-Garmendia, Brad Matsen, Ron O'Dor, Gemma Parkes, Mike Rivkin, Carl Safina, Vern Scholey, Mike Stokesbury, Dan Sulmasy, Craig Van Note, John Volpe, and Gail Morchower of the International Game Fish Association. My old friend and new agent Steve Wasserman brought me and this book back to Knopf, where the book and I benefited enormously from the efforts of my editor—and the editor of eight of my previous books—Ash Green. During the writing of this book and others—particularly *The Empty Ocean* and *Singing Whales and Flying Squid*—I would often consult Ransom Myers about the facts and figures of the massive worldwide depletion of marine fish species. I knew he was particularly interested in tuna, and I was hoping he would read the manuscript of this book for what I expected would be an insightful and helpful review. Ram Myers died in March 2007, and I never got to send him the manuscript. This would have been a much better book if he had had a chance to look it over.

In Port Lincoln, South Australia, where the future of the bluefin tuna may be written, I was aided and abetted by Neil Williams, Joe Puglisi, Hagen and Marcus Stehr, Morten Deichmann, Rob Staunton, Robert Stutzer, Thomas Marguritte, Barry Besold and the lovely Pamela Price-Besold, Jim Ellis, Captain Adam King and the crew of the *Sanchez-J,* and Captain Coppy and the crew of the *Bella Isha.* But it was Kiwi White—spotter, diver, historian, photographer, surfer—and my host during my flying visit to Port Lincoln and Ceduna in February of 2007, who arranged for me to talk to pilots, fishermen, fleet owners, and many others connected with the industry, making it possible for me to acquire an understanding of how this complex and fascinating business works.

Without Stephanie Guest's steadfast support and encouragement, this book would never have been written. Indeed, without Stephanie, I would probably not get anything done at all.

TUNA

One

PORT LINCOLN

Sanchez-J returning to Port Lincoln, South Australia, after feeding the tuna in the pens. Once a small fishing village, Port Lincoln has become a prosperous community, marked by expensive houses along the waterways.

FEBRUARY 13, 2007. At 5:30 in the morning I boarded the *Sanchez-J,* a tuna-feeding boat owned and operated by the Stehr Group, one of Port Lincoln's largest and most important tuna companies. Illuminated by the glaring spotlights on the dock and on the boat's bridge, blocks of frozen baitfish were being loaded aboard by crane and deposited on the broad, flat afterdeck, obviously designed for just this purpose. When I asked where the pallets of frozen fish came from, I was surprised to learn that they were from Monterey, California. I thought the Monterey sardine fishery died off around 1945, when John Steinbeck wrote *Cannery Row,* a book about the end of the sardine fishery. After the peak season of 1936–37, when 726,000 tons of sardines were taken, the population plummeted, not because of overfishing, but because of a sixty-year warm-water-cold-water cycle that causes the Monterey sardine and anchovy populations to fluctuate. The Cannery Row sardine canneries are gone (the Monterey Bay Aquarium was built on the site of one of them), but the fishery has been revived, mostly for the benefit of the tuna in the feeding pens.

Also lifted aboard the *Sanchez-J* were four rectangular aluminum containers, eight feet by four, and three feet deep, filled with unfrozen baitfish, called "pilchards" here, but actually sardines, mostly caught locally. In an hour we cast off and headed east, toward what would be the sunrise. As always on a voyage like this, my primary concern was staying out of the way of people who had a job to do, but I was also observing the lifeblood industry of Port Lincoln, the feeding of fish that would be transformed into the town's major export: southern bluefin tuna for Japan.

After an hour, we arrived at the first of the pens, some five miles offshore, east of Boston Island. The five pens clustered together are a part of Stehr Group's total of fourteen. They are round, floating corrals, 120 feet across, consisting of nets suspended from a ring of stanchions mounted on a circle of floats. Some fifty feet deep, the net corrals each hold between one thousand and two thousand half-grown southern bluefin tuna caught in purse seines in the Great Australian Bight to the west. They have been transferred at sea to towing cages, and after a one-knot tow that may take as long as three weeks, transferred again to the floating pens. These tuna are a precious commodity—and "commodity" is exactly the right word—and they are pampered and coddled to an extent that would embarrass a purebred Pekinese. Every day of the year, the baitboats make the five-mile journey to the pens at around six a.m., and then return to the PL marina to pick up another consignment of baitfish to feed the penned tuna. At two in the afternoon, they do it again. Stehr Group feeds sixty tons of pilchards a day to their tuna; over the season that adds up to 5,500 tons.

Two kinds of fish are given to these piscine gourmands. One is frozen blocks of footlong pilchards imported from California and also eastern Australia, Morocco, and even Europe. The solid blocks of frozen fish are lowered by crane into a special cage in the pens; as they thaw, the

Tons of sardines are brought out twice a day to feed the tuna.

tuna feed on them. When our boat has been made fast alongside the pen, the feeding begins. Just as your pet guppies know when you are about to sprinkle fishfood into their tank, the tuna are aware that the arrival of the baitboat signals a feed, and they pick up speed in their endless counterclockwise circling. Sometimes they break the surface with their pointed dorsal and tail fins, sometimes they roll on their sides to flash a gleam of silver, yellow, and blue and a curiously intelligent glance at the baitboat. The feeding begins as a man tosses one shovelful after another of pilchards into the pen, causing the water to boil with tuna rushing to get their share. So they do not land in a clump, the shoveler, with his back to the pen, tosses the fish over his shoulder in a graceful arc. Looking down on the feeding frenzy, I notice the bright yellow, horizontal keels at the base of the tail; these yellow markers are the most prominent features of the tuna zooming by underwater (and may be, says Jessica Farley of Australia's Commonwealth Scientific and Industrial Research Organisation [CSIRO], one of the characteristics that differentiate the southern bluefin from its northern relatives).

Except for the occasional glimpse of a fish breaking the surface, or when the light is right, glimpsing the dark-backed fishes swiftly passing underwater, we do not see the tuna very clearly. But another feeding frenzy is impossible to miss: hordes of seabirds have also learned that the arrival of the baitboats signals feeding time for them too, and watching them dive en masse as soon as a shovelful of baitfish hits the water, it would be easy to assume that the *Sanchez-J* traveled out here just to provide for the birds. Sometimes the birds work only with their conspecifics: clouds of the dainty silver gulls, with their soft gray backs and bright red feet, will rise as a coordinated flock and dive in unison as a spray of baitfish hits the water; sometimes the hovering and squawking terns, with their bright yellow bills, will dive as a miniflock; and the dark brown, long-winged muttonbirds sometimes dominate the feedings. But it is mostly a melee, with birds of every species fighting for each fish, occasionally stealing one from another bird before the catcher has a chance to swallow the fish. Ominously lurking over the smaller birds are the heavy-bodied, black-backed, gimlet-eyed Pacific gulls, whose modus operandi is stealing—they rarely exert themselves to dive for a fish, but wait until they spot a smaller bird that hasn't swallowed its fish yet, and harass it into dropping it so that they can snatch it in midair.

Silver gulls pick off the sardines tossed into the pens before the tuna have a chance.

One of the world's most extraordinary birds is in action here. The short-tailed shearwaters *(Puffinus tenuirostris)* make one of the longest migrations of any bird, flying annually from their nesting sites in the Aleutian Islands and Japan to South Australia, a round trip of twenty thousand miles. Known as muttonbirds here because the early settlers plucked them from their burrows and ate them when other food was scarce, they are, like many other seabirds, competent in the air, on land, and in the water. (It is only the flying seabirds and the ducks and geese about which that can be said. Penguins, masters of the water realm, are more than a little awkward on land or ice, and unable to fly at all.) Muttonbirds are not plunge-divers like the gulls and terns, but usually wait on the surface, like ducks, poking their heads underwater to spot a sinking baitfish. They will then dive to catch the fish; properly positioned, you can see them "flying" underwater in their pursuit. They usually resurface—looking perfectly dry because their feathers are waterproof—and sit on the surface like ducks, but occasionally they will do something that startlingly emphasizes their mastery of multiple elements: from underwater swimming they break the surface and keep right on going, passing through the water/air interface and taking flight.

Catching fish to feed fish is an ecologically unsound concept, but tuna are carnivores (more accurately, piscivores), and they would not eat food made from grain. Research is now being conducted to develop a grain + fish products + vitamins pellet that would be the equivalent of the "fishfood" sprinkled into a home aquarium, but the tuna need fat in their diet to produce the fatty meat that the Japanese prize so highly, and while grain might be used to fatten beef cattle, it doesn't work that well for tuna. (Curiously, the Kobe beef so highly prized by Japanese gourmets looks very much like the best bluefin tuna sashimi: rich red in color, marbled with fat to give it the desirable texture and flavor.) So for the moment, thousands of tons of small fish are being caught and exported to Australia to feed the tuna that will end up in the Japanese fish markets, and ultimately in high-end restaurants, where a piece of uncooked fish can sell for a hundred dollars.

Southern bluefin tuna, known technically as *Thunnus maccoyii,* and sometimes as SBT, are born in Indonesian waters. For the next three or four years they leisurely work their way around the west coast of Australia, crossing the Great Australian Bight en route to the east coast, where they disperse, some swimming east to New Zealand, others coming about and swimming west, again past southern Australia, all the way to South Africa. Some of the seven- or eight-year-old fish return to the Indonesian waters where they were hatched, to spawn and begin the cycle anew, but many will not make it that far. They will pass unscathed through Western Australian waters and the Great Australian Bight, but then they will find that their migration route has brought them into the perilous seas off South Australia, where boatloads of fishermen with nets are dedicated to keeping them from completing their instinct-driven journey. (There is no tuna fishing in Western Australia at all; the fish are not big enough, so the quotas were sold years ago to South Australian fishers.)

The first Australian tuna fishermen caught their quarry in an old-fashioned, labor-intensive way: they hooked them on a line and jerked them out of the water one at a time. For bigger fish—they can weigh upwards of four hundred pounds—two men would use a pole each, attached to a single hook. The hooks were barbless, so as the fish were yanked out of the water, the hook pulled out and the fish flopped onto the deck. (Later, a thick foam-rubber "blanket" cushioned the fish's fall, so it would not be damaged.) "Poling" was replaced in the early 1960s by purse-seining, where a huge net was deployed around a school of tuna and "pursed" to close it around the fish, which were then hauled up on deck. Then the Japanese introduced longlining, which consists of setting lines forty miles long, festooned with baited hooks, to catch almost everything. In the early 1960s, the Japanese fishery was hauling in about eighty thousand tons of southern bluefins per year, while the Australian pole-fishery accounted for another twenty thousand tons. Even to the fishermen, it was obvious that such massive pillage would reduce the stock out of economic existence, so Australia, Japan, and New Zealand agreed to set self-restrictive quotas in an attempt to keep this lucrative business going. But the profit margins for the Port Lincoln fishers fell so low that they had to mortgage their fleets just to stay afloat.

Born into an Italian-Australian fishing family that had originally settled in Sydney, Joe Puglisi arrived in Port Lincoln in 1958, just as the newly discovered king prawns of Spencer Gulf were hitting the market. Although he began his career as a shrimper, he could see that Port Lincoln's future was going to be defined by the Japanese shift from cooked fish to raw. After World War II, the development of large freezers in Japanese ships and smaller freezers and refrigerators in Japanese homes and restaurants made it possible for fish caught almost anywhere to be served up, and as if to demonstrate that their "fresh" was better than anyone else's, the Japanese developed an almost obsessive taste for raw fish, which they called sashimi. Australian fishermen, particularly those in Port Lincoln, labored long and hard to satisfy this lust for uncooked tuna.

South Australian fishermen brought the fish to the Port Lincoln docks, where Japanese freezer ships lay ready to carry the frozen carcasses to the Tokyo fish markets. As early as 1970, representatives of the Mitsubishi subsidiary Toyo Reizo were hovering around the Port Lincoln docks, trying to figure out a way to get the Australian bluefins to Tokyo. In 1971, with the support of the Japanese, Puglisi iced down a shipment of bluefin tuna caught by his boats and shipped them to Japan. Then he contracted to bring a Japanese freezer boat out to the fishing grounds in the Great Australian Bight, where the pole-caught tuna were transferred at sea; later he offloaded longlined tuna onto ice-filled Japanese reefers to reduce the damage to the fish and shorten the time from capture to market. Other fishermen soon joined the expanding tuna fleet of Port Lincoln; Croatians Dinko Lukin and Tony Santic, and the German-born Hagen Stehr, all brought drive and ingenuity to the burgeoning tuna industry of Port Lincoln.

The combined fleets of Japan and Australia wreaked havoc with the southern bluefin population, and it all but collapsed in the late 1970s. By 1977, the total catch for Port Lincoln was 9,000 tons. It fell off drastically in 1978, and again the following year. Too many fishermen and not enough fish, opined the Australian government, which initiated a quota system whereby the entire fleet would be allocated a total of 14,500 tons, to be divided up among all boat owners, their allocation determined by a complex formula based on previous catch history and production. Most of the total went to Port Lincoln fishermen, and while they seemed at first to be nothing more than a way to keep the fishery in check, the quotas would eventually prove to be the most valuable component of the industry. When the fish were "wild-caught," weather, chance, and fishermen's luck influenced the number of fish that could be brought in, but when "farming" was introduced, around 1989, even schools of small tuna could be corralled and fattened, and the tonnage quotas became worth their weight in gold—or tuna, which was actually worth more per ounce than gold at that time. The system (known as individual transferable catch quotas, or ITCQs) enabled fishers who chose to leave the fishery to sell their quotas and thus avoid the financial disaster that would have occurred if they just quit, and it concurrently enabled some companies to accumulate quotas of such value that a few of them could completely dominate the industry.

In his 1996 book *Tuna and the Japanese,* Takeaki Hori claims that the idea for "tuna ranching" originated with a Japanese tuna auctioneer named Hideo Hirahara, who went to Australia to find people to catch tuna and put them in some sort of a "fish tank," where they would be fattened for three to six months, killed, frozen, and shipped to Japan. He found tuna fisherman Dinko Lukin. Even though he was not the first to farm tuna, Lukin is surely one of the pioneers of the enterprise, but his claims to fame lie almost as much with his fathering Dean Lukin, a gold-medal-winning weightlifter at the 1984 Los Angeles Olympics, as they do with tuna farming. At first, the tuna were caught with baited hooks and transferred to floating pens, but this proved to be too harmful to the fish, so with Hirahara, Dinko claims to have devised the scheme whereby they would purse-seine juvenile tuna in the Great Australian Bight and tow them in underwater cages to the waters off Port Lincoln. Whoever thought of tuna ranching rescued a faltering fishery and transformed it into a multi-million-dollar business. Tuna farming launched the little village of Port Lincoln into the stratosphere of economic prosperity.

Kiwi White, a spotter for the South Australian Tuna Association since 1969, wrote this description of the basics of tuna ranching in a letter to me.

> The whole tuna industry now concentrates on catching fresh fish in the wild using purse seiners transferring the live fish to huge float-

Dinko Lukin in 1985, before tuna ranching was introduced to South Australia

> ing cages and then towing these cages with up to 100 tonnes of live fish hundreds of miles back to our home port. There are many problems with bad weather smashing the cages and huge tides that slow the towing boats down to less than a knot and it is not uncommon for boats to be actually going backwards. The fish are fed twice a day on pilchards (sardines) and each cage requires 3 or 4 divers to clean out the dead fish, repair any holes, etc. . . . The tuna boat operators now have a supply of fat healthy tuna in the bay so when the fish are needed in Japan a team is sent out to harvest kill and prepare these fish ready to be airfreighted. These fish command huge prices (up to $50 a kilo) and they have been the savior of the fishing industry in Port Lincoln, many people are now employed and the owners are building huge mansions, driving the latest Mercedes and in general spending large amounts around our town.

Tuna farming (sometimes known as tuna ranching) utilizes the same "feedlot" technology used for cattle, only it takes place at sea,

The basis for tuna farming is the great circular pen that can hold thousands of tuna. The fish are fed until they are large enough for the Japanese market.

mostly underwater. Young animals are herded into special enclosures, where they are fed until they reach the desired size and weight, at which point they are slaughtered for human consumption. Spotter pilots locate schools of tuna, and radio their location to the waiting "chum boats," which will throw small fishes overboard into the school, causing the fish to follow the boat. When the chum boat is close enough to the purse-seiner (Australians say "purse-seiner" as one word: "per-*say*-ner"), a net is shot around the school of fish and the chum boat, which is directed out of the net by the spotter pilot overhead just before the net closes, leaving nothing but fish behind. The net is pursed around the school, perhaps thousands strong, and when it is closed and the tuna are trapped, it is drawn alongside the tow-boat, to which is attached a large net "cage," open at the surface, that will be used like a floating corral to tow the fish to the pens off Port Lincoln. Many net hauls are transferred to a single towing cage, which can hold as much as 150 tons of fish. At the pens, the towing cage is brought alongside, and divers drive the fish into the pen, where they will remain for the rest of their lives. About 150 pens are clustered off Port Lincoln in Boston Bay and off Rabbit Island, each owned by a particular tuna company, which will feed the fish twice daily, every day of the year, until the time comes for them to be killed. Each pen contains anywhere from twenty to fifty tons of fish.

While it sounds as if it is a fairly routine sequence—spot fish, catch fish, tow fish, pen fish, feed fish, kill fish—tuna farming is often a dangerous and risky business. In April 1996, South Australia's industry was crippled by a fierce storm that caused the death of thousands of captive fish valued at more than $55 million. The fish, which were kept in floating pens and unable to escape the storm, were suffocated as their gills became clogged in swirling clouds of silt, excreta, and sediment. Between 65,000 and 75,000 tuna died, representing about half the population of Port Lincoln's nine farms in Boston Bay. The mass deaths were a serious setback, but evidently not a lasting one for the booming industry, which has grown at a phenomenal rate since the first experimental farm was established in 1991. The $100 million fish-fattening industry now comprises 60 percent of the Australian tuna industry's 5,200-ton annual quota—and will probably rise even higher.

Australian "tuna cowboy" wrestling a four-hundred-pound tuna onto the deck, where it will be killed

One unexpected outcome of the 1996 disaster, when most of the season's harvest perished, was the export of Australian expertise to Croatia. When Australians of Yugoslav extraction (like Dinko Lukin and Tony Santic) learned about the tuna farming off Port Lincoln, they realized that there were bluefin tuna in the Adriatic too, and Croatian and Australian businessmen formed a consortium to bring this new industry to Croatia. Aussie divers, net makers, and fishermen advised the Croatians on the intricacies (and profitability) of tuna farming, and by 1997 the Croatian version of Australian tuna farming was up and running. The bluefin tuna of the Mediterranean is not quite the same fish as the SBT—it is the ABT or Atlantic bluefin tuna—but it is similar enough that the same capture and farming techniques can be applied to both species. Italy, across the Adriatic from Croatia, soon initiated its own tuna farming, and by 2000 every country on or in the Mediterranean was catching tuna and putting them in offshore pens for fattening and sale to Japan. The eastern Mediterranean is one of only two known breeding grounds for the Atlantic bluefin, and dozens of countries catching hundreds of thousands of half-grown tuna is the surest way to drastically deplete the population.

Because you cannot easily count fish underwater—much of the school is too deep to be seen even by a spotter pilot—only rough estimates of the size of the school can be made. However, experienced spotters can calculate the size of the fish in a given school with remarkable accuracy, so they are able to advise the catchers as to whether the school is worth pursuing. Kiwi White says that he can differentiate seventeen-kilo fish from nineteen-kilo fish from the air. If the fish are small—less than twenty kilos, say—you can catch more in a single haul, but you will have to feed them longer to get them up to market size. Catching larger fish means fewer fish in the cage, but a shorter time to feed them, which means less of the expensive baitfish that are required to fatten them up.

Depending on various factors—market price, the value of the Australian dollar versus the Japanese yen, abundance of fish, size when caught—the tuna are kept in these fattening pens for several months or even a year. When they are considered big enough to be slaughtered, a platform is lowered into the pen, and divers in wet suits get ready to kill the fish. One diver shoves the fish up onto the ramp and another grabs it by the gills, and hauls it partially out of the water. Then, as a metal spike is driven into its brain, another metal rod is rammed down its backbone so that the dying fish will not tense up and thus spoil the tender meat. This "humane" method is employed in Australia and other locations where the fish are small enough (less than one hundred pounds) to manhandle. In Mediterranean ranches—such as those in Spain—where they grow to 250 pounds (120 kg) or more, a man cannot restrain a struggling fish that is much stronger and heavier than he is, and the tuna are herded into a corner of the pen and shot by riflemen standing on a platform above the pen. "Shooting fish in a barrel" indeed. In Maltese fish farming, the living fish is lassoed around the tail, hauled up out of the water and shot in the brain, or its throat cut so that it "bleeds out."

There are many ways for a southern bluefin tuna to make the long—but often remarkably fast—trip from Port Lincoln to Tokyo. Sometimes the dead fish are brought to factories in the town, where they are gutted, iced down—but not frozen—and packed in plastic-lined, coffin-shaped boxes for shipment by air or sea to Japan. Sometimes the fish are transferred directly to waiting Japanese reefers, where they will be gutted and iced down for a sea voyage to Japan; and sometimes they are loaded into huge bins—again iced but not frozen—trucked from Lincoln to Adelaide, and placed on a plane to Sydney, there to be transferred to a plane to Tokyo. Some of these tuna take only thirty hours to get from Port Lincoln to Tokyo, fulfilling the Japanese requirement for fresh fish. Because they have never been frozen, these thirty-hour fish are probably the most desirable at the Tsukiji tuna auctions, but even frozen fish from Port Lincoln's tuna farms, pampered and virtually hand-fed, also demand particularly high prices. So far, the record price for a single fish at the Tsukiji tuna auctions is $173,600.

In 1991, only a couple of years after the Australians set up the first "experimental grow-out project," Takahiko Hamano (a tuna-farming expert) and Yoshio Koga (a fisheries biologist) were in South Australia as part of a joint $1 million project involving the Japanese Overseas Fishery Co-operation Foundation (OFCF), the Australian government, and the Tuna Boat Owners Association of Australia (TBOAA).

In Roger Nicoll's 1993 article in the *Australian Fisheries Newsletter,* Hamano defined the objectives of the project: "Our number one objective is to develop fish for the wild fishery. We aim to produce juveniles and release them into the open sea to build up the natural resource . . . It is not for commercial purposes but for the international fishery. Farming is a secondary consideration." The tuna farms were a South Australian innovation, and it is possible—but not likely—that the Japanese really intended to raise southern bluefin tuna and "release them into the open sea." As defined, however, the project was almost circular: breed SBT in captivity, release juveniles into the wild, then catch the juveniles and put them into pens to raise them to adult size. Obviously, it would be less complicated to raise the tuna from eggs to adulthood in captivity, but that would require feeding them for eight years or more, so it would be cheaper to let them feed themselves in the open sea, recapture them, and feed them only for a couple of years.

The 1991 breeding experiment did not work out and no spawning behavior was recorded, but the Port Lincoln tuna farmers expanded their business exponentially anyway. (As we shall see, the profits made by at least one of the tuna-farming companies have encouraged them to try again to breed bluefins in captivity, but this time on a much larger scale.) Capturing half-grown tuna, fattening them, and selling them to Japan was a cash cow for tuna farmers, but a tragedy for tuna. Southern bluefins were so overfished that they became scarce, and some of the larger tuna-fishing companies floundered. Joe Puglisi, now a retired tuna-company owner (he sold Australian Bluefin in 2000 for $100 million), was the first commercial tuna farmer in South Australia. As the first president of the TBOAA, he has participated in the rise and fall (and rise again) of the economic roller coaster that is the tuna-farming industry. This genial, white-haired man claimed that others were taking credit for what he did, but without rancor he told me how he had developed the "towing cages" that make the industry possible. His company had gone into receivership in October 1992, when "we almost lost everything." With generous refinancing from local banks, however, Puglisi's company bounced back; within three months it was out of receivership. "We got off our asses and worked ourselves out of it," he said. "It is a good industry but also

The author (left) with Joe Puglisi in Puglisi's office, Port Lincoln, South Australia (February 2007)

tough. There are a lot of rewards if you are willing to work hard." In the 1980s he was almost alone in supporting the idea of quotas. "We were catching twenty thousand tons out of Port Lincoln, which was ridiculous," he said. "If there weren't some restrictions, we would have run out of fish and everybody would have gone out of business long ago."

Once on the brink of failure because they were fishing the wild southern bluefins out of existence, tuna ranchers are now awash in money. Penelope Debelle, a reporter for *The Age* (Melbourne), visited some of these millionaires in August 2006. She wrote, "The most outspoken of the tuna kings, German-born Hagen Stehr, likes to boast that when he sits down for his Thursday get-together, at a Port Lincoln café with his old mates, Sam Sarin, Mario Valcic, and Joe Puglisi, they are worth more than a billion dollars between them." Some of the money comes from the tuna farms, but the bulk of it stems from the individual ownership of quotas, parceled out to the tuna fishermen before the establishment of tuna ranching in 1990. "A few years ago," wrote Debelle, "one of the richest of the tuna kings, Tony Santic, reportedly sold 337 tonnes of his quota for $72 million, or $214,000 a tonne." Even these men recognize that the tuna-ranching boom cannot last—they will run out of fish again—and Stehr's "Clean Seas" company is investing heavily in projects designed to breed bluefins in captivity. In the meantime, the tuna millionaires continue to prosper, and Brian Jeffriess, spokesman for the Australian tuna fishermen/farmers, says, "This is not a dying industry. These people are not stupid; they are highly successful business people."

Port Lincoln is situated on one of the world's largest protected natural harbors, encompassing Boston Bay, which covers an area more than three times the size of Sydney Harbor. It was discovered by Matthew Flinders under his commission by the British Admiralty to chart Australia's unexplored coastline in the ship *Investigator.* Dropping anchor in Boston Bay in February 1802, Flinders named the spot Port Lincoln after his native Lincolnshire in England. Initially considered as the alternative site for the state capital of South Australia, Port Lincoln was rejected in favor of Adelaide because Lincoln lacked an adequate fresh water supply. (Highways running south to Port Lincoln are accompanied by above-ground pipelines that bring in the city's water.) Now the home of Australia's largest commercial fishing fleet, Port Lincoln has a thriving tuna-farming industry, but also aquaculture farms for kingfish, abalone, mussels, oysters, and experimentally, sea horses and lobsters.

In 1993 Joe Puglisi predicted that the tuna-farming industry would bring three hundred new jobs to Port Lincoln. He was off by an order of magnitude. Although the grain silos on Boston Bay still do a healthy business in loading barley and wheat onto freighters, tuna farming has turned Port Lincoln into a one-industry town. There are more millionaires per capita in Port Lincoln than in any other city in Australia, most of them owing their success to the southern bluefin tuna. New houses, apartments, stores, marinas, hotels, restaurants, and shopping centers are being built at an astounding rate for a town whose population hovers around 14,000. As proud as the chauvinistic Aussies are of their accomplishments, however, they remain deeply indebted to the Japanese, figuratively and literally. Japanese dealers buy the entire catch of the Port Lincoln tuna fleet, and Japanese investments in the various tuna farms provided the capital for the development and implementation of the sophisticated technology that powered the industry. As the Port Lincoln tuna farmers (now fondly referred to as "tuna barons") prospered, the Japanese were assured of an almost endless supply of the precious red belly meat of the bluefin tuna. It was a win-win situation—unless of course you were a fish.

I went out with the baitboats and flew with the spotters, but I didn't get a chance to ship out aboard a tuna-fishing boat. I did, however, get to see the movies. The film *Tuna Cowboys,* made by National Geographic Films in 2004, shows sixty-eight-year-old Dinko Lukin, the owner of a tuna-farming company (and a multimillionaire), at the helm of the tuna boat *D-Three* heading for the offshore fishing grounds. We see the dangerous activities being pursued 250 miles offshore in a raging storm as the nets threaten to break loose; the divers repair the nets in high winds and crashing seas; and the "cowboys" manhandle living, aggressive blue, mako, and bronze whaler sharks ("bronzies") to get them out of the nets. A thousand swiftly circling tuna can create a whirlpool that can suck the diver down and pin him to the bottom of the net, unable to get back to the surface. Watching Nick Pluker bulldogging the big fish in the net suggests why these divers are called cowboys, but otherwise the setting is mostly antithetical to one's ideas of a cattle drive. Cowhands rarely had to worry about building or repairing a floating corral on the high seas, and they probably didn't give much thought to man-eating sharks, but these dangers are necessary components of the major industry of Port Lincoln, and risks are offset by the potential rewards. "One haul," says the narrator, "might be worth $11 million on the open market, but they'll have to cheat death a hundred times to bring in the herd."

Tuna Wranglers, which was shot just after I left Australia in February 2007 and aired in the United States in August, stays with the nomenclature of the Wild West. The emphasis is on the courage of the Port Lincoln fishermen and divers as they brave the "notorious" Southern Ocean in their attempt to corral and bring home the most valuable fish in the world. This time it is the boats, nets, and personnel of the Stehr Group, with Marcus Stehr taking the *Saxon-S* out of Port Lincoln in the search for bluefin tuna to catch and tow back to the fattening pens. A huge "patch" of fish is spotted by Kiwi White from the air, and the *Saxon-S* hauls in 240 tons of fish, a catch so large that one net cannot hold it. The towing boat *Kingfish* has to spend two more weeks towing two nets bulging with $20 million worth of tuna back to Port Lincoln. We are told that it is vitally important to protect the tuna from sharks that have gotten into the nets, because "if a shark bites just one tuna, that's a potential loss of tens of thousands of dollars." There are not too many sharks this time, but the possibility of ripping the overloaded nets provides the tension in this film. At no point in either film is there a mention of the status of the southern tuna stocks; the cowhands are "bringing in the herd" from what seems to be an endless open range. We are told that the fish are destined for the lucrative Japanese sushi and sashimi market, but how they are to get there is not explained. The romantic image of cattle ranching does not include the slaughterhouse, and so it is with tuna ranching. Even when we see a diver seizing a fish and dragging it onto a platform at the side of the pen, the killing of the fish is not shown. It is not even mentioned. *Tuna Cowboys* and *Tuna Wranglers* are not about fish, and not even about fishing. They are about money.

The Australian tuna-ranching industry has prospered since its inception in 1991. In 2005 the southern bluefin tuna harvest reached nine thousand tons, the biggest to date. (This region is one of the most famous of all locales for great white sharks, and it is not unusual to find one of these "man-eaters" trapped in a tuna net.) Environmental groups have been lobbying for quotas, arguing that the very stock is threatened, but the tuna-ranching industry claims that a reduction in the catch would put people out of work, and besides, the status of the stock is fine. "The benefits of ranching [in Australia] have been numerous," wrote Roberto Bregazzi, "not only to the owners but also to regional employment and the spin-off industries such as tourism, transport and freight. The total export value of the industry has grown from $45 million in 1994 to $252 million in 2001—an increase of 56% in just eight short years." According to the 2006 report of the Australian Department of Agriculture, Fisheries and Forestry, the southern bluefin tuna industry now earns an estimated $300 million annually.

Two

A CELEBRATION OF TUNA

Superfish: the bluefin tuna *(Thunnus thynnus)*

YOU ARE A TUNA, wrote William MacLeish:

> Most fish are neutrally buoyant, equipped with gas-filled swim bladders that enable them to hang at a given depth, conserving energy until they need it. Your bladder system doesn't have that buoyancy. You are heavier than water, so you must fly in it. Some other fish, like the great white shark, do this too, but sharks are fixed-wing aircraft; their pectorals remain rigid. You have sweptwing design. You sight prey and become a bullet, your first dorsal fin, the triangular one, depressing into a slot in your back and your pectorals folding into grooves in your flanks. Your fusiform body stays rigid. Your new-moon tail, one of the most efficient of living propellers, picks up the beat. Only when you are near your target do your pectorals and first dorsal extend, giving you the control you need to feed well. At the last moment your gill

covers pop open, and the extra suction draws in bait from the sides as well as in front of you.

What exactly is a tuna? It is a fast-moving marine fish, plump in the middle and pointed at both ends, rather like an elongated American football, with a suite of fins (dorsal, second dorsal, pelvic, anal, pectoral) found in most bony fishes, and little finlets aft of the second dorsal and the anal fin. The tail fin—also known as the caudal fin—is *heterocercal* (equal-lobed), and is characteristically crescent-shaped (lunate). On either side of the base of the tail, tuna have one or two horizontal keels, as do the billfishes. Tuna are usually dark above and light below, with a color scheme of blue, green, and silver that befits their characterization as jet fighters of the sea. Inside the fish, however, it is even more interesting: tunas (and some billfishes) have a countercurrent heat-exchanger system that allows them to retain metabolic heat to stay warmer than the surrounding water. Warmer muscles mean that the fish can move more efficiently and quickly—think of an athlete "warming up"—which gives them a pronounced advantage over their prey. "Their sleek muscle-bound bodies," wrote marine conservationist Callum Roberts, "cut through water with effortless mastery, driven by a high crescent tail beating side to side in rapid staccato. Pectoral fins shaped like hydroplanes flick and twist in the unseen marine breeze, lending remarkable agility to such stiff-bodied creatures." Tuna are among the fastest fish in the ocean.

The genus *Thunnus* consists of eight species, which have been divided into two subgroups: the warm-water group *(Neothunnus),* which includes the yellowfin, the longtail, and the blackfin. These species possess both central and lateral heat exchangers (about which more later). The cold-water tunas are the bigeye *(T. obesus),* albacore *(T. alalunga),* and the three bluefins, *T. thynnus, T. orientalis,* and *T. maccoyii,* in which the lateral heat exchangers (retia) are particularly highly developed, and there are additional retia which elevate the temperature of their viscera, eyes, and brains.* The skipjack, *Katsu-*

*The first part of an organism's scientific name refers to its genus, and is always capitalized. The second part—the species name—always begins with a lowercase letter. Also, because scientific names are in a "foreign" language (usually Latin, Greek, or some combination thereof), they are italicized.

wonus pelamis, is not officially a tuna, having been placed in a separate genus, but whatever you call it, the skipjack is the most heavily fished tuna of all; it is the "light-meat" tuna sold in cans by the billions around the world. The warm-water tunas *(Neothunnus)* are considered to be more primitive than their cold-water relatives; the albacore, bigeye, and bluefins have developed characters that enable them to elevate their body temperatures to a level where they are able to dive deeper and swim faster in colder waters. (It might also have worked the other way: perhaps the warm-water tunas came later, and did not need to develop the means to heat their bodies so much.)

Large or small, cold-water or warm, all tunas have the same basic body shape, generally considered the most advanced design for moving efficiently through the water. The pointy front end would encounter even less water resistance were it not for the tuna's need to move water over its gills by keeping its mouth open. As Richard Brill and Peter Bushnell (2001) wrote, "tunas have high metabolic rates, and are obligate ram ventilators. They suffocate rapidly if prevented from swimming, so special care must be taken to ensure that ventilatory requirements are met during all stages of an experiment." Also known as "ramjet breathing," this means that tuna cannot stop swimming, for if they do, they will die. As James Joseph, Witold Klawe, and Pat Murphy wrote, "the tuna and billfish are the fastest swimmers in the world. Because they depend on their own motion to pass oxygen-rich water over their gills, the tuna can never stop swimming. When they are being as languid as possible, they must move a distance equal to their own length every second, which for a large tuna, is faster than a man can swim at top speed." (According to the *Guinness Book of World Records,* the fastest human swimmer hit 5.05 miles per hour.) Do they sleep? They have no eyelids, so they can't close their eyes, and because stopping is not an option, they might move autonomically through the water, "sleeping" as the water passes over their gills. Chuck Farwell was asked if he had ever noticed any change in the behavior of the yellowfins and bluefins in the Outer Bay tank at Monterey, and he answered, "At night the tuna seem to swim slower

---

You are *Homo sapiens,* and the northern bluefin tuna is *Thunnus thynnus.* Sometimes, for convenience, when several members of a single genus are listed, the generic name is represented by its first letter, e.g., *T. orientalis, T. maccoyii,* and so on.

and closer together. Attempts at feeding them during the night usually do not catch their attention. These are limited observations and of course made in 30- and 40-ft. diameter tanks."

A goldfish in a bowl opens and closes its mouth, gulping in the water that contains the oxygen it needs to breathe. (Fishes do not breathe the "O" in H_2O, they breathe oxygen dissolved in water. That is why fish tanks have to be aerated, and why some waters are higher in oxygen than others.) All fishes need oxygen, which is absorbed into the gills, but oxygen requirements differ dramatically from species to species, depending upon the fish's lifestyle. The amount of gill surface of tunas is up to thirty times greater than that of any other fish, and in some species, the oxygen-absorbing surface approaches that of the lungs of air-breathing mammals of comparable weight. To transfer oxygen from the gills to the bloodstream and thence to the other tissues, the heart of a tuna is, relative to the body weight, about ten times the size of that of other fishes. A tuna's blood pressure and heart rate are about three times higher. Another factor in oxygen absorption is hemoglobin, the pigment that actually transports oxygen in the blood. The hemoglobin concentration in the blood of tunas is almost as high as it is in humans. A higher oxygen intake contributes to the tuna's stamina, and enables these superfish to undertake long-distance migrations, incredibly deep dives, and sprint speed that would do justice to a cheetah.

Also at the end away from the tail—the cutwater, if you will—are the tuna's large eyes, which, unlike the eyes of many other fish species, are flush with the fish's head, to enable it to pass through the water even more smoothly. The fins all fold flat against the body, and the first dorsal can be completely retracted into a slot. The scales of most tunas are so tiny as to be almost invisible; the great fishes slip through the water like polished torpedoes. All tunas have the same fin arrangement: a spiny, depressible dorsal fin, a second dorsal matched on the ventral side by the anal fin, and pectoral fins, used for steering and lift, that range from short in the bluefins and skipjacks to the long, knifelike blades of the albacore and yellowfins. (In some adult yellowfins, the second dorsal and anal fins are elongated into sweeping scimitars, which are often bright yellow, giving *Thunnus albacares* its common name, and also its popular designation as the most beautiful of the tunas.) The finlets on the yellowfin are also yellow, but in most other species they are a nondescript gray or grayish blue.* Except for the yellowfin, tuna are not brightly colored fishes. With minor variations, they are steel-blue on the back and pearly, opalescent white below, sometimes with faint stripes or spots. Many fish species can change color, often dramatically, and there are a few records of tuna "flashing" their colors in breeding displays.

At the base of the tapered tail stock (the caudal peduncle) there is a horizontal keel that differs in size from species to species. Where the tunas, swordfish, and mackerel sharks have a single, ax-blade-shaped keel on each side of the tail fin, the marlins and sailfishes have two. Because bluefins, swordfish, and mako sharks are known to be fast swimmers, we assume that the large keels have something to do with speed—although we're not sure what that something might be. At a right angle to the keels is the ichthyological wonder that is the scombrid (and istiophorid) tail fin. It is always lunate—crescent-moon-shaped—and proportional to the size of the fish. A hand span would cover the tail of a bonito, but from tip to tip, the tail of a giant bluefin would tower over a small child. Coupled to the powerful musculature of the tail stock, this empennage can power the tuna through the water at speeds up to fifty-five miles per hour, and enable it to swim steadily for hundreds—even thousands—of miles. That has to be so; these fishes cannot stop to rest and live their entire lives in motion. The "engine" that propels most fishes is the tail fin, oscillated from side to side to propel the fish forward. Some fishes move their whole bodies in a sinuous motion, while others use their anterior musculature to move only their hind end, but the scombrids and tunas have evolved such a specialized way of swimming that it has been named for them. "The fastest cruising fish (tunas, billfish and lamnid

*Operating under the assumption that the finlets had something to do with swimming performance, Blake, Chen, and Kwok (2005) analyzed the swimming performance of yellowfin tunas with finlets and without. Measuring speed and number of tail beats of yellowfins in a tank at the Kewalo Basin research facility in Honolulu, they found that there was no significant difference in the fishes' performance, whether they had finlets or not. They did suggest, however, that "given that finlets and finlet-like structures are a recurring morphological design in fishes it is likely that they . . . [perhaps] function in improving swimming performance in high speed sustained swimming, rapid acceleration (burst swimming) and transient maneuvers." In other words, because yellowfins have finlets and yellowfins are fast swimmers, it seems likely that the finlets contribute something to fast swimming. Q.E.D.

sharks)," wrote Moyle and Cech in their *Introduction to Ichthyology,* "are often placed in a separate *thunniform swimming* category":

> These fish have a low-drag, fusiform shape and undulate a narrow caudal peduncle stiffened by a keel and a large slim, lunate caudal fin for propulsion. With its "swept-back" tapered tips and "scooped out" center, the lunate tail provides efficient power for fast underwater movement. Frictional drag is minimized by the high-aspect ratio design of the tail. By having a small surface area per span length or distance between the lobe tips (high-aspect ratio), propulsive force is maximized, and energy wasted on lateral displacement of water (and consequent creation of excessive vortex wakes) is minimized. The vertically large tail allows a greater mass of water to be accelerated to the rear, thus increasing forward thrust.

The fastest of all fishes is believed to be the sailfish, with a speed of 68 miles per hour. This record usually appears without attribution, but in *Swashbuckler of the Open Seas,* his book about sailfish, Jim Bob Tinsley wrote, "Between 1920 and 1925, Florida's Long Key Fishing Club conducted stop-watch experiments in an effort to determine the speed of the sailfish. The club and its records were destroyed by a hurricane in 1935, but many of the former members recall the experiments and the results. It was found that a hooked sailfish, timed with a stop-watch, could take out 100 yards of loose line in three seconds, a speed of about 68 miles per hour." But speed alone does not qualify a fish for greatness. In the preface to *Tuna: Physiology, Ecology and Evolution* (2001), we read:

> Watching a tuna swim, you can't help but be impressed by its power, grace, and speed. If you've tried to catch a tuna on a fishing line, you have undoubtedly discovered its tremendous strength. Its superbly streamlined shape reflects its ceaseless activity. Even the scientific family name of the tunas, *Thunnus,* comes from the Greek verb *thuno*—meaning "rush." Almost 40 years ago the real life of tunas was hinted at when a bluefin tuna tagged in the Bahamas was caught less than 50 days later off the coast of Norway, 6700 km distant. This migratory lifestyle and the extraordinary anatomical and

physiological features that permit it have interested observers since Aristotle.

There are more than thirty thousand recognized species of bony fishes, also known as teleosts. The vast majority of these are cold-blooded (ectothermic), which means that their body temperature is controlled via external means (from the Greek *ecto,* "outside," and *thermo,* "heat"). More specifically, fishes are poikilothermic, meaning that their internal temperatures vary, often matching the ambient temperature of their immediate environment. Thus the body temperature of a codfish swimming in water that is 40°F will be 40°F, and an angelfish in tropical water that is 75°F will have a body temperature of 75°F. (Lizards, snakes, and crocodilians are also poikilothermic, and depend upon the heat of the sun to warm their bodies.)

The opposite of ectothermy is endothermy, the maintenance of body temperature by metabolic means, regardless of the ambient temperature. You are an endotherm, as are all other mammals and birds. Many dictionaries define fishes as "cold-blooded aquatic vertebrates possessing gills and fins," so it is not a little surprising to find that there is a small group of bony fishes (and an even smaller group of sharks, which are *cartilaginous* fishes) that are "regionally endothermic," which Kathryn Dickson and Jeffrey Graham define as conserving body heat "by vascular countercurrent heat exchangers to elevate the temperature of the slow-twitch locomotor muscles, eyes and brain, or viscera, [which] has involved independently among several fish lineages, including lamnid sharks, billfishes and tunas." All of these endotherms, cartilaginous or bony, are large, active predators that range from the depths to the surface in pursuit of prey, and one species—the bluefin tuna—is the most hot-blooded fish of all.

When writing about tuna, it is almost impossible to control one's admiration for these marvelous creatures.* In the introduction to an

*Unless you're the fisherman-writer Pat Smith. In a 1973 *Sports Afield* article about the bluefin tuna, he wrote, "He's something less than beautiful. Squat, round-shouldered and overweight, he's physically incapable of the disciplined power aloft of the blue marlin—Nureyev of billfish. Or the brilliant Bolgeresque hoofing of the tarpon—freestylist par excellence. Or the astounding Wallenda-like flights of the mako—that birdfish that can jump clear of the surface some six times the length of its body. No, as if bashfully aware of his charmless form, he's like a fat man who won't dance, rarely showing himself above the surface, preferring instead to run silent, run deep."

North Atlantic bluefin tuna in a pen off the Mediterranean island of Malta

article in *The Journal of Experimental Biology,* not normally a site for the expression of wide-eyed enthusiasm, Graham and Dickson (2004) speak of the tuna's uniqueness:

> Tunas capture the imagination of comparative physiologists because of the functional differences distinguishing them from other fishes. Relative to most other species, tunas have a fundamentally different swimming mode, a radically different thermal biology, increased rate functions [e.g. standard metabolic rate (SMR), aerobic capacity, heart rate and gut clearance], and a markedly different cardiac physiology. Lamnid sharks, which are remarkably convergent with tunas, are the only other group with a similar ensemble of modifications.

If you were to look at a big tuna and a lamnid shark (a mako, for example) from above, you would find aquatic creatures of essentially the same design: sharply pointed prow; broad "shoulders" tapering to a narrow tail stock with horizontal keels; hooked, flared pectoral fins; and a thin, vertical tail. (You couldn't see it from above, but the tail of the tuna and the mako are similarly shaped: a crescent moon, with the upper and lower lobes approximately the same size.) If you were able to look down on the swimming fish, you would observe that only the posterior third contributes to the forward motion; the anterior two-thirds barely moves as the animal plows through the water. The tail fin, of course, located at the posterior end of the moving part, describes the largest arc. The mako and the tuna (let's say a bluefin) are, respectively, the fastest shark and one of the fastest bony fishes. They achieve these distinctions by using the same biological machinery: that lunate tail, coupled to some pretty fancy musculature to power it, and a complex system of countercurrent heat exchangers that raise the temperature of the fish to the point where it is considerably warmer than the surrounding water, making each of them an almost oxymoronic "warm-blooded fish." The shark has a cartilaginous skeleton, and is therefore not technically a *fish,* but regardless of this distinction, both of these water-breathing, vertebrate predators developed very similar modifications that enabled them to function at a different level—a higher plane, if you will—than most other fishes and sharks. Here is Carl Safina, poet laureate of the tuna, on its special modifications:

> What allows a tuna to generate such dangerously explosive thrust, merely by wagging its tail, is a package of natural adaptations that exquisitely integrate specialized muscles, specialized circulation, and specialized external design. Making the bluefin so unbelievably tough is a body thoroughly designed to penetrate cold, food-rich waters and rule as the top predator there; its muscles function more effectively in cold water than those of any other fish. Of more than thirty thousand fish species plying the world's waters, the bluefin is among the few that have developed the ultimate weapon of vertebrates: heat.

In the thunniform fishes and the lamnid sharks (mako, great white, porbeagle, and salmon shark) the convergence is much more than skin deep; it extends to the depths of the musculature and to the mechanical basis for forward propulsion. In a 2005 article he called "How Tunas and Lamnid Sharks Swim: An Evolutionary Convergence," Richard Shadwick wrote:

> This parallel between tunas and the lamnid sharks is just one of the many anatomical and physiological specializations shared by these two groups of highly active pelagic predators. The evolutionary convergence between them is so striking that in many ways these distantly related groups resemble each other more than they resemble their own close ectothermic relatives . . . Tunas and lamnids are among the most active apex predators, cruising the world's oceans, and both groups have evolved as specialists, in similar ways, for a highly active lifestyle . . . That evolution has reached this endpoint twice, in separated circumstances, is a testament to just how efficient this body design is, both inside and out, for high-performance swimming.

Most fishes have two distinct types of muscle: a large mass of white muscle on the flanks and a smaller band of red muscle running the length of the animal, just under the skin. The body of a fish (or shark) is divided into blocks of muscle called myomeres, each one separated by a sheet of tendon called a myoseptum. As Adam Summers wrote about myomeres in 2004, "In most fishes these blocks of muscle form a pair of W's, lain on their sides and arranged one above the other. In tunas and mackerel the points of the W's are extended into very long tendons that run much of the length of the body and insert into the tail . . . Despite the difference in shape, mako tendons play an identical role in shifting the force generated by the red muscle mass toward the back end of the fish." These long, powerful muscles engage the drive train of the tuna and the shark, powering short, quick flicks of the tail, which enable the fish to accelerate at astonishing speeds. Despite their convergent adaptations for speed, there is at least one major difference between the mako and the tuna: the mako has a jawful of the wickedest-looking teeth in the animal kingdom, "crisped like claws," as Hemingway described them, while the teeth of the tuna are small, conical, and arranged in a single row, designed for gripping fish that the tuna has already caught. A mako might attack a tuna, but never vice versa.* In the tuna and the shark, the red muscle

*In September 2004, just before a juvenile great white shark was to be introduced into the Outer Bay tank at the Monterey Bay Aquarium, the staff contemplated eliminating some of the larger bluefins from the exhibit because they feared that the tuna—which outweighed the little shark by

is located within the white muscle, closer to the vertebral column than the skin, which allows heat from the continuously activated white muscle to be retained in the core of the fish, leading to local warm-bloodedness.

Along with the lamnid sharks, the tunas have developed a circulatory and respiratory system that enables them to elevate their internal body temperature, and that of their brains, eyes, and viscera, to as much as twenty-five degrees higher than the water in which they are swimming. In 1966, Frank Carey and John Teal, both marine biologists at Woods Hole, Massachusetts, wrote "Heat Conservation in Tuna Fish Muscle," the first published discussion of this surprising adaptation. In *The Sargasso Sea,* Teal explained its significance (referring to himself in the third person):

> Tuna should never have a body temperature different from the water in which they swim, yet they have. Francis Carey, a physiologist at Woods Hole, with John Teal and other colleagues sought the answer to the warm-body question. After much pursuing and experimentation on tuna and makos, they discovered a "countercurrent heat exchanger" within the fish's muscles, similar to the structures used in heating and cooling devices such as refrigerators, in which heat passes between two fluids flowing in opposite directions. Equal and parallel flows of a liquid through a heat exchanger, one at 0° and the other at 100°, would come to equilibrium and both would emerge at 50° temperature. If one flow is reversed, the hotter will always be hotter, but only by a small amount at any one place along the exchanger. In the counterflow system, the fluid that enters cold will be almost 100° when it emerges, and the originally hot flow only very slightly warmer than 0° at the opposite end.

The *rete mirabile* (wonderful net) is a finely intermingled sheet of blood vessels that acts as a countercurrent heat exchanger, placing the veins and arteries close to each other. As explained by Frank Carey,

---

350 pounds—would behave aggressively toward the shark and injure it. The plan was never implemented, and the tuna and the shark coexisted peacefully for 198 days, after which the shark was fitted with an electronic tag and released back into the Pacific.

one of the scientists who discovered this miracle of bioengineering, the "warm, oxygen-poor venous blood courses directly alongside the cold, oxygen-rich arterial blood coming from the gills. Heat passes from the veins to the adjacent arteries and is then returned to the warm muscles. The blood that reaches the gills is already cooled, so it doesn't lose its heat to the surrounding water." The effect is to increase the temperature and thus the power of the muscle, and all those warm-blooded fishes are top predators. The swordfish, a deepwater hunter, heats up its brain and huge eyes, but the tunas and the mackerel sharks heat up their whole bodies. (The differences in the mechanisms have led biologists to suggest that the heat conservation strategy evolved independently in the sharks, tunas, and swordfish.) The additional heat gives an extra boost to already powerful muscles, permitting the bluefin tuna, for instance, to accelerate quickly, to plunge into deep, cold water to search for food, or to migrate from the tropics to the polar regions without losing the ability to elude predators or capture prey. The tuna's steady, powerful swimming sustains a uniquely high metabolism, which permits its extraordinary growth rate. This also places a heavy oxygen demand on the fish, requiring tuna to swim continuously in order to meet that demand.

Over a long and distinguished career, mostly at the Woods Hole Oceanographic Institution in Massachusetts, Francis G. Carey (1931–94) studied the physiology, functional anatomy, and behavioral ecology of large fishes, including the mackerel sharks, tunas, marlins, and the swordfish. In 1966 (with John Teal) he identified warm-bloodedness in tuna, and by 1973 he had incorporated the mackerel sharks and the skipjack and other tunas, as well as the wahoo, into the category of warm-blooded fishes, previously thought to be cold-blooded like other species. Carey did his original research on bigeye and yellowfin tuna caught by longliners in the Atlantic and the South Pacific, but later he began to study bluefins of the North Atlantic, first planting muscle transmitters into hooked fish, and later inserting primitive stomach transmitters to measure the temperature and movements of the living fish. Following the transmitters, he learned that bluefins prefer to swim in the warm upper layer, just above the thermocline (the border between the warm and cold water), but when they dive below the thermocline, they thermoregulate—

Frank Carey, the Woods Hole biologist who pioneered tuna tagging

they stay warm in cold water. Hemoglobin is the oxygen-binding protein in blood, while myoglobin serves the same purpose in muscles. With Don Stevens (1981), Carey observed that the red muscle of tuna contains a high concentration of myoglobin, and suggested that one advantage of being warm-bodied is that warmth increases the delivery of oxygen to the myoglobin of the muscles. More myoglobin, more oxygen; more oxygen, more power.

Then he made another remarkable discovery: the bluefin tuna can also raise the temperature of its brain and eyes. Giant tuna, ranging in size from four hundred to nine hundred pounds, were caught in pound nets off Provincetown, Massachusetts, and then shot in the head with a twelve-gauge shotgun. Carey and Scott Linthicum (1973) took the temperature of the brain of the fish with a thermometer mounted on a foot-long needle, and found that where the water temperature was 21°C, the brain was 27.1°C and the eye 30.2°C. They concluded:

> The tunas are fast-swimming fishes that feed on small, active prey. Bluefin tuna have warm eyes and brains which are maintained at

> fairly constant temperatures over a wide range of water temperatures. It seems likely that warming these organs results in improved vision and responses which would be useful to such an active predator . . . In migration these tuna pass rapidly from the warm, near 30°C waters of the Bahamas to northern waters which may be as cold as 5°C. Telemetry records show that bluefin may undergo 10–15°C changes in water temperature within a space of minutes when passing through a thermocline. Without control the bluefin could probably not function through such a temperature change.

Working with these great fish in the tuna pens at St. Margaret's Bay, Nova Scotia (see pp. 159–160), Carey, John Kanwisher, and Don Stevens learned that bluefin tuna also warm their viscera to temperatures 10°–15°C above the temperature of the water they are swimming in. At first they experimented with dead fish, but "the period of several hours between death and landing . . . was too great for meaningful temperatures to be obtained," so they trapped living bluefins, brought them alongside their boat, and had them swallow a temperature sensor folded into a mackerel. These tuna were owned by the fishermen who were planning to sell them to the Japanese, so only "noninvasive" techniques were used at first, so as not to injure or disturb the fish, but they had to be killed anyway, so the scientists could also take measurements of freshly killed fish. Because the digestive organs are insulated from the heat transfer of the musculature, the visceral warming "must be in part due to the heat released during hydrolysis of the food," and indeed, the largest digestive organ (the pyloric cecum) was the warmest organ of all. Visceral warming speeds digestion and allows the tuna to feed frequently when food is abundant. Even the process of eating—which would appear to be what the tuna was designed for—contributes to the unique suite of adaptations that sets the tuna apart from other fishes.

We can learn about the biology of a fish by taking its temperature, dissecting it and identifying which parts do what, or observing wild and captive specimens, but a significant proportion of the information we now have about the lives of tuna has come from tagging. Barbara Block of Stanford University, who has probably tagged more tuna (or overseen the process) than anyone else, tells us (in her chapter in *Tuna: Physiology, Ecology, and Evolution*) that the first published account of the acoustic tracking of tuna is that of Yuen (1970), who tagged skipjack in Hawaiian waters in 1969. Around the same time, Carey and Lawson (who published their account in 1973) were tagging bluefins off Nova Scotia. With Robert Olson, Carey tagged yellowfins in the Gulf of Panama in 1981, and their observations were published the following year.

Frank Jewett Mather Jr. (1868–1953), an art critic and professor who taught at Williams College and Princeton, was the father of Frank Jewett Mather III, who in his turn was the "father" of tuna tagging. Born in 1911 and trained as a naval architect, Frank III fell in love with fish and fishing, and devoted most of his professional life to research on what he referred to as the "large pelagics"—tuna, marlins, and sailfish. Even though he was affiliated with the Woods Hole Oceanographic Institution for fifty-five years, he regarded oceanographers as sissies, and is reported to have said, "Catching water is easy; catching fish is hard" (MacLeish 1989). He would ask fishermen to catch a fish (or catch it himself), affix a small tag, and release it. The tag asked anyone who recaptured the fish to contact Mather with the date, location, weight, and so on, so he could calculate how it had grown, or how long it had taken to get from place to place. He began tagging small bluefins in 1954 off Martha's Vineyard, and found that two of them had traversed the Atlantic to the Bay of Biscay. This was the first time anyone realized that bluefins crossed the ocean, and that therefore there were not distinct eastern and western stocks. Mather was the first to recognize the transatlantic nature of bluefin migrations; by 1976, eight adult bluefins had traveled from the Bahamas to Norway. From 1954 to 1972, he recorded 13,496 bluefins tagged, with 2,248 returns. The great majority were released and recaptured from Cape Cod to Cape Hatteras, but there were many releases off Cat Cay in the Bahamas, the scene of a sport fishery that targeted large tuna and billfishes (Mather and Mason 1973).

Frank Mather's tagging program was originally intended as a means of charting the growth of the fish and their migration patterns. But the operation took an alarming turn in the 1960s when the tag return rate increased sharply. To Mather, the meaning was clear: the bluefin population was in decline. As catch rates declined, an increase in size of the giants was noted, which, along with an absence of medium-sized fish, indicated an imbalance in the stock structure. "By 1968," wrote Dick Russell and Brian Keating, "Mather was convinced that the species was in desperate trouble. He began a campaign to save the bluefin before it was too late. Armed with his evidence, he told anyone who would listen—and even those who would not—that the time had come to manage the Atlantic bluefin stocks."

Although simplistic when compared to today's pop-up satellite tags, which can track a fish for a year, Mather's primitive program set the stage for today's sophisticated research on the habits and movements of the various tuna species. Since Yuen's early experiences with skipjack, six species of tuna have been acoustically tracked: Atlantic bluefin, Pacific bluefin, southern bluefin, yellowfin, albacore, and bigeye tuna. Intense scientific curiosity (and improved technology) have enabled researchers to track tuna into the depths, over, under, and through thermoclines, and across entire oceans, and we have been able to solve some of the mysteries of the heretofore secret lives of these deepwater adventurers. Although we still marvel at this fish whose internal engineering enables it to function in such a superlative fashion, researchers are beginning to get a better picture of the interior design of the tuna's body. The deployment of electronic tags that can record and archive data eliminates the need to track individual fish acoustically from a ship. The tags collect data on depth, light, and external and internal temperature every two minutes. They also use sunrise and sunset data with sea-surface temperature to compute a daily location, yielding tens of thousands of data points. And finally, the development of archival tags, miniature computers that can collect and store data for a year, enable researchers to study the lives of tuna in a way never before believed possible. A satellite tagging program begun in 1997 might lead to the discovery of a heretofore unexpected breeding area for bluefin tuna, beyond those already known in the Gulf of Mexico and the Mediterranean.

In a 1984 interview, Mather expressed his love of the bluefin: "There's always been a magic about them," he said. "What beauty they have! A perfect fish! They're tremendously strong and fast . . . Such staying power. I've seen people who thought for sure they'd hooked the bottom." Frank Carey simply said (in the Woods Hole magazine *Currents*), "I was in love with tuna." Many other scientists marvel at every aspect of tuna biology. At a 2007 seminar about bluefins, Barbara Block exclaimed: "A body temperature of 80°! Why they're practically mammals!" In swimming speed, musculature, vision, stomach-warming, brain-warming, and hemoglobin-binding, the tunas stand (swim?) head and shoulders above other fishes; they are the piscine definition of *sui generis.* It is now recognized that a third of the North Atlantic population makes a transatlantic crossing, doing it in less than sixty days. Some of them make multiple crossings in a single year.

They even have their own research facility. In 1994, with colleagues at the Monterey Bay Aquarium, Block, professor of marine sciences at Stanford University and tuna tagger par excellence, founded the Tuna Research and Conservation Center (TRCC), located at the Hopkins

Barbara Block (center) tagging a tuna in the Pacific

Marine Station in Pacific Grove, California. A joint collaboration between Stanford's Hopkins Marine Station and the Monterey Bay Aquarium, the mission of the TRCC is to promote cooperative ventures in which the basic biology and captive husbandry of tunas are studied. Current TRCC activities include the Tag-A-Giant (TAG) program,* which follows the routes of bluefins tagged off North Carolina; and the Tagging of Pacific Pelagics (TOPP) program, in which tuna are tagged, as well as other marine predators such as billfish, sharks, baleen whales, elephant seals, and large cephalopods like the Humboldt squid. Tagging has given researchers new insights into the life history of bluefin tuna, just as observations of wild and captive specimens have opened windows to the fascinating biophysics of tunas that set them apart from other fishes.

While some scientists are collecting data on the giants, others are looking at their smaller counterparts. At the Large Pelagics Research Lab at the University of New Hampshire, Molly Lutcavage, Prop., there is the converse of the Tag-A-Giant program, here known as "Tag a Tiny." Both sizes—in fact, *all* sizes—are helpful in the quest for understanding the ways of the tuna. For the LPRL, juvenile bluefins are caught off the coast of Virginia and implanted with tags that record daily relocation, depth, and internal and ambient temperature every four minutes. These pop-up satellite archival tags are actually small minicomputers that are programmed to jettison from the fish and "pop up" to the surface. Data is then transferred from the tag to an orbiting Argos satellite and e-mailed back the researcher. They also deploy hydroacoustic "pingers" that can be attached to the fish by a modified harpoon and that transmit information to a nearby vessel equipped with ultrasonic receivers attached to hydrophones. This research has revealed that bluefins in the Gulf of Maine may travel as much as thirty miles a day in search of food. Just as Block heads up a tuna research program on the West Coast, Lutcavage is her East Coast

*As of 2007, the Tag-A-Giant program has become the Tag-A-Giant Foundation. Founding mother Barbara Block has defined their mission as "committed to reversing the decline of northern bluefin tuna populations by supporting the scientific research necessary to develop innovative and effective policy and conservation initiatives. We will engage scientists, policy makers, fishermen and citizens to chart the course towards rebuilding sustainable populations of bluefin tuna in the Atlantic and Pacific Oceans." For more information, write to Tag-A-Giant Foundation, P.O. Box 432, Babylon, NY 11702, or visit info@tagagiant.org.

equivalent: the director of a program designed to collect information on the bluefin tuna, which has become one of the most intensively studied fish in the world. But when the actual fish—alive or dead—does not provide the answers to researchers' questions, they have to build their own.

"An Efficient Swimming Machine" was the title of a 1995 *Scientific American* article by Michael and George Triantafyllou. Was it about tuna? Yes, in a manner of speaking. The Triantafyllou brothers (both of whom are Ph.D.s from MIT in ocean engineering) wrote, "Over millions of years in a vast and often hostile realm, fish have evolved swimming capabilities far superior in many ways to do what has been achieved by nautical science and technology. Instinctively, they use their superbly streamlined bodies to exploit fluid-mechanical principles in ways naval architects can only dream about, achieving extraordinary propulsion efficiencies, acceleration, and maneuverability." But the Triantafyllous weren't writing a paean to piscine efficiency; the efficient swimming machine of their title isn't a fish at all, it's a robot. In 1995, David Barrett built the first mechanical tuna for his Ph.D. thesis at MIT. In *Smithsonian* magazine (2000), Douglas Whynott described the way Barrett's "RoboTuna" was constructed:

> It had a stainless steel spine of eight vertebrae, each connected with low-friction ball-bearing joints. The spine was flexed by means of pulleys and cables driven by small electric motors mounted outside the fish on a control carriage. The cables were equivalent to tendons, and the motors to muscles. Steel spines mounted with plastic ribs were attached to the vertebrae. This "flexible hull" was covered with a thin layer of foam rubber—the fish's flesh—and over that went a Lycra sock for the skin. The fish was suspended from the control carriage by a hollow streamlined mast through which ran the cables and sensor wires. There was a controller for the movement of the carriage and another for the fish, with a single data collection system. The total cost of construction was about $30,000.00

The next robot tuna was an autonomous (free-swimming, as opposed to cable-controlled) undersea vehicle, designed and built at the MIT Ocean Engineering Lab, and called "Charlie I." The idea was to replicate the movement of the tuna to see if the "efficiencies, acceleration, and maneuverability" could somehow be incorporated into underwater ship design. The MIT robot tuna was modeled directly on the body plan of a tuna, with a tapered, bullet-shaped body and a rigid quarter-moon tail fin. Almost all other autonomous undersea vehicles (AUVs) are powered by propellers (as are most ships), but the robot tuna moves like a fish, with flexions of its tail. From trials in a water tunnel the researchers observed that the tail's efficiency lay in the interaction of the vortices created by rapid flexing of the tail—and not the hind third of the fish's body—but the hydrodynamics of the real tuna were far beyond their electronic model. They acquired a bet-

A robot tuna in the MIT testing tank. The robot could not match the abilities of a real tuna.

ter understanding of how tuna move so efficiently through the water, but the Triantafyllous realized that their model never came close to the real thing: "The more sophisticated our robotic-tuna designs became," they wrote, "the more admiration we have for the flesh-and-blood model."

Then Jamie Anderson, a graduate student at Woods Hole and MIT, built an even more sophisticated robot tuna, which was closely modeled on a forty-inch yellowfin that had been caught off Long Island. It cost nearly a million dollars to construct, and she named it VCUUV, for Vorticity Control Unmanned Undersea Vehicle. Scaled up from the actual tuna, it was eight feet long, and had "65 pounds of lead-acid batteries, electric motors for moving the pectoral fins, and a hydraulic power system for operating the four links making up the tail, each of which is powered independently. There is a computer inside VCUUV that can be programmed with an 'umbilical cord,' and also a compass, depth sensor, leak detector, and inertial implementation" (Whynott 2000). All that fancy technology reinforced what they already knew: that the tuna is an engineering marvel whose micromanipulation of vortices has been honed over millions of years by the designer that makes everything work so well: evolution.

Like wolves, bluefins often hunt in packs, forming a high-speed parabola that concentrates the prey, making it easier for the hunters to close in. Tuna are metabolically adapted for high-speed chases, but as opportunistic (and by necessity, compulsive) feeders, they will eat whatever presents itself, whether fast-swimming mackerel, bottom-dwelling flounder, or sedentary sponge. A study of the stomach contents of New England bluefins by Bradford Chase revealed that the most popular food item (by weight) was Atlantic herring, followed by sand lance, bluefish, and miscellaneous squid. (In addition to these prey items, Chase also found butterfish, silver hake, windowpanes, winter flounder, menhaden, sea horses, cod, flounder, plaice, pollock, filefish, halfbeak, sculpin, spiny dogfish, skate, octopus, shrimp, lobster, argonaut, crab, salp, and fig and finger sponges.) In a similar—but much less rigorous—study, Valérie Allain of the South Pacific Commission Fisheries Program analyzed the stomach contents of yellowfin tuna caught around Papua New Guinea, New Caledonia, and French Polynesia, and found that the predominant food item was the purpleback flying squid *(Sthenoteuthis oualaniensis),* followed by flying fish, skipjack, surgeonfish, triggerfish, and various deepsea species. Tuna will eat anything they can catch, and they can catch almost anything that swims (or floats, crawls, or just sits on the bottom). By and large, they hunt by vision.

Do they see in color? Kerstin Fritsches and Eric Warrant (2004) realized that it would be more than a little difficult to devise a test that would answer the question, so instead they examined the eyes of bigeye and yellowfin tuna caught off Western Australia. (Also part of the experiment were swordfish and sailfish.) They found that the "major visual pigment of all these species is well suited to light in deep and shallow water," which means that the fish have the "hardware" necessary for color discrimination. Earlier, Loew, McFarland, and Margulies (2002) had measured medium- and short-wavelength pigments in yellowfin tuna and concluded that they were not monochromats (color-blind), but—at least in the juvenile stages—that they have the necessary optical equipment to respond to a "violet and blue spectral range but also a reasonable flux of green and yellow photons." Fritsches and Warrant (who have also worked on the even more specialized eyes of swordfish) were unable to say unequivocally that tuna see in color, but the tuna's eyes are certainly designed to do so. "We have made good progress," they wrote, "in establishing an electrophysiological set-up to study the further processing of the color signal. This will allow us to test spectral sensitivity, and hopefully collect more evidence for color vision in pelagic fish."

While dissecting a large bluefin at Wedgeport, Nova Scotia, in 1952, Luis Rivas of the University of Miami was surprised to discover a translucent "window" on the top of the fish's head, roughly between the eyes. Many deepwater shark species have a pale yellow spot on the top of the head, which has been identified as a "pineal window," used by the sharks to detect downwelling light. (In a wonderfully simple experiment, Gruber, Hamasaki, and Davis [1975] put a lit flashlight in the mouth of a dead lemon shark to demonstrate the translucent nature of this "window.") Although Rivas first detected this structure in the bluefin, subsequent investigation found it in all tuna species. Histological examination of the pineal organ *(epiphysis cerebri)* reveals that it is nearly as photosensitive as the retina, and can be affected by light levels as low as moonlight. However strange the idea of a "third eye" might appear, it seems appropriate for animals that frequent regions of low light, but because the organ has also been found in tunas, the discovery of deep-diving bluefins supports the contention that the "pineal apparatus" functions as a receptor that controls movement in response to light. As Rivas noted, "The evidence . . . suggests that tunas, and other related scombrid fishes, react to light by means of the pineal apparatus. In the open sea, light may be one of the factors controlling their vertical movements. The correlation between sun altitudes, light intensities, direction of rays, water transparencies, depth, etc., and the behavior of tunas under natural and experimental conditions remains to be worked out . . ."

When Scott Allen and David Demer put hydrophones in the Outer Bay tank at the Monterey Bay Aquarium and also in a holding pen at Maricultura del Norte, a tuna farm in Baja California, they recorded (and filmed) bluefin and yellowfin tunas making sounds that resembled coughing. A series of photographs shows "the animal's mouth open wide with its jaw bones extended and its abdomen expanded, then contracted abruptly," which they described as "cough-like behavior." As a rule, fish are not big sound producers, but there are some, such as drums, grunts, and croakers, that get their very names from the sounds they make. The drumming, grunting, and croaking noises are generated by a muscle close to the swim bladder, which causes the swim bladder to vibrate, with the bladder acting as a resonator and amplifier for the sound. The swim bladder in fishes is a sac inside the abdomen that contains gas. It is what is known as a "hydrostatic device," an organ that the fish can use to regulate its buoyancy. Most tunas have a small swim bladder—although the kawakawa *(Euthynnus affinis)* has none at all—and Allen and Demer concluded that "adult bluefin and yellowfin tunas are capable of generating sounds . . . possibly caused by the contraction of muscles around the swim bladder." The researchers described the conditions in the two tanks as noisy, and suggested that the experiments "should be repeated under more controlled and lower noise conditions before tuna vocalizations can be claimed with certainty." Tuna can do everything else better than any other fish; surely they ought to be able to make a better sound than a grunt or a croaker. But can they talk? Why not?*

In his 1982 *National Geographic* article, marine biologist Michael Butler said, "Of the 20,000 fish species, the family Scombridae is among the most advanced, renowned for speed and endurance. The bluefin marks the zenith of this evolution." Although there is no scale upon which evolutionary accomplishments can be measured—every living creature, after all, is here because it is well enough adapted to survive over time—the tunas are usually considered the pinnacle of piscine evolution, and the bluefins are the most advanced of all. Many accounts identify the northern bluefin as the largest of all bony fishes, but the unofficial record for a black marlin—it was not caught according to IGFA rules—is 1,806 pounds; and marlins are believed to grow even larger. In *Fishing the Pacific,* Kip Farrington wrote, "The largest fish I ever heard of was harpooned off Cabo Blanco—a 2,250-pound black marlin." Left to its own devices—a virtual impossibility these days—a bluefin tuna could probably grow bigger than that. It is the tuna in chief, the ne plus ultra of the tuna world: the biggest, fastest, warmest-blooded, warmest-bodied fish in the world. The largest of the tunas—and one of the largest fishes in the world—the bluefin is known to reach a length of twelve feet and a weight of three-quarters of a ton. Whatever a yellowfin or albacore can do, a bluefin can do it better, and do more of it.

*In the 2003 Disney cartoon *Finding Nemo,* the fish of the title is a brightly colored, orange-and-white clownfish, who, with all the other fish, sharks, and birds, speaks English. Only cartoon fish can speak English, but the sound production capabilities of the real clownfish *(Amphiprion clarkii)* were recently identified in an article in the journal *Science.* Parmentier, Colleye, et al. (2007) wrote that "clownfish are prolific 'singers' that produce a wide variety of sounds, described as 'chirps' and 'pops' in both reproductive and agonistic behavioral contexts."

Three

THE FRATERNITY OF BIG TUNA

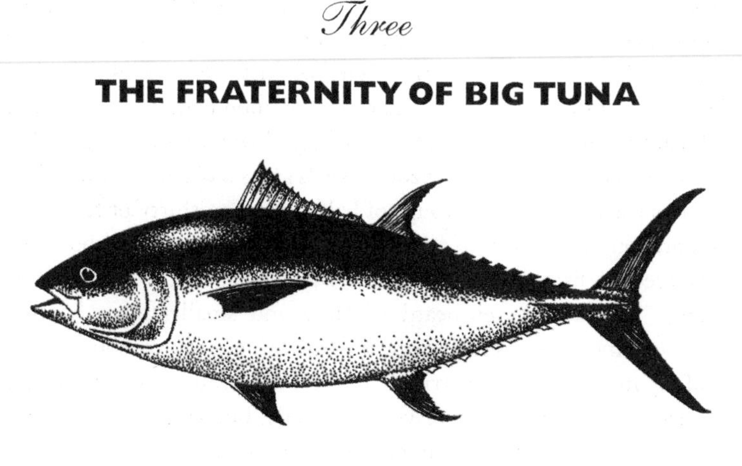

Bluefin tuna *(Thunnus thynnus)*

BLUEFIN TUNA (THUNNUS THYNNUS)

THE WORD "magnificent" is often applied to the bluefins, and it is exquisitely applicable. They are immensely powerful and fast, capable of speeds up to fifty-five miles per hour. They are mature at about the age of eight, and individuals have been known to live for more than thirty years. All tunas are shaped pretty much alike, but the bluefin is the paradigmatic tuna, exquisitely designed to move smoothly and quickly through the water. It is an elongated teardrop whose very design bespeaks the word "sleek." Its eyes are flush with its head, its fins can be tucked into slots, and the scales are small enough to be invisible to the naked eye. Its body is deepest near the middle of the first dorsal, and tapers to the narrow caudal peduncle before the insertion of the tail, the engine that powers the fish. The short pectoral fins are used for steering, and are usually flared out from the body, like the wings of an airplane. Like an oceanic missile, the bluefin tuna is a dark, metallic steel blue on the back, sometimes glittering with green iridescence. The flanks and belly are silvery white with faint spots and stripes. The gill covers are silver with pinkish undertones. The first dorsal fin is blue (giving the species its common name), the second is reddish brown, and the finlets are dusky yellow with dark edges. The tail fin is black, as are the horizontal keels at its base. Schooling fishes, bluefins usually travel in groups of half a dozen to thirty or forty, but they are sometimes seen in much larger schools. Most tuna schools are composed of fish of the same size; Bigelow and Schroeder (1953) wrote, "We have never heard of large and small tuna schooling together. And it seems that the very large fish usually are solitary."

Many marlin species, particularly the white, blue, and striped, are characterized by a pattern of vertical bands, but when excited, they "flash" these or other stripes in neon displays of blue, green, or purple, because they are either breeding, feeding, or struggling against a hook. When the breeding or feeding bouts are over, the fish return to their normal color, and if the flashing has been a result of a struggle against a fisherman, if and when the fish is gaffed and brought aboard, the colors fade as the fish dies. One doesn't think of bluefin tuna as "striped," but they are related to the marlins, and under some circumstances they have been known to flicker the lights on and off. Under the skin there are chromatophores that can expand or contract under the control of the central nervous system; it is these cells that are responsible for color change during various stages of excitement. Many squid and octopus species also have chromatophores under their skin, and can change color even more dramatically than fishes. It is now believed that cephalopods can communicate with one another through color change, and certain fishes may share that ability. For example, yellowfin tuna have been observed to change color rapidly during spawning, perhaps as a stimulant to the opposite sex. Captain Al Anderson, a professional fisherman out of Point Judith, Rhode Island, says that bluefin tuna feeding in a chum slick often become brightly variegated, spotted, or striped, and his book *(The Atlantic Bluefin Tuna: Yesterday, Today and Tomorrow)* includes two photographs of small tunas with a pattern of stripes that would rival that of a zebra. Not a zebra *fish*—a *zebra.*

Under stress, bluefin tuna can flash their "stripes" like a marlin. Nothing is more stressful for a fish than being hooked and yanked out of the water.

Bluefins spawn in April, May, or June in subsurface waters, but while spawning behavior has been observed in some of the smaller species, it has never been seen in bluefins. We do not know if they spawn once or many times in a season, of even if an individual spawns yearly. It is also not known whether individuals can spawn in the east and then in the west at different times. Mature male bluefins are somewhat larger than females, which is different from the arrangement in billfishes, where the largest ones are always females. (Macho big-game fishermen like Zane Grey and Ernest Hemingway went to their graves believing that all the big, tough, brave billfish and tuna they caught were males.) In the western North Atlantic, bluefins mature by age eight; after examining the ovaries of 501 female bluefins from the Mediterranean, Aldo Corriero and (fourteen) colleagues found that maturity for this "population" was reached between four and five years of age, and a length of fifty-four inches. Females release millions of eggs, which are fertilized by males.*

The spawning grounds for western Atlantic bluefins are in the Gulf of Mexico and the Mediterranean. (Spawning areas are identifiable by the presence of females with eggs ready to be broadcast, and the males with the sperm duct filled with spermatozoa.) Two of the Mediterranean spawning areas have been known for centuries: one just inside the Strait of Gibraltar, around the Balearic Islands, and the second from Sicily to the Gulf of Sidra (off Libya). It was not until well into the twenty-first century that a third spawning area was tentatively identified in the eastern Mediterranean between Turkey and Cyprus (Karakulak et al. 2004). Tagging results (such as Block et al. 2005) have shown that tuna have a definite "homing" instinct; bluefins hatched in a particular area return regularly to the place where they were born. But many of the tagged fish do not show up in any of the known spawning areas, which raises a tantalizing question for researchers: is there another breeding area for North Atlantic bluefins? Mather, Mason, and Jones suggested the possibility of a Gulf Stream spawning ground three decades ago, but confirmation has not been forthcoming.

Barbara Block and her colleagues (2001) reported the first descriptions of what they believe to be breeding (but not actual spawning) behavior in giant tunas. A unique behavioral repertoire recorded by the electronic tags suggests that the giant tunas breed at night in very warm waters, swimming up and down in the water column for hours. It is behavior such as this that can only be monitored by electronic tags; even if researchers were somehow able to follow the tuna in the dark, they could not see what the fish were doing. Most surprising is

*If even a small percentage of these eggs survived to adulthood, the oceans would be chock-full of tuna. That they're not points out the odds against survival for newborn tuna—and all other fish species that lay enormous numbers of eggs. For a long time, biologists believed that the fecundity of certain species of fish would enable even an overfished population to bounce back. In 1999, wrote Daniel Pauly, "Ransom Myers disposed of one of the central reasons why fisheries scientists had underestimated the impact of fishing and provided fisheries managers with over-optimistic advice." The North Atlantic cod, a species renowned for its egg production, turned the "millions of eggs" theory into the "millions of eggs" *fallacy*. Myers's research showed that, at low population sizes, the females of most commercial marine fishes produce only three to five viable young a year, in spite of the millions of eggs they may shed and are fertilized in the process. Even laying millions of eggs is not a safeguard against population decimation.

that the spawning period occupies less than a month in the tunas' yearly activities. Tuna spawn only when the appropriate water temperature is reached, which ranges from 75°F to 79°F, and for most species, spawning is restricted to warm, tropical waters with high salinity and strong current boundaries. For example, southern bluefins spawn only in a relatively small area off northwestern Australia in the eastern tropical Indian Ocean. Characterized as "reproductive broadcast spawners," mature females lay millions of eggs over a period of months. The eggs are tiny, spherical, and transparent, and contain a single oil globule to keep them buoyant. Only a very small percentage of the eggs survive. The larval tuna, which are two to three millimeters long, hatch in a few days, and begin feeding on plankton smaller than they are. The great majority of larval tuna are eaten by larger predators—it has been estimated that fewer than one in a million survives.

Bluefins begin life as big-headed, bug-eyed, finless creatures, less than a quarter of an inch long. They grow about two millimeters a day, quickly assuming a more tunalike shape, with a spiky first dorsal, and a forked tail that will eventually form the crescent-moon shape. For the first half-year of their lives, tuna grow rapidly, putting on between 8.5 and 10 percent of their body weight a month before leaving the spawning grounds for a life of constant travel. La Mesa, Sinopoli, and Andaloro (2005) calculated the age-to-weight ratio as

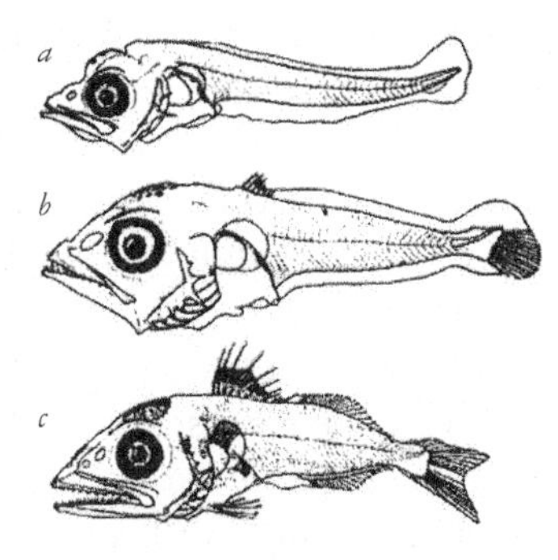

Bluefin tuna larval development: *(a)* 5.1 mm, notochord length; *(b)* 6.0 mm, standard length; *(c)* 8.5 mm, standard length

follows: at two months, the average weight was 168 grams (5.8 ounces); at three months, 429 grams (15 ounces); and at four months, 813 grams (28.4 ounces). "Our data indicate," they wrote, "that juvenile bluefin tuna have a very high growth rate in the first part of their life, reaching a weight of more than 1 kg [2.2 pounds] in four months." The growth potential for the larger tunas is amazing: from the time a giant bluefin hatches until the time it reaches full size, it will have increased in weight one billion times.

In their 1953 discussion of the fishes of the Gulf of Maine, Bigelow and Schroeder measured and weighed bluefins that had been caught off Ipswich, Massachusetts: 28 inches, 17 pounds; 34 inches, 40 pounds; 42 inches, 56 pounds; 60 inches, 144 pounds; 63 inches, 172 pounds; 66 inches, 188 pounds; 68 inches, 200 pounds; 88 inches, 516 pounds; 93 inches, 580 pounds. Bluefins segregate into schools based on size, often mixing with other species, such as albacore, yellowfins, bigeyes, or skipjack. After the first growth spurt, tunas grow slowly and reach immense size—thousand-pounders are not unusual—and estimates of the maximum age come in at anywhere from twenty to thirty years. In the late 1960s, when Frank Mather was beginning his release-recapture program, he tagged three tunas off New England, one of which was 78 inches long (6.5 feet) and weighed 275 pounds. In July 1979, thirteen years later, when the fish was recaptured, it was 111.6 inches long (9.3 feet) and weighed 633 pounds. Mather (1980) estimated that the fish was about eight years old when it was first tagged, which would have made it twenty-one at recapture. (By 1980, Mather's program tagged 17,822 bluefins in the western North Atlantic and retrieved 2,050 tags. Of the returns, 45 had made a transatlantic crossing.) Unlike the marlins and swordfish, where the very large specimens are always females, both sexes of bluefin tuna can grow to be giants.

It has been estimated that a tuna has a one in 40 million chance of reaching adulthood. During the early stages of their lives, tuna are susceptible to the predations of just about any fish (or squid, dolphin, seal, penguin, or pelican) that is bigger than they are, so almost anything can qualify as a predator of a small tuna. However, the only predators that can catch *adult* bluefins are mako sharks, killer whales, and of course, the world's ultimate predator, *Homo sapiens.* What does

Ken Fraser poses beside his 1,496-pound bluefin, caught off Auld's Cove, Nova Scotia, in 1979. It was the largest tuna ever caught, a record that has never been seriously threatened.

the tuna feed on? Almost everything. As warm-blooded fishes, they are always "warmed up," like runners before a race, and therefore have a distinct advantage over their colder-blooded, unprepared prey. Even so, creatures with such a high metabolic rate need a lot of food to stoke their furnaces. In their 2001 study of tuna metabolism and energetics, Keith Korsmeyer and Heidi Dewar wrote:

> Tuna white muscle differs from that of other fish, including other scombrids, in that in conjunction with a high anaerobic capacity,

> this tissue has an extraordinary aerobic capacity. Total aerobic capacity of the white muscle exceeds that of the red. This aerobic capacity is likely associated with the tuna's capability for rapid oxygen debt recovery. Rapid recovery, while also swimming at moderate aerobic velocities, will permit more frequent swimming bursts and can therefore result in a higher *average* swimming velocity.

In 1996, Molly Lutcavage of the New England Aquarium, along with Paul Howey of Microwave Telemetry in Columbia, Maryland, began to develop microprocessor tags that could be attached to Atlantic bluefins in an attempt to better understand their migration and biology. In previous tagging systems, the tag included only information about where and when it was inserted, so when the fish was caught, the most researchers could learn was how long the fish took to get from here to there. The new tags, however, logged the fish's temperature hourly, then averaged the temperatures and stored the figure. After sixty days, the tiny computer shuts down, an electric current is generated that corrodes the wires fastening it to the fish, and the tag floats to the surface, transmitting its data via satellite to waiting scientists. Barbara Block launched thirty-seven of Howey's pop-up tags, but then employed the "archival" tag, which is lodged in the tuna's abdomen and measures not only water temperature but also light, depth, and internal body temperature until the fish is caught and the tag is recovered. Among the more surprising findings revealed by the microprocessor tags was that tuna, long believed to travel and hunt near the surface, made regular feeding dives to 3,300 feet, and could maintain a body temperature of 80+°F in the ink-black, icy cold waters more than half a mile down.

By 2000, a total of 377 Atlantic bluefin tunas had been tagged along the East Coast of North America with both types of microprocessor tags, pop-ups and archival. In a report in *Science,* Block and her colleagues showed that the migratory behavior of the bluefin was far more complex than anyone had imagined. Rather than separating into eastern and western Atlantic populations, the tuna "mixed," which further complicated the already thorny issue of who has the right to catch which tuna. Moreover, this mixing did not occur just near the midline: some tuna were found to have traversed the entire ocean—some 1,670 nautical miles—in less than ninety days. Before the tagging data were analyzed, it was thought that perhaps 2 to 4 percent of the tuna might have crossed the ICCAT line (the 45th meridian) that separated the "American" tuna from the "European." ICCAT had established quotas on the assumption that there were two distinct populations that did not mix, but it now appears that while there are two distinct spawning areas—the Gulf of Mexico and the Mediterranean—the fish migrate across the Atlantic at will until they attain sexual maturity at eight to ten years, when they migrate to their particular spawning grounds. Both eastern and western populations feed at western foraging spots, but they separate for breeding. Moreover, it wasn't 2 to 4 percent of the bluefins that crossed the ocean; it was closer to 30 percent. Younger tuna from the east—the ones Block calls "adolescents"—travel to the western Atlantic until they reach breeding age, at which time they go back to the Mediterranean spawning grounds and are unlikely to return to North America.

"The natural history and migratory abilities of Atlantic bluefin tuna have fascinated mankind for millennia," wrote Block and her colleagues in the introduction to their paper in *Science.* This article discussed these very subjects, and some truly surprising information was revealed. Bluefin tuna tagged in the western Atlantic with electronic data-recording tags are mixing on their feeding grounds but moving to distinct spawning grounds in the western Atlantic and the eastern Mediterranean. Researchers discovered that Atlantic bluefin—considered overexploited in the western Atlantic since 1982—often are traveling throughout the entire North Atlantic and, in some cases, into the Mediterranean Sea. Tuna tagged in the western Atlantic in most cases resided for a year or more on western North Atlantic feeding grounds. While many fish remained in the west, some of them had migrated to the eastern Atlantic or Mediterranean. Individuals also migrated from the western Atlantic to the east and back again in the same year. Most of the fish examined showed at least one year of western residency, traveling between the Carolinas and New England and back again to the original release location. The fidelity to these two locations is associated with high primary productivity and represents bluefin feeding aggregations. A third aggregation site is near the Flemish Cap—a section of the Atlantic seafloor just east of the

Grand Banks off Newfoundland. (The Flemish Cap was the scene of the "perfect storm" that sank the swordfishing boat *Andrea Gail* in the book by Sebastian Junger.) Researchers also reported that the western-tagged bluefin travel to distinct spawning grounds in the Gulf of Mexico or the eastern Mediterranean.

Bluefin tuna migrate on a yearly cycle, arriving in the waters off the northeast coast of North America by June of each year and departing in late autumn. They can be found as far north as Newfoundland in the summer, and they are known to travel as far as 40 degrees south of the equator during the winter. Northern waters provide the fish with rich feeding grounds, where they can grow and store fat as an energy source for migration. Each year they return to their specific spawning grounds: the Gulf of Mexico or the Mediterranean. It was always assumed that there were two separate North Atlantic populations, one that spawned in the Gulf of Mexico and remained in the western quadrant, and another that spawned in the Mediterranean and swam around in the eastern part of the Atlantic. Starting in 1997, a research team led by Lutcavage satellite-tagged juvenile bluefins in New England and Canadian waters. The data they collected from the pop-up tags (meaning the tags were designed to pop *off*) revealed some surprising results: none of the tagged fish had appeared in either the Gulf of Mexico or the Mediterranean, the only known spawning grounds for bluefin tuna. It may have been nothing more complicated than that the timed release of the tags was insufficient to detect spawning, but it also might be that there is another breeding area in the central North Atlantic. If that is the case, the ramifications would be enormous. Management of the two North Atlantic stocks, already compromised by the tunas' disinclination to remain in their half of the ocean, would be more than a little complicated if a third breeding population appeared on the scene.

Of course, Lutcavage knew that such a possibility had been suggested by Frank Mather, who, with J. M. Mason and A. C. Jones, wrote in a 1974 ICCAT report on the distribution, fisheries, and life history of the bluefin tuna:

> There are less numerous indications of a secondary spawning area in the western North Atlantic at the northern edge of the Gulf

> Stream east of the Middle Atlantic States. This area is frequented mainly by medium (120–185cm) sized bluefin, and most of the ripe or nearly ripe fish captured there have been in this size range. More information on the spawning of these medium-sized bluefin is urgently needed.

From the moment Mather tagged the first bluefins in the 1950s, we have accumulated information on the lives of these fishes that has revised what we knew—or what we thought we knew—about *Thunnus thynnus.* Even though bluefins sometimes crossed the Atlantic Ocean, it was long held that each of the separate stocks was supposed to stay on its side of the 45th meridian, but tag returns showed that the tuna had no respect for this dividing line, and indeed, the Atlantic bluefin has now been shown to comprise a single, oceanwide metapopulation. The migratory patterns that we have managed to identify serve only to complicate the picture of the lives and loves of bluefins, and of course, our massive interference by capturing large numbers of these fish at one place or another has dramatically altered the picture. Fishes cannot migrate if they're dead.

At one time, wrote Brian MacKenzie and Ransom Myers in a 2007 study, "North Atlantic bluefin tuna *Thunnus thynnus,* used to migrate to northern European waters (Norwegian Sea, North Sea, Skaggerak, Kattegat, and Øresund) where it supported important commercial and sportfisheries. The species disappeared from the region in the early 1960s and [it] is now extremely rare." In Denmark in the early years of the twentieth century, where bluefin schools could be sighted not far offshore, fishermen first pursued these giant mackerel for sport. The first *thunfisk* caught in the Kattegat (the body of water that separates Denmark from Sweden) was taken in 1919, and by 1929 enough tuna were being caught there to motivate the construction of the first Danish tuna cannery. At the same time, there were also substantial French, German, and Norwegian commercial tuna fisheries, and smaller enterprises in Sweden and the Netherlands, suggesting that bluefin were present in significant numbers in western European waters. Sportfishermen from the United Kingdom were catching "tunnies" off the Yorkshire coast, and venturing to Denmark in the late 1920s to fish there. The catch statistics reported by MacKenzie and Myers would have been insufficient to decimate the population, but "these figures likely underestimate the true abundance and catches because many landings and sightings were not reported to fishery authorities." It seems indisputable that the large, concentrated, multinational fishery during the 1930s was responsible for the decline of the bluefin in the eastern North Atlantic.

When Mike Stokesbury's team tagged six large tunas off the coast of Ireland in 2003 and 2004, they were not sure where the fish would end up, but they didn't expect them to appear on opposite sides of the Atlantic. Two of the fish were tagged within ten minutes of each other off County Donegal on October 8, 2004, and according to an account in the *Edmonton Journal* (Munro 2007), "one of the giant fish, weighing close to 250 kilograms [550 pounds] showed up off the coast of Cuba eight months after it was tagged. The other ended up more than 5,000 kilometres [3,100 miles] away, in the Strait of Gibraltar." A third fish tagged at the same time "moved into the Mediterranean Sea, and was caught by a fishing vessel southeast of Malta" (Stokesbury et al. 2007). The tag data show that the fish were feeding off Ireland and heading for their spawning grounds, one in the Gulf of Mexico, the other (along with the one that was caught off Malta) in the Mediterranean. In a 2005 study, Block, Teo, Stokesbury et al. identified the two elements in the North Atlantic tuna population, "one using the spawning grounds in the Gulf of Mexico and another from the Mediterranean Sea . . . Bluefin tuna that occupy western spawning grounds move to central and eastern Atlantic foraging grounds," one of which is off Ireland. If they are not caught en route by the armada of international fishing boats that ply the North Atlantic, bluefin tuna return, like salmon, to spawn in the place where they were born.

In a discussion with the intriguing title of "Obstinate North Atlantic Bluefin Tuna," Phillipe Cury and his colleagues hypothesize a generalization of the concept of "natal homing." Earlier, Cury had postulated that

> a newly hatched individual memorizes early environmental cues, and that these later determine its choice of reproductive environ-

> ment. . . . Migration routes of north Atlantic bluefin are postulated to have been developed during and after the ice ages in the northern hemisphere and it is proposed that bluefin still return to ancestral spawning areas despite having expanded greatly their foraging and overwintering areas to the extent that the two populations may mix in these areas.

Beginning around 1962, a fleet of Japanese longliners began fishing for tuna, and even though the operation was successful at first, it was suspended in 1967. In their 2005 discussion of the population dynamics, ecology, fisheries, and management of the Atlantic bluefin, Jean-Marc Fromentin and Joseph Powers illustrate the catches for the 1960s, 1970s, 1980s, and 1990s; the map for the 1960s shows the largest catches off the southeastern United States and northeastern Brazil, but no catches whatever in the South Atlantic after that. The authors refer to this as the "Brazilian Episode," during which Japanese longliners caught five thousand to twelve thousand tons of Atlantic bluefin tuna (ABFT) in an area where they usually took tropical tuna. In Sakagawa and Coan's 1974 comprehensive review of Atlantic bluefin fisheries in the Atlantic, they wrote that for "Region V" (off northeastern South America), "a peak catch rate occurred in the mid-1960s, and thereafter dropped off sharply." In an interview that he gave to the BBC shortly after the publication of the paper he wrote with Boris Worm on the decline of large predatory fishes around the world, Ransom Myers said, "there were 200,000 large bluefin tuna removed from off the coast of Brazil in the first fifteen years of the Japanese longline fishery and in the last fifteen years off Brazil, with similar effort, the Japanese fishery has caught exactly zero fish." (In a letter to me in October 2006, Alberto Amorim, a Brazilian fisheries biologist, wrote, "We have been working with longliners based in Santos City, São Paulo state since 1974. In the 1980s, two southern bluefins were caught, but we did not see any more.") "Perhaps," wrote Fromentin and Powers, "the collapse of fishing . . . might have been due to overfishing on local (possibly isolated) populations of ABFT. [But] in the case of the Brazilian episode, questions remain as to how a long-lived fish such as ABFT could have been fished out within 6 years by a few number of longliners."* According to Matsuda, however, the Japanese fishery came back on line in 1976, with the leasing of two Korean longliners, but it did not reach operational capacity until 1983, when the fleet numbered eighteen vessels. Targeting mostly small bluefins, the Japanese took 1,160 tons in 1981, 2,660 tons in 1982, and reached a high of 4,778 tons in 1992–93 before quitting altogether. It appears that longlining for bluefins is no longer conducted off Brazil because the big fish are gone. (In the northwestern Atlantic, however, where the biggest bluefins can still be found, fishing continues apace.)

Long thought to consist of localized, discrete populations, the Atlantic bluefin is actually a single population, and has the "widest distribution of any tuna, and is the only large pelagic fish living permanently in temperate Atlantic waters" (Fromentin and Powers 2005). It now appears that the "giants" come to feed on herring and other small fishes off the coasts of North Carolina, Massachusetts, and Maritime Canada, and move on when the herring are gone. During the course of their wanderings, bluefins display a remarkable range of temperature preferences, from near-freezing waters when feeding (37°F or 2.8°C) to very warm temperatures when breeding (86°F or 29.5°C). Because we saw them there—sometimes feeding voraciously on those herring—we also assumed that bluefins were creatures of surface waters. Not necessarily. Pop-up tags revealed that these fish, more highly endothermic than any other fish in the ocean, regularly dive into icy black waters three thousand feet below the surface. (So too does the swordfish, also seen to frequent the upper, sunlit layer, but even more than the tuna, *Xiphias gladius* is a regular denizen of the deep.) Atlantic bluefins, bigeye tuna, and swordfish forage at heretofore unsuspected depths, to feed on the small creatures that make up the deep-scattering layer (a large assemblage of small fishes and

*Throughout the literature, there are many anecdotal mentions of the killing of 200,000 bluefins off Brazil, but I have not been able to locate a specific, responsible reference to this event. In his 1973 *Sports Afield* article on the bluefin, Pat Smith alluded to "the Japanese longliners who each year harvest no less than 240,000 tuna from South American waters alone"; and even Frank Mather (1974), normally a stickler for references, included the Brazilian story without revealing how he knew that "the largest bluefin catches (22,000–67,000 fish per year) were taken in the years 1962–1966, with the highest catch rates occurring off the easternmost parts of Brazil in the western North Atlantic." Because so many people refer to this event, it probably happened, but I'd be more comfortable if I could find something in the ICCAT files like "Catches of bluefin tuna off Brazil by Japanese longliners, 1962–1967, *Col.Vol.Sci.Pap.ICCAT,*" etc.

cephalopods that migrates daily toward the surface at night and descends during daylight hours) or perhaps as a means of cooling their overheated bodies. The energy required to maintain a high body temperature is costly, and the tunas cannot remain at depths indefinitely, but have to rise into warmer waters to restore their thermoregulatory balance. In a 1992 study, Holland et al. observed that bigeye tuna "can rapidly alter the whole-body conductivity to two orders of magnitude. The heat exchangers are disengaged to allow rapid warming as the tuna ascend from cold water into warmer surface waters, and are reactivated to conserve heat when they return to the depths." Bigeyes—and probably bluefins—can turn their heat exchangers on and off when necessary.

Nowadays, we tend to regard bluefins as engineering marvels, and at the same time as being among the world's most esteemed food fishes. But they were not always venerated; around the turn of the twentieth century, they were known as horse mackerel, and were considered a nuisance by herring or mackerel fishermen, because they sometimes followed the small fishes into the nets, and after eating their fill, tore their way out. Evidently the red meat was considered unappetizing then, for they were occasionally harpooned for oil when fish oil was a popular commodity. (Even the Japanese showed no interest in red-meat tuna; in *Tuna and the Japanese,* Takeaki Hori wrote, "In Edo [the former name of Tokyo] the samurai class abhorred tuna, considering it unclean and unfit to eat because it was seen as a vulgar item of food.") During the early decades of the last century, fishermen began harpooning the big ones, and landings in Maine and Massachusetts reached 94,000 pounds in 1919, 250,000 pounds in the 1930s, and 2 million pounds by 1948. (Note that these figures, which are taken from Bigelow and Schroeder's 1953 *Fishes of the Gulf of Maine,* are in *pounds.* In the discussion of commercial fishing later in this book, the not dissimilar numbers will be in *tons.*)

Bluefin are now fished around the world, but a large proportion caught off Maine, Massachusetts, or California is destined for Japan. North Atlantic bluefins use the Mediterranean as a spawning ground, and fishers from virtually every country with a shoreline on that inland sea catch the tuna there. Countries like Spain, Italy, Greece, Tunisia, Libya, Malta, and Turkey are now "ranching" bluefins. It is the Japanese sashimi market that sets the astronomical prices for these fish. They are the source of *toro,* the fatty belly meat that sells in Tokyo restaurants for the equivalent of $50 an ounce. When a big, top-quality tuna is caught in New England, Australian, or New Zealand waters, it may sell for $100,000 on the dock in Tokyo, and by the time it is served in a restaurant, its value may have increased tenfold. Because of these prices, bluefins have been overfished and their populations are threatened, but there have been few effective protection measures taken, because to do so would require unprecedented international cooperation.

To cooperate, of course, people need to know what fish they are cooperating about. Its scientific name (*Thunnus thynnus*) is used in scientific literature everywhere, but to fishermen and consumers around the world, it has vernacular names, many with a similar root. The scientific name comes directly from the Greek *thynnus,* which means to lunge or dart forward, and in Greek it is *tónnos* (actually, it's τόνος). In Latin, *Thunnus* means "tuna" and *thynnus* means "tunny"—the British still call tunas "tunnies." It is *atún aleta azul* (bluefin tuna), *atún cimmarón,* or *atún rojo* (red [meat] tuna) in Spanish-speaking countries such as Spain, Argentina, Uruguay, Chile, Venezuela, Mexico, and Cuba. French speakers know it as *thon rouge;* the Portuguese and the Brazilians say *atum;* and in Italy it is simply *tonno.* Germans call it *roter thun;* the Danes *thunfisk;* Norwegians *sjorje* or *thunfisk;* and Icelanders *túnfiskur.* To the Dutch it is *tonijn;* Russians say *tunets;* the Japanese refer to the bluefin tuna (whether caught in Japan, California, the Mediterranean, or New England) as *maguro, honmaguro,* or *kuromeji.*

To be so renowned in all those places, the bluefin has to be cosmopolitan in distribution, and indeed, *Thunnus thynnus* is one of the most widely traveled of all fishes. Fish tagged in New England have been recovered in Norway and even in Uruguay, demonstrating that these ocean rangers can easily (and repeatedly) cross the Atlantic Ocean. (Northern Pacific bluefins replicate these long-distance migrations; they have been known to cross the Pacific from Japan to California—and then turn around and do it in the other direction.) But, as John Gunn and Barbara Block point out in their study of tagging, "although most species of tuna are capable of long distance movements, there is growing evidence that for some species or populations, migrations are more the exception than the rule." Some bluefins might cross and recross oceans, while others may spend their adult lives on the same side, migrating only north-south for breeding.

At one time it was believed that there were two separate populations of Atlantic bluefins; one that spent time along the East Coast of the United States and offshore, and another that lived in the eastern Atlantic and the Mediterranean. In those simpler times, Americans thought they ought to be able to catch the bluefins of the western North Atlantic, and Europeans (along with Japanese, Chinese, and everybody else) believed that the European tuna (which spawned in and returned to the Mediterranean) were theirs for the taking. Block's tagging program showed that the population structure was infinitely more complex than was previously thought, and that the Atlantic bluefins paid very little attention to which side of the ocean they were supposed to be on. (An analogous situation occurs in the Pacific, where bluefin tuna born off Japan will sometimes swim to California, and if nothing interferes with their round-trip—think Mexican purse-seiners—they will swim back to Japan.)

For purposes of "stock management," ICCAT divides the Atlantic tunas into two groups based on their known spawning areas. One group spawns in the Gulf of Mexico and the Straits of Florida (the spawning areas of swordfish, too), and the other group spawns in the Mediterranean. It was never obvious to ICCAT that the Gulf of Mexico is open to the Atlantic and the Caribbean (except along the Gulf coast, of course), while the Mediterranean is a huge fish trap—hard to get out of once you're in—and that therefore the fisheries ought to be handled completely differently. (The first tuna "ranchers," at Ceuta in Spanish Morocco, took full advantage of this, and set their trap nets in the nine-mile-wide passage that is the Strait of Gibraltar, snagging the postspawning tuna on their way out of the Mediterranean.)

In response to dwindling catches, ICCAT's twenty-two member countries divided the North Atlantic into eastern and western sectors, each with its own quota. In 1991, when Sweden submitted a proposal to ICCAT that the bluefin be listed as "endangered," it was immediately voted down by the United States and Japan, two countries with a strong economic interest in catching tuna. (Japan consumes 36 percent of all tuna—of all species—caught in the world, and the United States follows with 31 percent.) Conservationists, fishermen, and bureaucrats continued to draft proposals and position papers, while the tuna populations plummeted and the prices rose. As John Seabrook wrote in 1994, "One reason that the price is so high is that there are so few of them left in this part of the ocean, and one reason that there are so few of them is because the price is so high." If someone is willing to pay $173,000 for a fish, a lot of fishermen will be looking to be the lucky one to cash in.*

The tuna of the eastern zone are managed under a strict annual quota set by the European Union, while those of the western Atlantic have been overseen under strict catch quotas since 1995. (As we shall see, neither of these quotas includes farmed tuna.) Nevertheless, in both areas, stocks have fallen dramatically: there has been an 80 percent decline in the eastern (European) stock over the past twenty years, and a 50 percent drop in the western Atlantic population. ICCAT's persistence in treating the "eastern" and "western" populations as separate stocks will only mean the continued and inexorable decline of Atlantic bluefin tuna. Because the fish of the western sector have been so heavily fished, they have received more of the benefits (such as they are) of ICCAT protection. The recent quotas of 3,000 tons in the western North Atlantic (compared with 32,000 tons in the east) will not benefit the western population if the fish are regularly venturing into the more heavily fished eastern regions. Until ICCAT recognizes that the bluefin are a single, mixed population, the numbers will continue to plummet.

North Atlantic breeding populations are estimated to have fallen about 90 percent in the last twenty years. As with all fish populations, exact counts are impossible, so there are vast gaps between the high estimates made—to no one's surprise—by the fishermen, and the low estimates made by those who would protect the tuna from overfishing. From dock to cabinet ministry, there have been endless discus-

*Writing in 1995, Sylvia Earle could not have suspected that six years later, a 440-pound bluefin tuna would sell on the dock in Tokyo for $173,600. She wrote, "But even at a price of $100,000, $1,000,000, or in fact, any price, duplicating such a fish is presently beyond human know-how. Who, I wondered, notices the cost to the ocean, now missing creatures that have taken perhaps twenty years to reach their Tokyo destination?"

sions about solving the problem at every level, but few protective measures have been taken.

Fisheries management is a fine and noble goal, but there has to be something left to manage. During the 1960s, bluefin catches peaked at about 35,000 metric tons, but less than a decade later, overfishing sent the catch plummeting to less than half of that figure, and a 1964 total of 20,000 tons in the western Atlantic fell to 6,100 tons in 1978. The collapse of the New England tuna fishery has been comprehensively documented, most eloquently by Carl Safina in his 1997 *Song for the Blue Ocean,* but where the big fish were before they arrived off Georges Bank is still a mystery. The same is true for the massive schools of tuna that every year entered the bottleneck of the Strait of Gibraltar and swam into the Mediterranean, the functional equivalent of a gigantic fish trap. Somehow, the size, speed, and range of the great bluefins have kept much of their life history hidden from the prying eyes of researchers. Through advances in tags and tagging techniques, we have been able to follow individual bluefins in the Atlantic and the Mediterranean, and we are just beginning to get an idea of where they go and when.

How and where they are fished is also an issue. Unlike some other tuna fisheries, bluefin in the Mediterranean are predominately harvested by purse-seiners. This type of gear has relatively little bycatch; but the illegal use of spotter planes makes it so efficient at locating and capturing schools of tuna that hardly any escape. Recent years have also witnessed tremendous modernization of the Mediterranean's purse-seine fleet with larger, faster vessels outfitted with the latest in fish-finding technology. These vessels are so effective that traditional fishing grounds in the western Mediterranean have been depleted and abandoned in favor of high densities of spawning fish in the eastern part of the sea. In a 2006 ICCAT paper, Jean-Marc Fromentin forwarded the novel suggestion that because large females produce larvae that have a much higher survival rate than first-time spawners, a way to further protect the stock would be to refrain from taking the largest fish. (Males and females look alike and reach the same size.) This makes sense, but unfortunately, it is exactly the opposite of what the fishermen are doing who hunt the "giants" for the Japanese sashimi market.

At the heart of the problem is the rapid development of commercial tuna-farming operations throughout the Mediterranean. Beginning as early as 1991, these farms are not true aquaculture operations that produce fish from spawning individuals, but are more accurately termed "fattening operations," where tuna captured by purse-seine vessels are transferred to floating cages. Tuna are caged anywhere from several months to several years, during which they are fed to increase their fat content and improve the color of the flesh in order to better meet Japanese market standards. The cages are nothing more and nothing less than tuna feedlots.

Why is this a problem? With roughly ten to twenty-five kilograms of baitfish necessary to produce one kilogram of tuna, the fattening operations are grossly inefficient. In addition, the number of illegal, unregulated, and/or unreported catches has increased dramatically because purse-seine vessels no longer need to land their catch at port, but instead transfer live tuna to cage operations directly at sea. Indeed, even fish that are not destined for fattening are also directly transferred to floating cages called "tuna hotels," where they are slaughtered at sea and processed or blast-frozen. Some estimate that up to 50 percent of the Mediterranean's total bluefin catch may be illegal. Ultimately, the WWF study reports that the increase of illegal catches, exacerbated by the popularity of tuna farms, may have generated a total catch in excess of 45,000 metric tons. This catch is 40 percent above the 32,000 metric-ton quota set by ICCAT, which itself is 23 percent higher than the total quota recommended by scientists. The harvest of Mediterranean bluefin tuna stock is 63 percent higher than what the best science suggests.

Bluefins come in three varieties: the northern *(Thunnus thynnus),* which breeds in the Gulf of Mexico and the Mediterranean and cruises the North Atlantic; the Pacific *(Thunnus orientalis),* which breeds in the waters of northern Japan and crosses the North Pacific; and the southern *(Thunnus maccoyii),* which breeds in the waters north of Australia and south of Indonesia, and can be found throughout the Southern Ocean. All three look very much alike, except that the northern version grows largest—particularly in the waters of eastern Canada and New England. Because the three species do not mingle or interbreed, they are mostly differentiated by geography.

PACIFIC BLUEFIN TUNA (THUNNUS ORIENTALIS)

A will-o-the-wisp species, *Thunnus orientalis* seems to fade in and out of existence. When Jordan and Evermann classified the "Giant Mackerel-Like Fishes, Tunnies, Spearfishes and Swordfishes" in 1926, they recognized eight species of *Thunnus.* They were *Thunnus thynnus,* of the North Atlantic and the Mediterranean; *T. saliens* of California; *T. coretta* of the Caribbean; *T. subulatus* of the West Indies; *T. secundorsalis* of the western Atlantic, north to Nova Scotia; *T. phillipsi* of New Zealand; *T. orientalis* of Japan and Hawaii; and *T. maccoyii* of Australian waters. Some of these names have been found to be invalid, or the species combined with others ("synonymized"), but still with us are the northern bluefin *(T. thynnus),* the southern bluefin *(T. maccoyii),* and *T. orientalis,* which the authors call "Kuroshibi," "Maguro," or "Black Tunny." Jordan and Evermann wrote: "This species is clearly different from any of the other forms referred to Thunnus. It is rather common in southern Japan, and occasionally seen in Honolulu. None of the other species has been definitely noted, either from Japan or Hawaii. The finlets are yellow instead of blue; dorsal and anal lobes high, the pectoral rather short, reaching two-thirds distance to anal . . . Belly with 12 obscure pale cross-bars of grayish silver, narrower than the interspaces, replaced above and below by round spots, alternating with the bars." As *T. orientalis* is one of the species that Jordan and Evermann recognize as valid, it seems odd that it would soon begin to disappear from popular accounts.

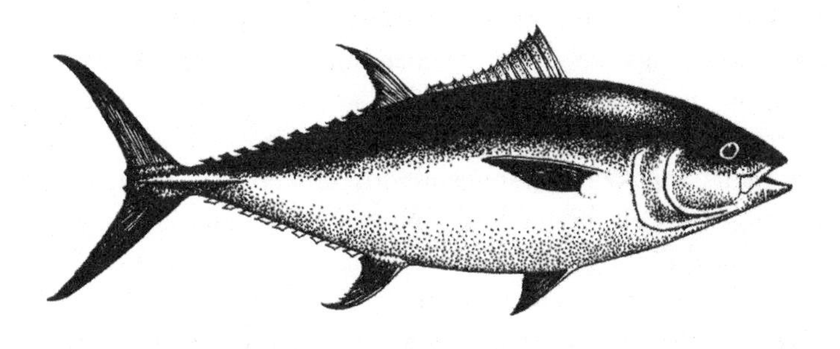

Pacific bluefin *(Thunnus orientalis)*

But it keeps appearing—sometimes under pseudonyms, sometimes nameless—in the fishing literature. Zane Grey, who was born in 1875, began his big-game fishing career in the waters of the coast and islands of Southern California, where large tuna weighed about one hundred pounds. To members of the Avalon Tuna Club, hundred-pounders were "blue-button" fishes; capturing one allowed you to wear the coveted lapel button that proclaimed you a master tuna fisherman. Grey concedes that it took him five years to catch his first blue-button fish, but in 1919, he wrote, his "big tuna weighed 117 pounds, 114, 111, 109 and 109," and his brother R.C. caught a 149-pounder. Grey never bothered with scientific names, but in discussing the fish caught off Catalina Island, he said, "there are two kinds of tuna, yellow-fin and blue-fin." The one-hundred-pound California "blue-button" tunas were large enough, but to Grey, they were probably nothing more than smaller versions of the eight-hundred-pounders that he had heard about off Nova Scotia. And yet, right there in Jordan and Evermann's catalog of giant mackerel-like fishes is *Thunnus saliens,* "the Leaping Tuna of California," which they call (probably after reading Zane Grey's accounts) "the most appreciated of our game fish."

In 1939 Ralph Bandini (1884–1964), another member of the Avalon Tuna Club, published *Veiled Horizons: Stories of Big Game Fish of the Sea,* in which he also paid homage to the tuna of California. But, he wrote, "while the tuna about which Catalina has grown are small compared to those of the North Atlantic and the North Sea, nevertheless, in the Southern California channels, as well as elsewhere, are very, very big tuna. A few of us who have had the good fortune to get into these huge fish have returned sadder and wiser by reason of our experience." The fish were so big that Bandini did not believe they could be caught, "not out here in the Pacific, at least. I doubt that the man is born or the tackle made capable of stopping and landing one of those giants. I make this statement with full knowledge that astoundingly large fish have been taken in Nova Scotia, Maine, and the North Sea . . . Grey, Farrington, Mrs. Farrington, Peel, and dozens of others have taken great tuna. How they do it, I don't know—unless there is a difference between our big tuna out here and those of the Atlantic and the North Sea." The really big tuna of the California channels were probably the species known today as *orientalis*—which Jordan and Evermann say "reaches a weight of 750 pounds." Not quite the half-ton of Nova Scotia tuna, but a respectable fish nonetheless. We know that *orientalis* breeds off Japan, and that it is the object of the local Japanese tuna fishery. So why does the species' name keep coming and going?

In Francesca La Monte's 1952 *Marine Game Fishes of the World,* it is omitted altogether, and replaced by a generic "bluefin tuna," found throughout the North Atlantic, the Mediterranean, and (maybe) Japanese waters. Under the heading "tuna" in Fitch and Lavenberg's 1971 *California Marine Food and Game Fishes,* only the albacore (*Thunnus alalunga*) is mentioned, and of the others, "albacore, bluefin and bigeye tunas prefer temperate waters, but the yellowfin tuna is a tropical species. None lives off our coast on a year-round basis." Not a bluefin in sight. *T. orientalis* is absent from James Joseph, Witold Klawe, and Pat Murphy's 1988 publication *Tuna and Billfish—Fish Without a Country,* and the North Pacific distribution usually given for this species is here assigned to the northern bluefin:

> Subtropical and temperate waters of the north Pacific, in the Western Pacific; in the western Pacific present also in the tropical waters extending from the Ryukyus to northern New Zealand, eastern Australia, and eastern Tasmania, rare in the southeastern Pacific Ocean off Chile, sporadic in the eastern Indian Ocean; south and North Atlantic Ocean and the Mediterranean and Black Seas, recently rare in the latter.

The range of the southern bluefin is given in *Fish Without a Country* as "Subtropical and temperate waters of the southern region of the Indian, Pacific, and Atlantic Oceans." The species received a new lease on life in 2001, for when Bruce Collette, Carol Reeb, and Barbara Block discussed the systematics of the tuna and mackerels, they said, "Bluefin tunas extend into temperate waters of the North Atlantic *(T. thynnus)* and North Pacific *(T. orientalis).* The southern bluefin *(T. maccoyii)* has a distribution pattern . . . in the Southern Ocean." In fact, the northern bluefin ranges across the entire North Pacific, migrating from California to Japan and back. (The $173,000 fish was one of these, caught off Aomori Prefecture in northern Japan in 2001.)

In the *FAO Species Catalogue, Scombrids of the World,* generally accepted as the authoritative last word on mackerel, bonito, and tuna taxonomy (as of the 1983 edition, anyway), the northern bluefin *(Thunnus thynnus)* is said to exist "as at least 2 subspecies, one in the Atlantic and one in the Pacific." The range map for this species is cross-hatched from Japan to Alaska and Baja California in the Pacific, and from Nova Scotia and northern South America across the North Atlantic to northern Norway and the Mediterranean. Collette and Cornelia Nauen, the authors of the catalog, identify the "two subspecies" as "*T. thynnus thynnus* (Linnaeus) in the North Atlantic and *T. thynnus orientalis* (Temminck & Schlegel) in the North Pacific."

The species now known as *Thunnus orientalis* is the Pacific northern bluefin tuna, found primarily in the North Pacific from Japan to California (spawning in the vicinity of Japan), although specimens have also been taken off South Africa, Western Australia, New Zealand, and the Galápagos Islands. This is obviously the species that might be confused with *T. maccoyii,* so Robert Ward, Nicholas Elliott, and Peter Grewe (1995) collected samples of skeletal muscle and liver from this species and *T. maccoyii,* and found that the two could be differentiated by allozyme analysis (allozymes are forms of enzymes that differ in amino acid sequence from other forms of the same enzyme) and also by mitochondrial DNA analysis. They concluded:

> Northern bluefin tuna [*orientalis*] may have a wider range than previously accepted . . . Southern bluefin tuna are far more abundant in Australian waters than are northern bluefin tuna, and the incidental catches of northern bluefin have been allocated by Australian management to the southern bluefin quota. In Japanese markets, the few Australian tuna that have been identified as northern bluefin have sold for a high price. Both marketing and management would be aided by unequivocal species identification.

The three species—southern bluefin, northern bluefin, and Pacific northern bluefin—have long been held to be separate. The true northern bluefin does not visit the South Pacific, so there is little chance of finding one off Australia. As for the other two, the visible differentiating factors appear to be yellow versus blue caudal keels and yellow versus blue finlets. Too bad the name "yellowfin tuna" is already taken (or "preoccupied," as taxonomists say)—it would make things so much easier. Everything depends on where you catch your bluefin. In the North Atlantic and the Mediterranean it will be *thynnus;* off Australia it will be *maccoyii,* and just about everywhere else in the Pacific, from California to Japan, it will be *orientalis.*

Species that can wander thousands of miles into an area already occupied by a very similar species might hybridize under certain circumstances. One way to identify distinct species is through the analysis of the mitochondrial DNA (mtDNA). In 2006, scientists at Japanese laboratories (Chow, Nakagawa, Suzuki, Takeyama, and Matsunaga) examined the "intra and interspecific nucleotide sequence variations of rDNA first internal spacer (ITSI)" of all eight species of the genus *Thunnus.* They reported that "the Atlantic and Pacific northern bluefin tunas *Thunnus thynnus* and *Thunnus orientalis* recently

Pacific bluefin, caught off California and photographed in the Outer Bay tank of the Monterey Bay Aquarium. The ridge above the outstretched pectoral fin marks the slot into which the fin is tucked when the fish is moving at speed.

proposed to be distinct species, were found to share nearly identical ITSI sequences, well within the range of intraspecific variation." Sienen Chow, the lead author, explained what that meant:

> As you noticed, my latest paper seems to be inconclusive. *T. thynnus* and *T. orientalis* apparently have been separated for a long time. There are several morphological differentiations between the two morphs, and mtDNA is very different. This may be enough to determine these two to be completely different species. And Dr. Collette has described them to be two nominal species . . . and many tuna related scientists followed Collette's description . . . However, there are Pacific mtDNA types in the Atlantic population, and Atlantic mtDNA in the Pacific population. Furthermore, we have not observed any distinct differences in nuclear genome between the Atlantic and Pacific bluefin tuna. The point of my latest paper is that I can not say these two morphs are different species but also I can not say these two are the same species.

It is possible that the Atlantic and Pacific bluefin tuna species are separate; it is possible that they are a single species; and it is also possible that both populations are hybrids of two previously separated species. (That, after all, is one of the ways new species develop.) For the most part, however, large bluefin tuna found in the North Pacific are probably *Thunnus orientalis.**

It's a whole lot simpler in New Zealand: the bluefins found around those islands are *Thunnus maccoyii.* Smith, Griggs, and Chow (2001) compared the mtDNA of southern bluefins (the predominant species taken in New Zealand waters) with that of *T. orientalis* to ascertain

*Why does the scientific name of a particular tuna make any difference? Zane Grey certainly didn't care, and Japanese auctioneers aren't concerned if the fish that is going to be turned into sashimi is *Thunnus thynnus, Thunnus maccoyii,* or *Thunnus orientalis.* The differences are important to commercial fishermen and fisheries biologists, however, because it is necessary to know which species you are dealing with when you are trying to assess the status of a given population, to report on migration routes or patterns, or to breed a particular species. You certainly don't want to be classified as merely a generic *Homo sapiens*—you have a first name, a last name, maybe a couple of middle names, and a genealogy that you can trace back for generations. Likewise, maybe tuna should pay more attention to their family tree. Leighton Taylor, an ichthyologist turned winemaker, once claimed to be a member of the "American Miscellaneous Society's Committee to Inform Animals of Their Scientific Names."

whether the Kiwis were catching one species or two. If it was only *T. maccoyii,* it would indicate that it was a "single-species fishery," and therefore within New Zealand regulations, but if some *T. orientalis* were also taken, the fishermen would be operating outside government restrictions. Furthermore, if the DNA of the two species was different, it would enable researchers to "determine if there were reliable field characters for identification of Pacific bluefin tuna on commercial vessels." Where most of the specimens of the Pacific bluefin *(T. orientalis)* had a bluish mottled pattern posterior to the anal fin, and blue speckling above the pectoral fins and sometimes on the head, the SBT had none. In the New Zealand fishery, more than 17,500 southern bluefins *(T. maccoyii)* were examined between 1990 and 2000; during the same decade only 59 Pacific bluefins *(T. orientalis)* had been confirmed, less than 0.3 percent of the total. It is now clear that the New Zealand bluefins are *Thunnus maccoyii.*

Even though it breeds close to the islands of Japan, making it susceptible to a local fishery, the Pacific tuna population appears to be the only one of the three bluefin species that has not been severely overharvested. There is currently no management of bluefin in the North Pacific, a cause for worry given the status of the other populations. At present we know little about the movements of bluefin in the eastern Pacific, especially the larger animals. Based on the available information, we can conclude that bluefins appear to spawn only in waters off southern Japan, and that late in their first year some individuals swim east. Data suggest that when the sardine population off Japan diminishes, bluefin leave the area. Off the coast of Mexico and California, bluefin are most abundant in the summer and fall months, after which they seem to disappear.

Except in the fishing literature, there is not much information on the Pacific bluefins off California, or, for that matter, off Japan. In his 2001 chapter on tuna conservation, Carl Safina wrote:

> Pacific northern bluefin tuna are known to breed (based on larval distribution) only between the Philippines and northern Japan, and in the Sea of Japan. From those only known origins, they inhabit much of the temperate North Pacific, east to North America. In the eastern Pacific Ocean, nearly the entire catch of bluefin tuna is

> made by purse seiners fishing not far offshore from California and Baja California. Catches of Pacific northern bluefin tuna were as high as 35,000 mt in 1956, and as low as 9,000 mt in 1990. But as recently as 1995 fluctuating catches went as high as 24,000 mt. Though recruitment has been judged not to have declined significantly, IATTC (1999a) states, "fishing has greatly reduced the abundance of mature [northern] bluefin in the Pacific Ocean."

In the *National Tuna Fisheries Report of Japan as of 2004,* Miyabe, Ogura, Matsumoto, and Nishiwaka discuss the Japanese fisheries for YFT (yellowfin), BET (bigeye), SKJ (skipjack), ALB (albacore), SWO (swordfish), MLS (striped marlin), and BUM (blue marlin), but the report contains no mention at all of BFT. In the section on "Pacific Bluefin Tuna" in *Historical Trends of Tuna Catches of the World* (Miyake, Miyabe, and Nakano 2004), we are finally able to learn a little about the Japanese bluefin tuna fishery in the western Pacific. Because no larvae have ever been reported elsewhere, we assume that *T. orientalis* spawns only in the area between the Philippines and southern Japan, and also in the Sea of Japan. Juveniles move northward from the spawning grounds when they are about twenty centimeters (eight inches) in length—at which time they become susceptible to the southern Japanese troll fishery. Some fish undertake transpacific migrations at one or two years of age, and if they are not caught by Mexican fishermen off Baja, they might recross the Pacific and return to Japan (a trip of 6,700 miles), but some wander into other parts of the ocean, making appearances off Hawaii, New Zealand, and southern Australia. Miyake, Miyabe, and Nakano again:

> The major fishing grounds for Pacific bluefin are located in the middle latitudes of the North Pacific between 20°N and 45°N in the west, and between 23°N and 45°N in the east. In the northwestern Pacific, catches have been made . . . in the waters from eastern Taiwan to northeastern Japan, including the East China Sea between Japan and the Republic of Korea and the Sea of Japan. In the eastern Pacific, bluefin are caught off Baja California and southern California, by purse-seine and sport fisheries. [So *these* are the big bluefins caught off Catalina!]

Since 1952, the total catch for the whole Pacific Ocean has fluctuated between ten thousand and thirty-five thousand tons, peaking around the 1980s, dropping off in the 1990s, and climbing again at the beginning of the twenty-first century. By far the largest proportion was taken by Mexican purse-seiners in the eastern Pacific, but there is also a Japanese longline fishery, as well as purse seines that target skipjack, yellowfin, and Pacific bluefins. The bluefins are not the most important product of this fishery by weight, but they are worth more because of the sashimi market. Unsurprisingly, the Japanese dominate this fishery, but the Taiwanese took nearly three thousand tons in 1999.

The warming of central California's Monterey Bay has evidently attracted large numbers of Humboldt squid *(Dosidicus gigas),* previously found only in the tropical waters off Peru and Mexico. Juvenile Humboldt squid are a known prey item of tuna, but overfishing the Pacific bluefin has effectively removed the squid's predators, allowing the cephalopods to proliferate and colonize an entirely new region. As Louis Zeidberg and Bruce Robison put it in their 2007 article, "This geographic expansion occurred during a period of ocean-scale warming, regional cooling, and the decline of tuna and billfish populations throughout the Pacific." The orange-red Humboldt squid, also known as the jumbo squid, reaches a length of eight feet, which is far less than the sixty-foot length of the giant squid *(Architeuthis),* but *Dosidicus* is an aggressive, powerful, pack-hunting predator, known to attack divers, which may be yet another, totally unexpected problem tied to global warming.

SOUTHERN BLUEFIN TUNA (THUNNUS MACCOYII)

The southern bluefin tuna *(Thunnus maccoyii)* is a large, fast-swimming, pelagic fish, found throughout the Southern Hemisphere mainly in waters between 30° and 50° South, but only rarely in the eastern Pacific. Some consider it a subspecies of the northern bluefin, which is technically known as *Thunnus thynnus.* The northern variety, found in all the temperate and tropical waters of the world—except the high southern latitudes—looks very much like the southern, but it gets much larger. (The IGFA record for a southern bluefin is 436

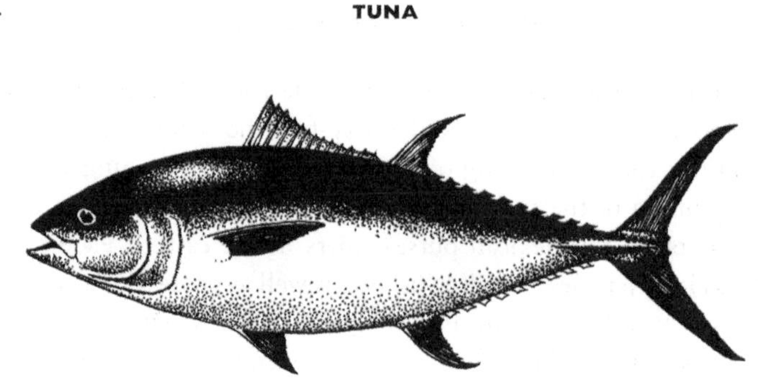

Southern bluefin *(Thunnus maccoyii)*

pounds; the record northern was a thousand pounds heavier.) Except for size, though, the difference in the two species of bluefin tunas is minuscule. In Jordan and Evermann's 1926 discussion of the "Mackerel-Like Fishes," we find that *Thunnus maccoyii* can be identified by "insertion of the dorsal fin *before* the last ray of the second dorsal, under the third or fourth ray from the last; pectoral nearly half length of heads; finlets 8/8, blue and brown, without yellow." Other descriptions, such as that found in the 2006 IGFA record book, say that the difference between the northern and southern bluefin is the number of gill rakers; the southern has 31 to 40, while the northern has 34 to 43. (Gill rakers are bony, comblike projections which point forward and inward from the gill arches; they act as coarse filters to prevent large debris from entering the opercular cavity and damaging the gills, and also to keep prey items from escaping between the gills.) If that is the differentiation factor, what species is a bluefin that has 35 to 40 gill rakers?

In 1926, Jordan and Evermann probably believed that they had resolved the southern versus northern bluefin controversy, but it would appear that not everybody agrees with their solution.* Many authors describe *T. maccoyii* as just a smaller version of *T. thynnus,* and

*Just for the record, this is what they wrote:

{T.} thynnus (Europe): Second dorsal falcate, higher than first, about 2 in pectoral. Insertion of anal below last ray of second dorsal. Pectoral fin short, about 7 in length to base of caudal; depth slightly more than length of head; finlets 9/9, not yellow; young with pale streaks.

{T.} maccoyii (Australia): Insertion of anal before last ray of second dorsal, under third or fourth ray from last; pectoral nearly half length of head; finlets 8/8, blue and brown, without yellow.

in McClane's 1965 *Fishing Encyclopedia* we read, "It now seems fairly certain that the bluefin tuna is a single, worldwide species, but subspecific variation is still in need of further study." When we look up "bluefin tuna" in Ayling and Cox's *Guide to the Sea Fishes of New Zealand* we find that the fish is indeed *Thunnus maccoyii,* but the description—particularly the size ("One of the largest of the bony fish, attaining a maximum length of over 2.5m and a weight of over 400kg")—is for the *northern* bluefin. In another New Zealand guide (*Marine Fishes of New Zealand,* by Larry Paul), there is this: "The northern bluefin is very difficult to distinguish; the colour (black) of the tail base keel is one feature, and the pectoral fin shorter (usually less than 20% of body length to tail fork)." In *The Marine and Freshwater Fishes of South Australia* (1973), there is a good description of *Thunnus maccoyii,* followed by this remark: "This species is very common in South Australia in large schools during the summer months, and it gives promise of providing a good fishing industry. In Queensland and northern Western Australia, it is replaced by the northern bluefin tuna *(Kishinoella tonggol).*" And Harold Vaughan's *Australian Fisherman's Companion* (1981) lists *Kishinoella tonggol* as the northern bluefin tuna, and *T. maccoyii* as the southern, describing the latter as "the largest of the tuna family, growing better than 136kg [300 pounds]. Dark blue to blue-green on the back above the lateral line, and silvery below. The edges of the fins are yellow."

Jessica Farley at CSIRO's Pelagic Fisheries and Ecosystems branch in Hobart, Tasmania, described how the species could be differentiated:

> Most people can't tell the difference between the two, and even fishers sometimes get it wrong. I haven't seen too many NBT so I'm no expert, and we genetically test tissue samples to be sure. However, one of the main distinguishing features of SBT is its yellow median caudal keel (in both juveniles and adults). It's black in adult NBT, and I've read that it's "semitransparent" in juvenile NBT. Some say you can also tell the difference from the colour/pattern of lines and spots on the side and belly but I'm not convinced.

Southern bluefins can reach a weight of over four hundred pounds and measure more than seven feet in length. They mature at ten to twelve years of age, and if undisturbed, may live forty years. Breeding takes place from September to April in warm waters of the Indian Ocean, southeast of Java—the only known breeding area. Mature females produce several million eggs in a single spawning period, and those that survive to juvenile size migrate down the west coast of Australia. During the southern summer months (December–April), they tend to congregate near the surface in the coastal waters off the southern coast of Australia and spend their winters in deeper, temperate oceanic waters. After age five, southern bluefins are seldom found in nearshore surface waters, and follow instinctive migratory urges that take them to New Zealand and southern South America to the east and South Africa to the west. Wherever they go, southern bluefins are targeted by dedicated fisheries.

In the 1950s, Japanese fishermen, expanding their fisheries to feed a country ravaged by war, discovered previously unknown aggregations of large tuna south of Indonesia in the Indian Ocean. The Australians found the same tuna in their own waters, but unlike the Japanese, they had no tradition of offshore fishing, and it took a while for them to get geared up. Offshore trawling began in the late 1930s, but the same war that devastated the Japanese fishing fleet also sidelined Australian boats. In *Fish and Fisheries of Australia* (1953), Theodore C. Roughley wrote:

> Various species of tuna have been known for a long time to occur on the Australian coast but little information was available concerning their abundance until 1936 when I investigated their occurrence on the south coast of New South Wales. It was then made plain that the southern bluefin tuna (or southern tunny as it was known at that time) occurs regularly on the western half of the New South Wales coast in large shoals, principally during the months of September, October and November.

We can learn about the variations in finlet color from the scientific literature, and about the history of the fishery from the historians, but probably the best account of what actually went on in the Port Lincoln fishery can be found in Colin Thiele's *Blue Fin,* a novel about the adventures and misadventures of a fifteen-year-old boy named Snook

Part of the South Australian tuna fleet, in Port Lincoln in 1985. Tuna are still being caught from boats, but now they are brought to pens where they will be fattened like cattle in a feedlot.

Pascoe who goes fishing with his father. Thiele (1920–2006) was the author of a hundred books, mostly about his home state of South Australia, some of which were set in Port Lincoln, the home of the Australian tuna fleet in 1969, and now the scene of the tuna ranches that replaced the offshore fishermen. In the 1960s, most of the tuna caught off Port Lincoln were destined for the cannery. Here's what it was like when Snook worked the line:

> This was the conveyor line down which each poor tuna had to travel on his last slow journey. Cooked and skinned, his blue sheen gone, his grace and beauty destroyed, he went along the line like the lump of flesh that he was. And then the picking started. A hundred women, ranged in long rows like harridans on either side of that final funeral path, ripped at him with nimble fingers. They stripped off the dark meat and whipped out the bones until only

> the pure flesh, rich and faintly pink, went off to its rendezvous with the zombie columns of cans. Here it was doled out by another mechanical henchman—fifteen ounces, or six and a half ounces, or even a midget bite of three and a half ounces, according to the can it was destined to meet. And so on at last to the final corroboree of cans and cartons where cardboard sheets reared up and folded miraculously into shapes to swallow their portion, seal and label themselves, and bind their sides with wire so that forklifts could trundle and dump and carry them into the vast haste of business.

Blue Fin was also made into a movie in 1978, starring Greg Rowe as Snook and the German actor Hardy Kruger as his father, Bill Pascoe. As much a historical document as the book, the film incorporates footage of actual pole fishing in South Australian waters, and thus serves as an important record of fishing techniques now relegated to history.

"The Australian fishery," continued Roughley, "has developed principally in New South Wales and in South Australia, though it is known that fish of much larger size than those taken on our inshore grounds are caught by Japanese longliners operating in the Tasman Sea and the eastern Indian Ocean." An American company based in Samoa decided to try to catch tuna in Australia using the kind of pole-and-line fishing that had proven so successful in California, and they succeeded (eleven tons of tuna in four days), leading to the development of the South Australian fishery, principally out of Port Lincoln. Most of the tuna caught by the Australians were being sold to the Japanese, not only because of their desirability as sashimi, but also because Australians were not particularly fond of fish. Indeed, in the late 1980s, a concerted effort on the part of the government was required to wean the Aussies—particularly men—from a diet of beef and lamb.* In 1986, this note appeared in *SAFISH,* the newsletter of

*Evidently, things haven't changed that much. On the website of the Australian Consumers Association in 2006, we read: "In Australia we eat on average only 26 grams of fish per person per day (compared with 157 grams of meat). According to nutrition experts we should be eating a lot more. This is because fish contains omega-3 fats, which we don't get much of from other foods. There's strong evidence that omega-3s help to prevent heart disease and strokes, and they may also help to prevent some cancers. The National Heart Foundation recommends eating at least two fish meals a week. All fish is good, but some types (fatty fish) are especially high in omega-3s—though unfortunately these

Blue Fin, the 1978 movie based on the novel by Colin Thiele

the South Australian Fisheries Association: "In a study of 800 men over a 20 year period, Dutch scientists have found strong support for the idea that the more fish you eat, the less likely you are to suffer from coronary heart disease . . . the chance of coronary heart disease was reduced by half in those who ate one or two fish meals a week."

As early as the 1950s, in the northeast Indian Ocean, the Japanese were catching southern bluefins, which proved to be so valuable that they soon expanded their operations to South and Western Australia. The tuna were frozen at very low temperatures (–60°C), and either unloaded at intermediate ports and shipped to markets in Japan, or

no longer include canned tuna. Skipjack and yellowfin are now used instead of southern bluefin tuna and recent research by the CSIRO has found that these species have only about half the levels of omega-3s found in southern bluefin."

shipped directly to those markets. From 1988 to 1995, a number of Japanese longliners (also catching yellowfin tuna, bigeye tuna, albacore, swordfish, and marlins) entered into a joint-venture agreement with Australian companies, but it ended when the Australians switched over from pole-and-line fisheries to purse-seining (Robins and Caton 1998). The Australian component of the fishery now uses purse seines to enclose a school of fish, and then tows them to waters near the Australian mainland, where they are transferred to floating cages anchored to the ocean floor. The tuna are then fattened for several months and sold directly to Japanese markets as frozen or chilled fish. Because of the high fat content of their flesh, premium prices can be obtained for *toro* in Japan. From 140 tons in 1992, Australian tuna farmers boosted production to more than 4,700 tons in 1998. The total value of the southern bluefin fishery is now estimated to be about AUS$1 billion.

Until recently, the only countries fishing the southern bluefin were Australia, New Zealand, and Japan. The Australians used pole-and-line, purse seines, and trolling gear, but the Japanese employed only longlines. In the 1950s the annual catch of southern bluefin was twelve thousand to fifteen thousand tons, and the fish was used mostly for canning. (Of the southern bluefin, Vaughan's 1981 *Australian Fisherman's Companion* says, "Although sought by the canning industry, its flesh is too meaty and strong to be eaten fresh.") When Japanese longliners adopted supercold freezers and started fishing for the sashimi market, the red meat of southern bluefins acquired the same market value as that of their northern cousins, and in 1960 and 1961 the catch ballooned to some eighty thousand tons a year. Fishing that intense cannot last, and by the early 1980s the annual catch had dropped to forty thousand tons. Heavy fishing pressure resulted in a significant decline in the numbers of mature fish and the catch began to fall rapidly. It became apparent that the bluefin stock was at a level where some kind of management and conservation was required, and as of 1985, Australia, Japan, and New Zealand began to apply strict quotas to their fishing fleets, and in May 1993, the voluntary management arrangement between the three countries was formalized with the creation of the Convention for the Conservation of Southern Bluefin Tuna (CCSBT). Other fishing nations, such as Korea, Taiwan, and Indonesia, were active in the southern bluefin fishery, which reduced the effectiveness of the members' conservation and management measures, and of course new countries trying to take advantage of the rich Japanese market caused the stocks to decline again. As a matter of policy, the CCSBT encouraged the membership of these countries; on October 17, 2001, the Republic of Korea joined the commission, and Taiwan's membership became effective on August 30, 2002. Indonesia was accepted as a formal cooperating nonmember on August 2, 2004.

Not long after the formation of the CCSBT, the organization was called upon to adjudicate a conflict between the original signatories of the convention, with Japan on one side and Australia and New Zealand on the other. Because the southern bluefin is a highly migratory species, it passes through the exclusive economic zones of several countries in the Southern Ocean. The commission was formed because it was recognized that the stock was declining precipitously, and action was necessary by the countries that were actively fishing on the only population. Instead of cutting back, however, Japan began a no-limit "experimental" fishery in 1998, and took 1,400 tons over their assigned quota. In protest, Australia banned Japanese fishing boats from visiting Australian ports. (Not unlike "Scientific Research Whaling," the Japanese experimental tuna fishery was nominally conducted for scientific purposes, but somehow the meat ended up in fish markets and supermarkets.) This effectively shut down the lucrative arrangements between Japan and Australia for Japanese freezer ships to pick up the frozen tuna (farmed and wild-caught) that the Australians were consigning to Japanese markets. Australia and New Zealand requested an injunction under the International Tribunal for the Law of the Sea (ITLOS) and the tribunal suggested binding arbitration. An agreement was reached in May 2001, whereby Japanese ships were once again allowed into Australian ports, and the Japanese would substitute a limited Scientific Research Program for their unrestricted "experimental" fishery.

Although management restrictions under the CCSBT have been in place for more than ten years, the southern bluefin biomass has continued to decline. The parental biomass is currently estimated at less than 10 percent of the 1960 level, well below what is considered a biologically safe level. Annual catches for countries such as Korea and Indonesia are probably on the order of 2,500 tons, and China is expanding its distant-water fishing fleet. According to the World Conservation Union's Red List of Threatened Animals, the southern bluefin tuna is "critically endangered," which is defined as "facing an extremely high risk of extinction in the wild in the immediate future." In a 1997 TRAFFIC report on the southern bluefin tuna fishery, Elizabeth Hayes wrote, "The collapse of significant fisheries in the past several decades has indicated that it is possible to fish marine species to commercial extinction and that great economic suffering occurs as a result. If the warning signs are ignored, this possibility exists for SBT."

In 2001, in the journal *Population Ecology,* Japanese fisheries scientists Mori, Katsukawa, and Matsuda published a "recovery plan for an exploited species, southern bluefin tuna." They recognized that the stock was heavily depleted in the 1980s and that the spawning stock biomass (SSB) was also in decline. Using some complex computer models, they ran ten different "scenarios" that incorporated natural mortality rates and fishing mortality rates, to conclude that if the spawning potential (SP) is increasing, "we do not need to worry too much about future reproduction because the decline of SSB may be temporary," but if SSB is increasing and SP is decreasing, "we need to be careful because this indicates that recruitment has not succeeded well and that SSB will decrease in the future." Their "recovery plan" consisted of reducing fishing mortality by 30 percent. The scientists should have checked with the fishermen, who were catching a great many more fish than they were reporting, thereby *raising* the fishing mortality.

In August 2006, an Australian investigation revealed that Japanese tuna boats, along with Taiwanese and Thai fishers, had been secretly taking more than twice the CCSBT quotas for the past twenty years, and bringing the fish into Japan without reporting the actual catch. The discrepancies were first noticed by Brian Jeffriess of the Australian Tuna Boat Owners Association when he saw "irregularities" in Japanese records of the fish caught in the Great Australian Bight. In an article in the *Sydney Morning Herald* (Darby 2006), Richard McLoughlin, managing director of the Australian Fisheries Management

Authority, was quoted as saying, "Essentially the Japanese have stolen $2 billion worth of fish from the international community, and have been sitting in meetings for the past fifteen years saying they are as pure as the driven snow. They are virtually killing off the stock." Under the CCSBT, Japan had agreed to a six-thousand-ton annual quota, but investigators learned that they had been taking twelve thousand to twenty thousand tons every year for twenty years and hiding it.

The Japanese allowed only Japanese observers on their boats, so there was no one to report the excessive tonnage or the misrepresentation of the species; some of the fish being delivered to Tokyo fish brokers as "bigeye tuna" were actually southern bluefins. Also, tens of thousands of tons of SBT were never entered in the Japanese public auction system, but sold directly to retailers instead. "This is a scandal of epic proportions," said Nicola Beynon of Humane Society International; "if proven, this will make clear that Japan cannot be trusted to exploit marine resources within internationally agreed quotas. They must never be given a commercial quota for whales." Benyon was making reference to the 2006 events at the International Whaling Commission, where Japan had stacked the voting to allow them to control future quota levels for endangered whale species. By this time, whalemeat had become an insignificant element in the Japanese diet, but tuna is the very lifeblood of Japanese cuisine, and it appears that they are willing to do anything to ensure a steady supply.

Following the revelations of Japanese overfishing, a CCSBT meeting was held in October 2006 at the Japanese tuna port of Miyazaki. Australian delegation leader Glenn Hurry announced that the 178,000 tons of bluefin taken by the Japanese were actually worth $8 billion, not the $2 billion that had been previously estimated. The Australian delegation proposed that the Japanese annual quota, which was at 6,065 tons, be reduced to 3,000. At the meeting, Brian Jeffriess announced that Japan had agreed to a (non-Japanese) observer aboard every tuna boat, a boat-to-market paper trail, and regulations over transshipping at sea. It is obvious that ten years of flagrant disregard of quotas and regulations by Japanese has been a major factor in the decline of southern bluefin stocks. It remains to be seen whether reduced quotas will enable the stock to recover. (Or if the Japanese intend to abide by the new quotas.)

But long before the Japanese had developed an obsessive craving for bluefin tuna flesh, the great fish had established itself—or more accurately, had itself established—as one of the most important of all game fishes. Toward the end of the nineteenth century, the red meat of the horse mackerel was considered fit only for dogs and cats, but American sportsmen realized that some very large fishes were swimming not too far offshore in places such as New Jersey and Nova Scotia. It is easy to imagine that one or another of these sportsmen, often looking for ways to fill their otherwise nonproductive days, thought it might be interesting to see if they could find a way to bring one of these monster fish to the gaff. As far as we can tell, they had no intention of eating any part of their catch. With the capture of the first giant tuna, they were originating the new pastime of fishing for sport.

Four

SPORT FISHING FOR TUNA

Michael Lerner (center) and fishing companions with twenty-one bluefins caught in an eleven-day stretch in Nova Scotia waters, September 1935

AROUND 1496, Wynkyn de Worde, one of the first printers in England, printed the *Treatyse of Fysshynge with an Angle,* which was based on even earlier treatises on "fysshynge." Izaak Walton published *The Compleat Angler* in 1653, inspiring a vast number of his fellow Englishmen to take up fishing. But no matter how much they enjoyed the sport, they still ate the fish they caught. Since Walton's day, however, the art and science of fishing often took precedence over the number and size of the catch that the fisherman brought home to feed his family. Thus did salmon and trout fishing—particularly in England—develop into pastimes suitable for gentlemen, along the lines of fox hunting or bird shooting. The conquest of a fox or a pheasant might require a certain degree of skill and courage, but for the most part, what were needed were a few dogs, a proper kit, and privacy and space enough to engage in these patrician pursuits.

Another field usually restricted to the aristocracy or to those officers posted to exotic locales was big-game hunting. Much of this sort of thing was dedicated to the acquisition of a large set of horns or antlers to display over the mantelpiece, but hunters with powerful weapons also took aim at tigers and leopards in India; lions, leopards, rhinos, and buffalos in Africa; and almost everything larger than a woodchuck in North America: bison, bighorn sheep, moose, elk, every kind of deer, and of course the predators, such as bears, wolves, and mountain lions. If there was ever any justification offered for the slaughter of these animals, it was a sort of hunters' "manifest destiny": the privileged classes were entitled to shoot everything with legs or wings, and if the targets happened to be predators, so much the better, the hunters were making the world safer for farmers and their livestock. If and when the predators preyed upon people—think of "man-eating" lions and tigers—well, it was clearly the duty of the hunters to rid the world of these malicious carnivores.

No fish ever threatened a farmer's livestock, and with the possible exception of some of the larger shark species, none threatened people either. But because some fish were so large and so powerful that their capture required more than a little skill (and often a lot of expensive equipment, sometimes including a big boat), the idea of big-game fishing was born. In his discussion of the origins of the very exclusive Cat Cay fishing resort in the Bahamas, Philip Wylie explained, "the day came when men who had once exulted over a three-pound trout taken on a rod and reel began to think three hundred pounds was no great shakes for hefty relatives of the same species of tackle. This new sport was dashing, daring—and expensive. And for those who could afford it, there was a need of a base suitable to the tastes of the elite." The idea of catching fish that you had no intention of eating—fishing for sport, in other words—is a very recent development, and probably to some extent based on the great billfishes. They certainly are edible, but fight takes precedence over fillets, and the idea of eating a thousand pounds of fish might be a little daunting to any but the most intrepid (or hungry) angler.

Obviously, "big-game fishes" had to be big—many of them could exceed a thousand pounds—but they also had to be "game," that is, prepared to put up a valiant fight to escape capture. This conspicuous reluctance to be reeled in often took the form of spectacular, repeated leaps out of the water, an exaggerated, large-scale version of the fight a hooked salmon might put up. The pursuit of large fishes on an adversarial basis began in the early decades of the twentieth century, where the goal was the conquest of an opponent worthy of an intrepid and well-outfitted fisherman. (The idea of fishing for food was eschewed; some of these big-game fishes were far too large to be eaten by anything but a small village.) The big-game fishes include the larger tunas—bluefin and yellowfin—and all the billfishes: the marlins, sailfish, spearfish, and of course the broadbill swordfish.*

In Van Campen Heilner's 1953 history of saltwater fishing, we learn:

> The first sportsman to test the quality of Nova Scotia tuna . . . was Thomas Pattilo, a schoolmaster, who tackled them from a dory in Liverpool harbor in 1871. He took thirty-two fathoms of ordinary codline, wound it on a swivel reel of some sort, fashioned a hook of steel "three-eighths inches thick, eight inches long, with a three inch shank," and sallied forth in an ordinary fisherman's dory with a single companion. He was soon hooked to a monster tuna which towed his boat across the harbor and hurtled it merrily into a fleet of herring netters, swamping one and creating havoc with the rest. One of the irate herring fishermen cut the line and thus Pattilo's first tuna got away.

On Pattilo's second attempt—presumably with the same gear—he brought to gaff a six-hundred-pound bluefin. In 1908, Commander J. K. L. Ross came to Cape Breton Island in pursuit of these monster mackerel, and although he managed to hook twenty-two of them during the season, he lost every one. The following year, he hooked one that towed him all over the harbor for nineteen hours before "the exhausted sportsman admitted defeat and cut his line."

*Over time, the category of big-game fishes has come to include sea basses (which can weigh upwards of 800 pounds); tarpon (which are spectacular jumpers and can weigh 250 pounds); wahoo (among the fastest of fishes; maximum weight 185 pounds); dolphin fish or mahi-mahi (maximum weight 85 pounds); and several species of large sharks, including the mako (probably the most acrobatic of all game fishes; record 1,221 pounds); the tiger shark (record 1,780 pounds); and the great white, which can weigh as much as a full-grown buffalo.

Sailors and fishermen who found themselves off the coast of Seabright, New Jersey, in the early decades of the twentieth century were aware that there were some large fish there. They just couldn't figure out how to catch them. Heilner, who fished there between 1912 and 1916 (with Zane Grey, among others), wrote:

> We fished off Seabright out of bank skiffs, sitting on camp chairs, and we gave those giant horse-mackerel some of the finest collection of hooks and lines you ever saw. We could hook them, and that was about all. The rest was fun for the horse-mackerel . . . They looked as big as hogsheads there under the sterns of our dories and some of them ran over a thousand pounds. I saw a fish there one day so big it scared me. I was afraid to put a bait over for fear he'd take it.

On September 13, 1915, wearing a primitive harness, Jake Werthheim hauled in the first New York Bight "horse mackerel," a monster

Zane Grey, one of the first of the big-game fishermen

of 286 pounds. Seven years later, Christian Feigenspan landed a 407-pounder, and that American record stood until Francis Low boated a 705-pounder in 1933. The big fish then seemed to disappear from New Jersey waters. According to Al Anderson's history of the tuna fishery, fishing for giants off New Jersey did not resume until the 1980s. Captain Al Ristori caught a 1,022-pound fish in 1980 that held the New Jersey record for a little more than a year, when Roy Parsons landed one that weighed eight and one-half pounds more. The current American record holder is Jim Dempsey, who reeled in a fish of 1,140 pounds off Galilee, Rhode Island, in 1981.

Zane Grey, considered the most macho of big-game fishermen—at least until Hemingway came along—had been christened Pearl Gray, after the color Queen Victoria had made popular in 1872, the year of his birth. He grew up hating his given name, changed the spelling of "Gray" to "Grey," and renamed himself "Zane" after his mother's family, the Zanes of Virginia, who had founded Zanesville, Ohio. The author of popular western novels in the 1920s and 1930s, Grey was a passionate big-game fisherman, spending most of his not inconsiderable royalties (his books sold 13 million copies) on fishing trips, boats, and gear. In addition to books like *Riders of the Purple Sage,* he also wrote about his fishing experiences, and in *Tales of Swordfish and Tuna* (1927) he described his battles with giant bluefin tuna, first in California waters and later off Nova Scotia. Grey admired the tuna so much that his writing approximated the fury and excitement of a tuna feeding frenzy. Here is an example, written in 1925:

> The giant Nova Scotia tuna, if he struck a trolling bait, would demolish the tackle and jerk the angler overboard. I have no idea what a splash he would make, but it would be tremendous. Blue-fin tuna, at Catalina, hitting a bait attached to a kite, make a thrilling strike. A bulge appears behind the skipping bait, then the tuna dives. Smash! He makes a round cracking circle of foamy water. The blue-fin hitting a trolled bait—as was the method before kite-flying—makes the same kind of a strike, only magnified in every way, and of course the angler gets the full benefit of that powerful smash.

With Captain L. D. Mitchell, an Englishman who worked in the fishing tackle department of Abercrombie and Fitch in New York, Grey headed for Nova Scotia in 1924. Near Liverpool, with Mitchell accompanying him in a twenty-foot skiff, he hooked a bluefin and fought it for five hours: "The tuna heaved to the surface, he rolled and gasped, lunged out his huge head with jaws wide and black eyes staring—a paralyzing sight for me. Then he wagged toward to the bow, his wide back round and large as a barrel, out of the water." Mitchell called it the "gamest tuna I ever saw or heard of." Brought to the dock, it measured 8 feet 4 inches in length, 6 feet 2 inches in girth, and weighed 684 pounds. In the prose for which he was famous, Grey described the vanquished fish:

> He was built like a colossal steel projectile, with a deep dark blue color on the back, shading to an exquisite abalone opal hue toward the under side, which was silver white. He blazed like the shield of Achilles. From the edge of his gill cover to the tip of his nose was two feet. He had eyes as large as saucers. His gaping mouth was huge enough to take in a bucket. His teeth were like a strip of sand paper, very fine and small. The massive roundness of his head, the hugeness of his body, fascinated me and made me marvel at the speed he had been capable of. What incalculable power in that wide tail! I had to back away to several rods' distance before I could appreciate the full immensity of him.

Later, Grey hooked another Nova Scotia monster, and described the chase in prose that could have come straight out of *Riders of the Purple Sage:* "Blue Island seemed a mountain, green on top, black at the sea line, a bleak jagged precipitous shore against which the great swells burst ponderously. The white spray shot high. I saw the green swells rise out of the calm sea and move in majestic regularity to crash and boom into white seething ruin . . . The feeling of the sea under me was something at the moment to take heed of. If I had not been hooked to what must be a gigantic tuna, I would have grown panic-stricken." This time his brother R.C. and his son Romer were along, and Romer cried out, "Must be a whopper! Don't work too hard, dad. Don't let him get away. Don't give him any rest." The "whopper"

Zane Grey gazes in awe at his (then) world's-record 758-pound bluefin tuna, caught off Nova Scotia in 1924.

weighed 758 pounds, and Grey rightfully claimed the world's record for bluefin tuna. As he described it:

> I was struck dumb by the bulk and beauty of that tuna. My eyes were glued to his noble proportions and his transforming colors. He was dying and the hues of a tuna change most and are most beautiful at that time. He was shield-shaped, very full and round, and high and long. His back glowed a deep dark purple; his side gleamed like mother-of-pearl in a lustrous light; his belly shone a

> silver white. The little yellow rudders on his tail moved from side to side, pathetic and reproachful reminders to me of the life and spirit that was passing. If it were possible for a man to fall in love with a fish, that was what happened to me. I hung over him, spellbound and incredulous.

Grey was not only among the first to write about tuna fishing, he was among the first to capture these great-hearted fish. Another trailblazer was Michael Lerner (1891–1978), a successful businessman who turned an early and avid interest in hunting and fishing into a mature scientific avocation. Michael was one of seven children of Charles and Sophie Lerner. With his father and brothers he founded Lerner Stores, a nationwide chain of women's clothing shops, but he left the business that provided enough money for him to pursue his passion for catching very large fish. Michael and Helen Lerner visited the island of Bimini in the 1930s, and by 1936 they had founded the Bahamas Marlin and Tuna Club, with Ernest Hemingway as the first president. Around this time, Lerner's association with the American Museum of Natural History began, as he would donate his prize catches, often mounted in lifelike poses, to become museum exhibits. As a trustee, he led seven American Museum expeditions (always where there was good fishing) to locations such as Cape Breton Island (Nova Scotia), Bimini, Australia, New Zealand, Peru, Chile, and Ecuador.

Lerner was also among the first anglers to seek out the giant tuna of Nova Scotia, pioneering the rod-and-reel fishery for giant tuna there in 1935. In eleven days in September 1935, he caught eleven that ranged from 86 to 450 pounds. He was also the principal organizer of the International Game Fish Association (IGFA), which held its first meeting at the American Museum of Natural History on June 7, 1939. Also in attendance were Van Campen Heilner and museum ichthyologists Francesca La Monte, John Nichols, and William Gregory. Lerner agreed to "personally underwrite all the expenses of the new organization, a level of support he would continue to maintain until his resignation as IGFA President in 1961" (Rivkin 2005). Almost alone among the company of big-game fishermen, Lerner never wrote a book about his accomplishments—which were signifi-

Helen and Michael Lerner stand next to a day's catch in Bimini.

cant in both ichthyology and angling. It would be up to others, such as Mike Rivkin, who wrote a history of the IGFA in 2005, to record Lerner's fascination with the giant tuna.

Fishing off Long Island (New York) in the 1920s, Lerner caught a seventy-pound bluefin, and was so impressed with its brawn that he was prepared to devote a good part of his life to the pursuit of big, powerful fishes—including *Thunnus thynnus.* In the early thirties, Heilner had set up a primitive fishing camp on the island of Bimini, less than a hundred miles east of Miami, and invited Mike and Helen Lerner to join him. (The Lerners would eventually build a house there, which was subsequently turned into the Lerner Marine Laboratory.) In 1935, Lerner's tuna fishing off Wedgeport, Nova Scotia, was successful beyond all expectations; in eight days he caught twenty-three fish that weighed a total of 5,536 pounds. Francesca La Monte was one of the ichthyologists on his expedition, and in her children's book *Giant Fishes of the Open Sea,* she described Lerner's fishing style at Louisburg:

> He was accustomed to using the best equipment, including a well-outfitted cabin cruiser, but here at Louisburg, things were different. He had to do his fishing from a small dory. It was an awkward and dangerous arrangement. His swivel fishing-chair had to be crudely attached to the bottom planking of the boat. This clumsy, low-lying craft pitched and rolled in the rough water, but Mr. Lerner expertly handled his rod and reel and made some very fine catches.

In *Angling and War,* the engrossing story of big-game fishing before and during World War II, Rivkin describes Lerner's reaction to the Nova Scotia giants:

> Thunderstruck by the quality and quantity of the giant tuna, Lerner immediately called on his old friend Kip Farrington to drop everything he was doing and come up to Wedgeport. Upon arriving, Farrington was awed not only by the fishing but also by the spectator sport it had quickly become. Hundreds of locals on dozens of boats would come out to watch the curious scene of a crazy American angler trying to conquer a fish more than thrice his size with little more than kite string. If that wasn't hard enough, the tidal flow was so blistering that boats had to run their engines at full speed ahead just to remain stationary in the current. When a fish was finally hooked, it would often tow the small boats for miles.

La Monte did not mention (but Heilner did) that Lerner's dory was towed into position by motorboat. When he hooked a tuna, a guide in the motorboat cut him adrift, and he had to battle the fish from the dory. To be in a small dory attached to an angry thousand-pound bluefin tuna is not everybody's idea of relaxation, but Lerner loved the thrill of the chase (which was often the piscine equivalent of a "Nantucket Sleigh Ride"), and also the moment of triumph when the exhausted fish was brought to the gaff. Many big-game fishermen did not think that their "sport" was a contest; some "anglers" just wanted to demonstrate that they could bring in a very big fish, even if they had to recalibrate the odds. Heilner tells the story of six men who fought a 792-pound tuna for sixty-two hours, taking turns on the rod, and being fed from an accompanying yacht during the ordeal.

We are used to seeing photographs of modern big-game fishermen strapped into "fighting chairs" on the rear deck of big, expensive fishing boats. Indeed, a major portion of fishing magazines such as *Marlin* and *SaltWater Sportsman* are devoted to ads for these luxury craft, from sixty to seventy feet in length, powered with 1,800-horsepower (or more) twin diesel engines. Most of the ads say "Price on Request," but these boats are for sale for one, two, three million dollars, or even more. Imagine how Michael Lerner or Kip Farrington would feel about a $3-million, 1,800-horsepower fishing boat: they fished from rowboats. Lerner often fought the fish alone, but Farrington had one man to row, a "fishing guide" to locate the fish, and himself (or his wife) to catch it. These newly minted big-game fishermen would sometimes position the double-ended dory amidst a netted school of herring, and drop a bait overboard. If tuna were feeding on the herring that escaped from the nets, they were likely to take the bait, which was also a herring, but with a hook in it. (As Farrington wrote, "If it were not for the presence of these nets, the tuna, in all probability, would not be there, or if they were, it would not be possible to hook them.") The intrepid angler, sitting low on a thwart in the skiff, would then fight a fish that might weigh six hundred pounds. "In August, 1936," wrote Farrington, "Mrs. Farrington fought a 493-pound tuna for ten hours and thirty-five minutes." This was before the IGFA established the rule book for game fishing, and in Nova Scotia, the anglers and guides did not hesitate to shoot the fish, pull them in hand over hand rather than with a rod (a process known as handlining), turn the rod over to others while they took a rest, or even catch a tuna that had been trapped in a net. Farrington "doubted that ten tuna were legitimately caught on the south shore of Nova Scotia after Zane Grey took his 758-pounder."

When people like Farrington, Lerner, and Heilner were fishing for giant bluefins, they incorporated the tunas' fondness for herring into their techniques. "The easiest and most popular procedure," wrote Heilner in 1937, "is to take a station near one of the herring nets set in every inlet by the inshore fishermen. Very often the angler's boat is moored to one of the nets; in fact many guides set herring nets espe-

Socialite sportsman Winston Guest fighting a giant bluefin off Wedgeport, Nova Scotia, in 1935. Back then, all a fisherman needed was a rod, a harness, and someone to row the dory.

cially for the use of their parties, well-anchored and conspicuously buoyed." The guide would then ladle out chum, which consisted of small pieces of herring or mackerel, and even an occasional whole herring, which was supposed to entice the fish into gobbling up the angler's actual bait: a herring with a big hook in it.

In 1937, Lerner and Farrington organized the International Tuna Cup Matches (later known as the Sharp Cup, for the winner's trophy donated by Alton B. Sharp of the Eastern Steamship Line), to take advantage of the spectacular tuna fishing off Wedgeport, Nova Scotia. Yearly contests followed to see which team could catch the most fish and amass the greatest poundage of bluefin tuna over a three-day period, the winners determined by a complex point system. The first year, only teams from the British Empire and the United States competed, but in the 1938 tournament, which had been moved to Liverpool (Nova Scotia), a Cuban team joined the fray, caught twelve fish, and won the cup. The tournament was returned to Wedgeport in 1939, but after teams from the United States, the British Commonwealth, France, Cuba, and Belgium had assembled, on the day fishing was to begin, war was declared in Europe, and the competition was canceled. The event resumed in 1947, and again Cuba won, this time with only six fish, but with a total weight of 3,397 pounds. Capturing a half-ton of fighting Nova Scotia tuna was no easy task; Rivkin describes the process:

> Once a line-shy bluefin was finally induced to take the hooked bait, all parties and the boat would spring into action. As the line fell from the bamboo outrigger, the captain would gun the boat away from the fleet while the angler struck the fish hard. The initial run was often breathtaking in its speed as the hooked giant headed towards open water. As the battle commenced, a crewman would pour water over the reel to avoid overheating as he kept the seated angler positioned front and center to his rapidly departing line. Finally, the fish would settle down and a tug-of-war would begin. If the tackle didn't fail, if the angler was resolute, and if the fish managed to avoid all the entanglements on its way out to sea, a weary but happy crew would eventually make its way back to Wedgeport for a well-earned celebration.

By the early 1950s, sportsmen from Argentina, Chile, Brazil, Mexico, Venezuela, the Netherlands, and "Scandinavia" had signed on. The 1949 Sharp Cup went to the United States, with a total of nineteen fish weighing 7,591 pounds. Although the number of teams was on the increase, the number of caught fish began to drop. The 1953 winner was Mexico, with only one fish, a 728-pounder. The tournament was covered extensively by the media, and all sorts of celebrities converged on Wedgeport during the annual event, which was held in early September. Among those whose photographs appeared in newspaper articles in Canada and abroad were President Franklin D. Roosevelt, Bernard Baruch, Amelia Earhart, Gene Tunney, Kate Smith, and of course the crème de la crème of the big-game fishing fraternity, including Zane Grey, Kip Farrington, Van Campen Heilner, Alfred Glassell, Lee Wulff, and Michael Lerner. In 1956, teams from Britain, Argentina, Germany, Cuba, and the United States caught a grand total of four fish between them; the British Empire won, with one

648-pounder. According to the Wedgeport Museum website, "*Les matchs sont annulés en raison de la rareté du poisson*"; there were simply not enough fish to make a contest out of catching them. The tournament was moved to Cape St. Mary in 1965 and the competition extended from three days to five, but by 1968 only two bluefins were caught by five teams. The event limped along until it became clear that they were running out of fish; the Sharp Cup competition ended quietly in 1970.

In 1949, sportfishermen from Copenhagen established the Scandinavian Tuna Club to sponsor a tuna tournament in the Øresund, the narrow strait between the eastern Danish island of Sjælland and Malmö, Sweden. Tournaments were held every year until the *thunfisk* disappeared in the 1960s. The club still exists, but according to its website (www.tunaclub.dk), the absence of tuna means that an award is given every year to the member catching the largest fish, of any species, anywhere in the world (MacKenzie and Myers 2007).

The Canada Tuna Cup now takes place at North Lake, Prince Edward Island, which bills itself as the Tuna Capital of the World. (The women's world record bluefin tuna, a 1,170-pounder, was caught there in 1978 by Dr. Colette Perras of Montreal.) Anyone who can afford the $150 registration fee can enter, and there are four anglers to a boat, each boat representing a province of Canada. Like the old Sharp Cup, points are awarded for the total poundage taken, with bonus points for the boat that lands the largest number of fish, as well as for the largest single fish. The contest is run under IGFA rules that govern the line, reels, hooks, and leaders. If a fish is harpooned, lanced, shot, or otherwise mutilated, it will be disqualified from the competition. The prizes are the cup itself, plaques, and a $300 award for the boat captain whose crew catches the largest fish.

Ernest Hemingway never wrote a nonfiction book about fishing, but he did write the introduction to Kip Farrington's *Atlantic Game Fishing,* in which he bemoaned the lack of "sportsmanship" in the activity and criticized those anglers who spend vast amounts of money in pursuit of world records. He wrote, "Seriously though, it is a grand sport, but it needs some simple and decent rules if it is to continue competitively. It will be all right with me. I would like to go back to fishing for fun and take a day off and go snapper fishing over by the concrete ship." Hemingway truly admired the great billfishes, and his experiences with marlin and swordfish were woven into his novels *The Old Man and the Sea* (1952) and *Islands in the Stream* (1970). According to Farrington, who also wrote *Fishing with Hemingway and Glassell,* Hemingway "was one of three men who turned down the honor of fishing with the U.S. Tuna Team after the international matches were begun in 1937. He never fished up there at all." However, from his boat *Pilar,* off Bimini in 1935, Hemingway landed two bluefins weighing 310 and 381 pounds. George Reiger wrote that "Hemingway evolved a theory for the successful capture of an unmutilated tuna. From the instant the fish took the bait, he argued, you simply had to fight the animal like there was no tomorrow. He believed that

Ernest Hemingway, proud conqueror of a large tuna

once any fish 'understood' that it was dealing with a superior force, then the job of landing it became half as hard."

In his forward to *Fishing the Atlantic,* Philip Wylie wrote that the author, S. Kip Farrington, was

> a super-angler, endowed with an oak back for the big ones and a violinist's fingers for light tackle; he is probably the foremost journalist of the ocean seas and "all that in them is." He is a journalist whose appetite and memory for fact is as voracious as any fish's appetite. Sometimes, listening to Kip talk, I think he must have a pipeline to the sea itself—for he appears to know immediately, as if by telepathy, the length, weight, line, bait, boat, and mate for every whopper taken . . . Kip is a serious fisherman.

In his fishing books (he also wrote about railroading), Farrington demonstrates his almost obsessive attention to detail—perhaps supplying more than anyone who wasn't there wants to know. He even wrote a book that he called *The Trail of the Sharp Cup: The Story of the 5th Oldest Trophy in International Sports.** Of the 1947 tournament, he wrote:

> During the first day's competition Major C. H. Frisby, V.C., the great British angler from London, took a 681-pounder, and Louis Mowbray, Bermuda's leading angler, came through with a 614-pounder. This fine sportsman was captain of the winning British Empire team in 1937, and is also a great ichthyologist. Thorvald Sanchez caught a 696-pounder for the Cuban team and also had another strike while drifting.

With his wife Chisie, Farrington fished almost everywhere—and wrote about it at length in his numerous books. In the Atlantic, he searched for (and usually found) tuna off Nova Scotia, Maine, and

*The other four trophies are the Davis Cup (tennis), the Walker Cup (golf), the Stanley Cup (hockey), and the America's Cup (sailing). That none of the other competitions involved killing animals did not seem to have occurred to Farrington, who obviously regarded big-game fishing as a sport comparable to tennis or golf. He wrote: "The nations competing for the Sharp Cup have directly challenged the fish, not each other. Trying to beat the fish has led to a much more friendly match than if they were hitting a golf or tennis ball against each other."

Massachusetts; in the Pacific, he went to Chile, Peru, Panama, Mexico, Hawaii, New Zealand, and Australia, mostly in pursuit of marlins and swordfish. Occasionally he thinks about tuna, however, and in the middle of a discussion about the black marlin of Cabo Blanco, Peru, we find this anomalous note:

> Big-eye tuna are marvelous fighters and very beautiful. This is the fish we had been thinking of as the yellowfin, but now we are told it has different livers and that's how it is identified, even though it has a big eye. John Olin tells me that one he lost was one of the toughest tuna he's ever fought. He really slayed them with half a dozen in a week taken in Nova Scotia, all weighing over 600 pounds.

There are, of course, all kinds of tuna in the Pacific, and when Farrington gets to Hawaii, he says it "was the Allison tuna that primarily attracted me to Honolulu, as it is by all odds, the hardest fighting member of the tuna family besides being one of the most beautiful fish that swims . . . The Allison tuna is called *ahi* in the islands, which means "ball of fire," and I consider this a perfectly appropriate name." (*Ahi* actually means "yellowfin tuna" in Hawaiian, and "Allison tuna" is just an old name for *Thunnus albacares,* said to be named by Lewis Mowbray for James Allison, founder of the Miami Aquarium.) Farrington provides the details of catching them:

> The presence of the Allison tuna is easily discovered, for they are wont to feed and play on the surface . . . Birds often betray the presence of the fish but a fast boat is required to catch up with the school. Hawaiian anglers use different varieties of their native feather baits made of bright-colored chicken feathers and are particular as to the kind of feathers and type of head of which the jig is made. On some days these tuna are also taken on the traditional Japanese feather jig . . . Allison tuna can be taken on 24-thread line with a 16- to 20-ounce rod but I believe that 39-thread is more practical when used with a 22- or 23-ounce rod, unless the angler has a great deal of time to spend fighting the fish. Feather jigs can be procured in Honolulu and a light grade of stainless cable leader is generally used with a 9/0, 10/0 or 11/0 hook.

Chisie Farrington in Nova Scotia in 1946, with a 678-pound bluefin, the Ladies' World Record

Selwyn Kip Farrington Jr. will be remembered for his voluminous descriptions of big-game fish and fishing, but he leaves another legacy, the tuna tournament, resuscitated today in a somewhat modified fashion. In 1958, a group of local fishermen organized the Rhode Island Tuna Tournament, to celebrate what they believed to be the "Tuna Capital of the World." Captain Al Anderson, now based in Galilee (as is the tournament), wrote: "In the late 1960s and 1970s, the RITT grew to be the largest tournament of its kind in the world. Contestants annually numbered over 300—some years over 400—and boats entered from an eight-state area. With the return of the fleet in the late afternoon, thousands of spectators would jam the bulkheads and docks hoping for a glimpse of a giant bluefin." In 1965 Jim Cullen brought in an 804-pounder, and in 1977, Tom Crafford broke that record with a fish that came within ten pounds of a thousand. Crafford's 990-pounder is the largest fish ever caught in the Rhode Island tournament. The RITT continues today, but with a 50 percent drop in participation from its heyday due to a reduction in the numbers of tuna to be found off the Rhode Island coast. Increasingly, emphasis is being placed on "tag and release" fishing, both in the tournament and in sportfishing for tuna in general. (Always ahead of the curve, Anderson has written a book called *Game Fish Tag & Release.*)

Nowadays, there are tournaments in almost every American city that is situated on the ocean, with fishermen eager to match their skills against the wily tuna (and one another). For the most part, these tournaments are run "Calcutta" style, which means that every fisherman (or every boat) contributes to a pool of entry fees, and the total is dispensed as daily jackpots and prizes. One of the largest is not exactly in America, but at the tip of the Baja California peninsula, at Cabo San Lucas. In 2005, the "Los Cabos Tuna Jackpot Tournament" divided $447,600 between 135 boats in overall and daily jackpots. The winner got $241,800 for a 92.4-pound yellowfin. At the Sturdivant Island Tuna Tournament in Maine, some of the money goes to charity (they have given away $175,000 over seven recent years), but there are cash prizes totaling $12,500, and a $25,000 incentive to the fisherman who lands the Maine State Rod and Reel record bluefin. The annual "Make-A-Wish Tuna Challenge" donates *all* of its proceeds to the Make-A-Wish Foundation; in 2006 they donated $300,000 to grant the wishes of children in the San Diego area. There are tuna tournaments in Maine, California, Virginia, Oregon, Massachusetts, Maryland, New York, New Jersey, and elsewhere. There are any number of fishing tournaments that are not dedicated to a specific kind of fish, but these are all *tuna* tournaments.

Catching tuna is relatively easy as fishing goes; because they are voracious feeders, tuna tend to strike at almost anything that is dragged in front of them. Here's the entry on fishing for tuna from A. J. McClane's 1965 *Standard Fishing Encyclopedia:*

> Bluefin tuna are caught chumming at anchor, chumming while drifting, and trolling . . . Menhaden "soup" and cut mossbunkers are most effective for attracting and holding tuna . . . In the chumming method, the angler ordinarily strikes his tuna by hand. The rod is placed in a holder in the fighting chair, and the reel is left in free spool with the click engaged. Generally, two lines are fished, one with a bait about thirty feet below the surface and the other on the bottom. When a tuna is felt, the angler strikes his fish with the line, then takes his place in the fighting chair. The reel brake is engaged to the correct drag, while the crew hauls anchor.
>
> Trolling is employed when tuna are plentiful. Schools are sighted in the clear Bahamian waters, for example, while migrating north. They may also be visually located in the Maritimes as they "fin" or "push water." For trolling, live or dead mackerel, mullet, and squid are commonly used baits. In the trolling method, the angler does not hold the line but strikes with the rod. The angler holds the rod, while seated in the fighting chair. The bait is dropped back a considerable distance (approximately one hundred yards) so the boat wake will not spook the fish.

In New England waters in recent years, another method of tuna fishing has been successfully employed, usually at night. A variation on the theme of chumming, where a mash of ground fish and blood is trailed behind the boat, "chunking" consists of tossing over chunks of freshly cut baitfish or sometimes live fish or eels, some of which have hooks embedded in them. Bluefin and yellowfin tuna seem to be able to identify bait objects by the rate at which they fall through the water, so wily fishermen pack the baits with styrofoam "peanuts" to compensate for the weight of the hook. In a 2007 article on chunking for yellowfins, Peter Barrett wrote, "Despite a downward tick in yellowfin numbers overall, smart anglers still score a good catch on the night shift—especially if they take advantage of the more innovative chunking techniques and the superlative advances in braided lines and smaller, powerful reels."

About a century ago, in places like Nova Scotia and Seabright, New Jersey, as the first tuna fishermen struggled to land a five-hundred-pound fish, they believed they had invented a new "sport."

As fishermen-writers like Grey and Farrington conveyed the thrill of the chase to amateurs and even told them where and how to find the fish, recreational tuna fishing was born. Big, powerful, and capable of putting up a hell of a fight, the bluefin tuna was on the way to becoming a game-fishing icon. The highly publicized Sharp Cup attracted international "sportsmen" (for which read: millionaires), and while not everyone could afford that kind of time and equipment, many fisherman tried to catch the mighty bluefin on a reduced budget. The Sharp Cup served to whet the appetites of sport fishermen up and down the East Coast of North America. But unknown (and unforeseen) factors, such as a shortage of fish, changes in water temperature, oceanic currents, and availability of food, led to a decline in the population of big Atlantic bluefins, and recreational fishing declined accordingly. Still, fishermen occasionally hauled in a giant.

Early on, the commercial value of bluefin was about a nickel a pound, so sport fishing was almost exclusively recreational—that is, if fighting a half-ton fish for several hours is one's idea of recreation. Trophy tuna that were not kept for personal display or consumption were sold to pet food producers. The red-meat tunas are not inedible, but until the Japanese popularization of raw fish, most people simply didn't consider bluefin meat particularly desirable. However, the 2006 Massachusetts Division of Marine Fisheries guide to the bluefin suggests that "the bright red flesh is excellent on the grill after marinating in your favorite concoction . . . Bluefin is even better eaten raw, using wasabi and soy to garnish and spice the ancient ritual." The "horse mackerel," once considered a junk fish, suitable only for cats and dogs, has not only managed to implant Japanese cuisine into the American diet, it has also blurred the distinctions between commercial and recreational fishing. Now any fishermen, even those just out for a day's sport, can sell their catch on the dock to a waiting Japanese buyer. The bluefins—especially those classified as "giants"—suddenly represented big money to commercial as well as to the sport fishermen. Most big-game anglers invest significant amounts of money in travel, gear, boats, and other paraphernalia used in the pursuit of the giant black or blue marlins, but fishing for big bluefins off, say, Massachusetts, is probably the only recreational fishery in which one might expect to be compensated handsomely for a day's catch. (The

eries came to New England ports such as Gloucester and Provincetown. These buyers were particularly interested in fish that were more or less undamaged, and therefore paid a higher price for those that were harpooned or caught on rod and reel than they did for purse-seined fish, because those caught in a net would thrash around and beat themselves to death, which ruined the quality of the meat. Anderson notes that "muscle contraction not only produces heat, which is retained and carried by the blood to all parts of the body, but waste products of activity as well. Combined over time they work to reduce the freshness and hence the flavor of the muscle tissue."*

With the exception of the International Game Fish Association in Florida, recreational fishers don't have a lobby, but commercial fishermen are extremely well represented in the corridors of Washington, where various government organizations—such as the National Marine Fisheries Service (NMFS)—decide who can catch how many fish and where they can catch them. The NMFS regs for recreational fishers are so complicated that it is easy to imagine these fishermen throwing up their hands and leaving the tuna to the commercial fishers, which might be exactly what ICCAT is lobbying for. In a 1993 article, Martin Garrell wrote that Joe McBride, skipper of the Montauk charter boat *My Mate,* decided that he needed a "tuna lawyer" to figure out which tuna he could take for his parties during the runs between July and October. For example, a party of six was entitled to two tuna per angler in each of the two school sizes. A small-school tuna was a fish between 26 and 45 inches, weighing about 14 to 66 pounds; a large-school tuna measured more than 45 inches and weighed between 66 and 135 pounds. The boat was also permitted to retain one medium tuna—a small medium weighed between 135 and 235 pounds, a large medium between 235 and 310 pounds—plus

*In the introduction to *The Atlantic Bluefin Tuna: Yesterday, Today and Tomorrow* (1990), Anderson says, "as they age and put on weight, they appear to further develop their blood circulatory mechanism that enables them to maintain body temperatures above the cold environment." But otherwise, I can find little discussion in the book of the warm-bloodedness of tuna, even though the early studies on that subject had been published long before his book appeared. Beginning with the studies by Frank Carey in the 1970s, Barbara Block, Richard Brill, Kathryn Dickson, and any number of other scientists published detailed descriptions of the elevated body temperature of tuna and other fishes. This is a curious oversight on Anderson's part, because the muscle tissue of bluefin tuna is *defined* by the endothermy of the fish, and while "thrashing" might produce excess heat, the fish's complex circulatory system produces more than a little heat in the first place.

Captain Gary Cannell (far left), aboard *Tuna Hunter* out of Gloucester, Massachusetts, hauls a 920-pound "giant" bluefin. Notice the horizontal keels at the base of the tail.

eleven school fish. For recreational fishers, the 2006–07 NOAA classifications for bluefin tuna were as follows:

Young School: 27″ fork length; 14 lbs. round weight (head and fins intact)
School: 27″–47″ fork length; 20–35 lbs. round weight
Large School: 47″–59″; 35–44 lbs.; 66–135 lbs. round weight
Small Medium: 59″–73″; 135–235 lbs. round weight
Large Medium: 73″–81″; 235–310 lbs. round weight
Giant: 81″ +; 310+ lbs. round weight

If they can decipher the regulations, anglers still go after bluefin tuna in the North Atlantic, and there are any number of boats that will take them out. But despite the party-boat captains' declarations that there are tuna all over the place, and ICCAT's belief that the stock is at maximum sustainable yield, numbers of bluefin tuna are at an all-time low. It will not help the diminished population if fish of any size are removed, either in large numbers by purse-seiners, as individuals by harpooners, or one at a time by weekend anglers. Even though it appears that the multitude of nontournament tuna fishermen in New England might have an effect on the tuna populations, they probably don't make much of a dent. The denting, as it were, comes from commercial fishermen, who, using every conceivable method to catch tuna in large numbers for the Japanese market, have brought low the populations of northern and southern bluefin tuna. Both are now considered endangered species.

COMMERCIAL TUNA FISHERIES

In the Sicilian *mattanza,* the tuna are hauled to the surface in the last of the net rooms and then killed.

THE ANCIENT GREEKS and Romans fished the Mediterranean Sea for albacore, skipjack, and of course the great bluefin. In *Halieutica,* his hexameter poem on fishing, Oppian, the second-century A.D. Greek poet, described fishing for tunnies:

> The pursuit of the Tunny is commonly designated as "big fishing" by the people of Italy and Sicily, and the places in which they are in the habit of storing their huge nets and other fishing gear are called "big-fishing tackle stores," for they wish henceforward to segregate the huge Tunny into the class of "big fishes." And I learn that the Celts and the people of Massalia and all those in Liguria catch Tunny with hooks; but these must be made of iron and of great size and stout. So much then for Tunnies in addition to what I have already said earlier on.

Fishing with traps is perhaps the oldest known type of "commercial" fishing—that is, catching large numbers of fish to feed more than the fisherman and his family. The Phoenicians, who lived along the shores of the eastern Mediterranean some thirty centuries ago, are believed to have built traps that consisted of palm-tree branches that were stuck in the sand to guide the tuna into shallow water where they were beaten to death with clubs or stabbed with spears (Sara 1980). Almost completely enclosed by Europe and Asia Minor to the north and Africa to the south, the Mediterranean is open to the North Atlantic through the Strait of Gibraltar on the west, and to the Black Sea by the narrow passage known as the Bosporus. The average depth of the Mediterranean is 4,900 feet, but there are extensive areas of the sea floor that are below 6,800 feet, and the deepest recorded point is 17,275 feet (about 3.27 miles) in the Hellenic Trough, west of the Peloponnesus. It is 2,400 miles from the Strait of Gibraltar to the shore of Lebanon, and the longest north–south distance (from Trieste to the shore of Libya) is about 1,000 miles. The Mediterranean's 1,450,000 square miles constitute the largest fish trap in the world; it is easy for the fish to get in, but very hard for them to get out.

For thousands of years, fish trappers have taken advantage of the sea's enclosed nature. Improved and modified over time, the individual traps consist of a complex of nets that are held up by means of floats and anchored to the bottom with weights. They form a series of one-way passages that funnel tuna into a final chamber, where they can be killed. Raimundo Sara's summary history of trap fishing in the Mediterranean identifies those areas around Spain, the Balearic Islands, Corsica, Sardinia, Italy, Sicily, the Adriatic, Greece, and Turkey where trap fishing for tuna has been practiced for thousands of years. The surface circulation of the Mediterranean essentially runs counterclockwise, moving eastward along the coast of North Africa, passing through the narrow Strait of Sicily, and circling back westward in northern waters toward the Strait of Gibraltar and the open Atlantic. Because migrating tuna tend to swim close to the surface, they follow this circulation pattern, until their journey is interrupted by fishermen who lay traps for them. (By the twentieth century, fishermen would catch the fish off North Africa as well, but that is part of the tuna-farming story, which comes later.)

In her book *Mattanza* (an Italian word that means "the killing"), Theresa Maggio chronicles the long history of *tonnara* (tuna fishing), from the time of the Carthaginians and Phoenicians five centuries before Christ, right up to present-day Sicily. "The bluefin were to ancient Mediterranean peoples what the buffalo were to the American Plains Indian," Maggio wrote, "a yearly miracle, a reliable source of protein from a giant animal they revered, one that passed in such numbers that the cooperation of an entire tribe was needed to kill them, and preserve their meat. Around the Mediterranean the migrating bluefin was a staple food for entire civilizations." Oppian's *Halieutica* described the second-century *tonnara:*

> Dropped in the water are nets arranged like a city. There are rooms and gates and deep tunnels and atria and courtyards. The tuna arrive in great haste, drawn together like a phalanx of men who march in rank: there are the young, the old, the adults. And they swim, innumerable, inside the nets and the movement is stopped only when there is no more room for new arrivals; then the net is pulled up and a rich haul of excellent tuna is made.

The *mattanza* of the village of Favignana, located on one of the Egadi Islands off the western tip of Sicily, is the most famous of recent *tonnare,* and one of only two remaining. A complex arrangement of gigantic nets is deployed along the tuna's expected route, and the fishermen wait for the fish to swim into them. The nets are hung from ropes, "arranged like a city" as they were in Oppian's day, and stretched to the bottom by massive anchors. "The trap is oblong," wrote Maggio, "except for a widening at the shoulders that makes it look like a widening coffin. It is divided into seven rooms by net walls with gates in them . . . It is fifty meters squared." The last "room" the tuna will ever see is the *camera della morte*—the chamber of death—the only room with a net bottom, which the *tonnaroti* raise by hand from seventy-five-foot-long open boats.

In *The Silent World,* Jacques Cousteau describes his experience in a tuna net in Tunisia, where the death chamber is called a *corpo:*

> The noble fish, weighing up to four hundred pounds apiece, swam around and around counter-clockwise, according to their habit. In

Around 1967, the Spanish surrealist Salvador Dalí painted *Tuna Fishing.* When Dalí lived in Cadaqués, a fishing village on the Costa Brava, he had ample opportunity to observe the net fishery, as depicted in the background, where men are stabbing fish and hauling up the nets, and in the left foreground where a man stabs a bluefin tuna.

contrast to their might, the net wall looked like a spider web that would rend before their charge, but they did not challenge it. Above the surface, the Arabs were shrinking the walls of the *corpo,* and the rising floor came into view . . . The death chamber was reduced to a third of its size. The atmosphere grew excited, frantic. The herd swam restlessly faster, but still in formation. Their eyes passed us with almost human expressions of fright.

My final dive came just before the boatmen tied off the *corpo* to begin the killing. Never have I beheld a sight like the death cell in the last moments. In a space comparable to a large living room tunas and bonitos drove madly in all directions. It took all my will power to stay down and hold the camera into the maddened shuttle of fish. With the seeming momentum of locomotives, the tuna drove at me, head-on, obliquely and crosswise. It was out of the question for me to dodge them. Frightened out of sense of time, I

At Favignana, the struggling fish are gaffed as the nets are raised.

> heard the reel run out and surfaced amidst the thrashing bodies. There was not a mark on my body. Even while running amok the giant fish had avoided me by inches, merely massaging me with backwash when they sped past.

The nets are raised and the struggling fish are gaffed as they come to the surface. Cousteau: "The fishermen struck at the surfaced swarm with large gaffs. The sea turned red. It took five or six men whacking gaffs into a single tuna to draw it out, flapping and bending like a gross mechanical toy. The boats rocked with convulsive bleeding

Seeing a bluefin tuna gaffed with spears is like watching a Thoroughbred racehorse being hacked to death with an ax.

mounds of tuna and bonitos." In 1957, at the height of the fishery, 7,480 tuna were killed in the Favignana *tonnara.*

Films of these sleek, graceful creatures being gaffed are heartbreaking. One moment they are on what Cousteau called their "honeymoon," and the next they are thrashing in a panicked melee as heavy steel hooks are smashed into their bodies, and they are then hauled ignominiously from the only element they have ever known into the one where they will die. Bluefins are among the most powerful and beautiful of the oceans' top predators, and seeing them gaffed is like watching a Thoroughbred racehorse being hacked to death with an ax. Tuna of all kinds are among the world's most popular food fishes, and people are no more interested in how they die than they are in visiting a terrestrial slaughterhouse. It might change our attitude toward tuna fish sandwiches or seventy-five-dollar pieces of sushi if we realized that tuna are wild animals that happen to live in the ocean and therefore cannot be viewed like herds of zebras or wildebeest, or packs of wolves. They are the oceans' nobility, described by Carl Safina as "half a ton of laminated muscle rocketing through the sea as fast as you drive your automobile . . . among the largest and most magnificent of animals."

Tuna fishing was once one of Sicily's most important and profitable industries. Until the first decades of the twentieth century, coconut-fiber nets over a mile in length were deployed by the hundreds, but diminishing numbers of tuna and market laws that have made this technique more capital-intensive have left only about ten *tonnare* (tuna fisheries) in the Mediterranean; only Bonagia and Favignana remain in Sicily. What was once a source of pride (not to mention the primary source of income) for entire communities has now turned into a tourist attraction, providing a few makeshift jobs in a social context poor in prospects, and kept alive by the obstinate will of the remaining *tonnaroti.* The canneries in Sicily are closed; almost all of the tuna caught in Favignana are shipped to Japan. There is a rumor in Favignana that the gaffing will be eliminated and the fishermen will simply wait for the tuna to die in the nets, because the gaffs make too many holes in the flesh. "Once," wrote Theresa Maggio, "the tuna snares thrived in Algeria, Corsica, Tunisia, Malta, Dalmatia, and Turkey. In Portugal they were called *armações;* in Spain, *almandrabas;* in France, *madragues.* The cause of abandon: insufficient fish to make a profit. Once there were tonnaras all over Sicily . . . Gone, all gone."

Other than the trap fishers, the earliest commercial fishers probably speared their prey, or worked from a canoe or a dory with a baited hook on the end of a line. A variation of this method is employed today in tropical Pacific and Atlantic waters to catch the smaller tuna species, such as albacore and skipjack. Dories are not used, but the fishermen stand at the rail of a fishing boat, attracting schools of tuna by chumming, and holding bamboo poles with unbarbed, unbaited hooks. The tuna are yanked out of the water by the large crew of fishermen, thrown over their heads onto the deck, killed, and stored belowdecks in freezers. Some technically advanced countries have replaced the fishermen with "jigging machines" that perform the same function, thus cutting down on labor costs.

When the Japanese developed monofilament fibers that could be used in open-ocean drift netting in the mid-1970s, they introduced the most durable and destructive method of fishing ever devised.

Large-scale, high-seas drift nets were first used in the North Pacific by fleets from Japan, Taiwan, and South Korea. Free from a connection with any boat, drift nets are set with floats at the top and weights at the bottom, so that they move passively in the water and trap anything that swims into them. Traditionally, these were small nets used in coastal waters to catch dense schooling fish, like herring, but with the introduction of light synthetic netting, drift-net fishing underwent a major change. The nets could now be used on the open seas where they are very effective at catching wide-ranging species such as tuna and squid. Barely visible in the water, these nets are also devastatingly effective at catching all other wildlife in their path. The boats set as much as forty miles of drift net each, totaling some *forty thousand miles* of drift net every night—enough to circle the earth one and a half times. Because of the huge bycatch of marine wildlife in these nets they have been labeled "walls of death"; to date, hundreds of thousands of whales, dolphins, seabirds, sea turtles, sharks, and other nontarget species have been killed "incidentally."

Early in the fishery, big tuna were caught on hook and line, with chum being thrown in to excite the voracious fish into biting at anything and everything in the water. Fishermen stationed along the rails with heavy rods and unbaited, barbless hooks yanked the heavy fish over their shoulders, hoping to have them land on deck—which they often didn't. When Jordan and Evermann wrote *American Food and Game Fishes* in 1902, the bluefin wasn't considered a big-game fish in the North Atlantic. They reported: "The fishermen about the Gulf of Saint Lawrence sometimes take the horse mackerel by means of steel hooks tied to solid lines and baited with herring . . . The fishing is quite exciting, although tiresome and requiring a good deal of skill, as in the efforts of the fish to escape they pull with such violence as to endanger the lives of the fishermen by dragging them overboard."

Some time later, when American Museum of Natural History ichthyologist Francesca La Monte was writing *Giant Fishes of the Open Sea* (a book for children), she observed:

> The giant tuna fishes are caught in several ways by commercial fishing companies. One way is with nets that are set sometimes over a stretch of two miles. Longlines or set-lines, which are lines contain-

> ing many baited hooks provide another method. A third method is fishing with a rod and reel from the deck platform of a tuna clipper. The deck platform runs along the side of the big boat. As the boat is guided into a mass of tunas, the fishermen stand in a long line on the platform, rhythmically baiting, casting and hooking fishes. When a fish is hooked, the fisherman swings the heavy fish up and over his head so that it falls free of the hook and lands on the deck behind him.

Sportsmen who seek to do battle with giant bluefin arm themselves with a heavy rod, a sturdy reel wound with line that will not snap when a five-hundred-pound fish hits the lure, and a strong back. Rod-and-reel sportfishing is one man versus one fish. Commercial pole-and-line fishing is practiced along the same lines, but there are more men working to catch more fish. Although it is labor-intensive, pole-and-line fishing is still a popular way of catching large tuna.* The modern version developed during the early twentieth century, when the Japanese switched over to larger pole-and-line vessels capable of traveling to any ocean where tuna occurred in fishable quantities. The vessels, which carry live bait in tanks of circulating seawater, can freeze their catches and stay at sea for three or four months. The greatest growth in pole-and-line fishing occurred in Southern California in response to a rising demand for tuna following the introduction of canning in the early twentieth century. It was in this fishery that the "tuna clipper" originated: a pole-and-line vessel capable of packing up to six hundred tons of frozen tuna, carrying large quantities of live bait, and staying at sea for many months. "Their great weight and strength," wrote Robert Morgan (1955) of the tuna, "often make landing by one man with a line impossible . . . and therefore, each hook is operated by two and sometimes three men." There are now regions where a similar technique is employed, but the

*Pole-and-line tuna fishing can show up in the oddest places. In *Pinocchio,* the 1940 Walt Disney cartoon, when the puppeteer Gepetto is trapped inside Monstro the whale, he tries to catch the fish that the whale (an evil amalgamation of sperm whale and baleen whale, with plenty of teeth and ventral pleats) has chased down. As the fish cascade into the whale's gullet, which is waterless so Gepetto (and Figaro the kitten) can breathe, we see that they are supposed to be tuna, typical shape, finlets and all. Gepetto lands them by the deckful, tossing them over his shoulder one at a time, until he is stopped by the realization that one of the creatures he has just hauled in is Pinocchio.

hooks, lines, and jigs are mechanized, and there are no fishermen, just a battery of rods, bobbing and yanking tuna out of the water and onto the deck. Surprisingly, in terms of tonnage of tuna captured, pole-and-line fishing, like longlining, takes about 14 percent of the world catch.

We knew that there were large tunas in the Northern Hemisphere and in the Southern, but only recently have fishermen and consumers cared very much about the difference—if indeed there was one. Ichthyologists, of course, worried about such matters, and in 1926, Messrs. Jordan and Evermann, two of the best-known fish biologists of their time, subdivided the genus *Thunnus* into eight separate species, distinguishable—to them, anyway—by minute variations in the length and placement of the dorsal, anal, and pectoral fins, and by the areas in which they were found. Of these "species" only *T. thynnus, T. maccoyii,* and *T. orientalis* are currently recognized: the northern and southern bluefins and the Pacific bluefin. In his 1951 *Fish and Fisheries of Australia,* however, T. C. Roughley added *Kishinoella tonggol,* which he identified as the northern bluefin tuna—north of Australia, that is. It is therefore the northern southern bluefin tuna.*

In Roughley's Australia, the citizenry didn't think much of tuna as a food fish. He observed:

> Practically all the tuna are canned. Canned tuna is regarded as a luxury fish in the United States. In the fresh state it is not favoured, the oil being strong and unpleasant to the palate, the fish is therefore always cooked before it is canned in order to remove some of the natural oil which is replaced in the can by a bland vegetable oil, such as cottonseed, peanut, or soya-bean oil. Treated in this manner, the tuna develops a consistency and flavour not unlike chicken,

*There actually is a species of tuna known in Indonesia as *tongol;* sometimes called the longtail tuna; its scientific name is *Thunnus tonggol.* It is typically tuna-shaped, but not as hefty as the large bluefins. Found in the western Pacific and the Indian Ocean, its tender, white meat is usually canned by Thai and Indonesian concerns, for sale as canned white-meat tuna. Maximum length is 52 inches, and the heaviest one ever caught on rod and reel weighed 79 pounds, but Thai fishers usually land fish in the 3–5 pound range. Whole Foods' "365 brand" canned tuna is usually tongol, which is described thus: "The meat is often very firm and light in color though usually darker and oilier than albacore. Many countries allow tongol to be labeled as 'white' tuna but the U.S. does not." (In the United States, only albacore can be labeled "white-meat tuna.")

and it is frequently referred to in the United States as "chicken of the sea."

Because tuna was reaching new levels of consumer popularity in the United States and Europe, the International Commission for the Conservation of Atlantic Tunas was established in 1969, at a conference in Rio de Janeiro. Some thirty species of large pelagic fishes are of direct concern to ICCAT: Atlantic bluefin, skipjack, yellowfin tuna, albacore, and bigeye tuna; swordfish; billfishes such as white marlin, blue marlin, sailfish, and spearfish; mackerels such as spotted Spanish mackerel and king mackerel; and small tunas like black skipjack, frigate tuna, and Atlantic bonito. According to its website, "ICCAT is the only fisheries organization that can undertake the range of work required for the study and management of tunas and tuna-like fishes in the Atlantic. Such studies include research on biometry, ecology, and oceanography, with a principal focus on the effects of fishing on stock abundance. The commission's work requires the collection and analysis of statistical information relative to current conditions and trends of the fishery resources in the convention area. The commission also undertakes work in the compilation of data for other fish species that are caught during tuna fishing ('bycatch,' principally sharks) in the convention area, and which are not investigated by another international fishery organization." Probably the species to which ICCAT has devoted the most time and energy is the Atlantic bluefin tuna.

Longlining, one of the most ruthlessly efficient of all fishing techniques, consists of a single line that may be one hundred miles long (the distance from New York to Philadelphia), supported along its length by floats. The lines are hung with thousands of hooks baited with live or frozen baitfish, and deployed in an area where a particular species is being sought. Hanging from the longlines are as many as three thousand hooks on "branch lines" that are dangled in the water and can be adjusted to fish at depths ranging from 180 to 500 feet. It can take up to eight hours to set the net and twelve to retrieve it. The gear is passive, in that it captures whatever fish happen to take the bait. The target species takes the bait, but so does every other kind of fish in the area. If tuna are the object of the fishery, sharks, billfishes, and any other fishes that are caught are often discarded, it being too much trouble to separate them. In 1995, an estimated 1,500 shy albatrosses *(Diomedea cauta)* out of a breeding population of 8,000 were caught in longlines and killed. Probably the most heinous use of longlines is in sub-Antarctic waters, where fishers for the Patagonian toothfish (known as "Chilean sea bass" in restaurants) are scouring the waters for their target species, but killing hundreds of thousands of other fishes, whales, seals, dolphins, and as many as 150,000 seabirds annually. Longlining accounts for about 30 percent of the world catch, including most of the billfishes taken commercially. The largest longline fleets are those of Japan, followed by those of Taiwan, China, and South Korea.

In the western North Atlantic, the purse seine and the longline changed the institution so much that by 1964, the combined landings from these two fisheries had skyrocketed to twenty thousand tons. As it became apparent that some sort of conservation measures were necessary to protect the great fish, ICCAT was born. For control purposes, the commission considered the Atlantic tuna as two stocks, separated by the 45°W meridian in the North Atlantic, with a dogleg around the bulge of South America to 25°W in the South Atlantic. Until the 1980s, most commercial tuna fishermen avoided taking the really big fish—bluefins can reach 1,500 pounds—because they were just too difficult to handle, but when the Japanese sushi market opened its bottomless maw, the larger fish became a specific target. As the prices rose, it became almost impossible to convince people *not* to fish for tuna. The giant bluefins were under siege. In a 2005 article about the sustainability of the western Atlantic tuna fishery, Clay Porch wrote:

> Like other tunas, western bluefin have a broad geographic distribution, but they also have a tendency to concentrate in well known spawning and feeding grounds where they are very conspicuous and easy to catch. The bluefins' plight is exacerbated by its popularity as a game fish and the exorbitant prices paid by the Japanese sashimi market; spotter planes and other expensive services remain cost effective even at very low densities. Moreover, bluefin are often taken as a bycatch of the vast longline fleets spread across the Atlantic.

As detailed in Carl Safina's *Song for the Blue Ocean,* the New England tuna fishers believed that the scientists were wrong in their estimates of the population, and that the stock was in fine shape. A Maine spotter pilot named Tim Voorheis showed Safina a picture of a school of bluefins taken from the air, and tells him, "You can count one thousand fish on the surface . . . I could see schools of medium-sized fish like this for fifteen miles around me. And they probably stretched all the way to Canada . . . Otherwise you would never know, in God's world, that there were that many fish out there." And most tuna fishermen believe the same thing. The New England Fishery Management Council was created in 1976 to enable American tuna fishermen to compete with foreign fleets—particularly the Japanese—that they believed were catching all their tuna. The New Englanders lobbied (successfully) for the passage of the Fisheries Conservation and Management Act, which extended U.S. control from twelve to two hundred miles offshore, guaranteeing the dominance of American fishers. The act was ostensibly designed to protect the fish stocks, but there was so little oversight that instead of protection, the result was massive overfishing. As Safina wrote, "Putting billions of dollars worth of fish in front of the industry and asking them to police themselves was not terribly realistic."

The conflicts between regulators and fishermen were clearly delineated in a 1994 National Audubon Society film, *Hunt for the Giant Bluefin.* Safina, who founded the Audubon Society's Living Oceans program in 1990, is one of the bluefin's greatest champions, and is presented here as the voice of reason, trying to find a compromise position between the two extremes: saving the fish or saving the fishermen. He recognizes that New England commercial fishermen are creatures of the lucrative Japanese market, and that these men will not be talked out of their bonanza by deskbound government bureaucrats "who have never even seen a fish." NMFS spokesmen acknowledge the precipitous decline of the tuna populations, but the fishermen claim that they are seeing (and catching) just as many fish as they always did. One fishermen says, "Every year, we're seeing more fish than the previous year. If we're killing them all, where are these fish coming from?" NMFS scientist Brad Brown says its because the fishermen are going where they know (or hope) the fish can be found, while the sci-

In Mexican waters, the superseiner *Jeannine* hauls in the huge purse seine. The helicopter is used to locate the schools of tuna.

entists are going to those places where the fish aren't—or where they used to be.

Safina proposes that the bluefin be listed by CITES (the Convention on International Trade in Endangered Species), which listing, he says, would mean it could not be shipped from one country to another, cutting the New Englanders out of the Japanese market completely. (Every attempt to list the bluefin by CITES has failed.) Safina also suggests that the fish caught by recreational fishermen not be sold to Japanese buyers at the Massachusetts docks. This may not seem like it would make a big difference, but there are a lot of fishermen catching a lot of tuna. And, he says, "stopping sportfishermen from making a profit does not affect a lifestyle." In fact, it might even bring sportfishing back to its origins, which did not include selling your catch to the highest bidder at the dock. His suggestion seemed to fly in the face of every American's right to fish for whatever he wants, and to sell it to anybody willing to buy, so it never went anywhere, either. Acknowledging the American entrepreneurial spirit, the film never condemns the fishermen who have historically been responsible for the bluefin's decline, but it is clear that the Audubon Society was deeply worried about the status of the Atlantic bluefin tuna.

In his book *Ocean Bankruptcy (World Fisheries on the Brink of Disas-*

ter), Stephen Sloan tells the story of a commercial fishing expedition gone horribly wrong. In August 1993, south of Block Island (Rhode Island), a spotter pilot for the purse-seiner *Connie Jean* reported a large school of yellowfin tuna at the surface. The *Connie Jean* dispatched the net boat, and the school was surrounded. "As soon as the Connie Jean pulled the net next to her port side and began to lift," writes Sloan, "it was immediately apparent to all observers nearby that the net contained not yellowfins, but very large bluefins, and the catch was illegal because the [bluefin] season had not yet opened." Seven of the large bluefins were hauled on deck, but the rest of the catch remained submerged in the net and died. The fish would have been worth a fortune to the captain of the *Connie Jean,* but because he could not get permission to land them before the season officially opened, "he sent the entire catch to the bottom for crab food." Sloan, who was a renowned fisherman, a member of the U.S. delegation to ICCAT, and the founder of the Fisheries Defense Fund, brought suit against the *Connie Jean*'s owners for illegal practices, but before the case came to court, the owners agreed to a settlement, which at Sloan's insistence consisted of a $10,000 donation to the Helen Keller Fishing Club for the Blind, in Brooklyn, New York.

Gerry Abrams is the founder and president emeritus of the East Coast Tuna Association, an organization, recounts Safina, "solemnly dedicated to making a dollar." Abrams is one of several people who claim credit for arranging to ship fresh bluefin tuna from New England to Japan for a sushi connection. Abrams believes the tuna population is as healthy as ever; the reason not so many are being caught is that they are becoming smarter and not coming to surface where they can be seen—and harpooned. "The National Marine Fisheries Service," says Abrams, "is perpetrating a deliberate lie that bluefin have become scarce. They refuse to believe anything the fishermen tell them." Abrams wants ICCAT to lower bluefin catch quotas, but as Safina notes, "In 1981 the commission's scientists concluded that the western Atlantic bluefin population was depleted and that catches 'should be reduced to as near zero as feasible.' The commission's managers responded by setting an annual catch quota of 1,160 metric tons, ostensibly for 'scientific monitoring.' The fishing industry complained, and in 1983, the 'scientific quota' was summarily raised to

2,660 metric tons." Abrams's association fought for higher quotas for fishermen and succeeded in persuading Congress to pass a law prohibiting NMFS from setting lower quotas than those set at ICCAT's annual meetings. That means that ICCAT commissioners, including those from Japan, were establishing quotas within U.S. waters.

For more than two decades, ICCAT considered the biomass which existed in the mid-1970s to be the maximum sustainable yield (MSY) level for the western Atlantic bluefin tuna stock. But since 1975, the population has declined by an additional 88 percent. According to ICCAT's latest stock assessment, the biomass (total weight of fish) of this stock had decreased to only 12 percent of that needed to produce the MSY. Thus, in just four decades, the population was driven from a healthy level (over three times the MSY level) to one just above extinction. This occurred because of unrestrained fishing sanctioned by ICCAT. The population is being held at this precarious level by continued overfishing allowed by ICCAT, ostensibly to provide scientific monitoring information on its status. The tuna of the eastern zone are managed under a strict annual quota set by the European Union, while those of the western Atlantic, targeted by American fishermen, have been supervised under strict catch quotas since 1995. Nevertheless, in both areas, the stocks of bluefin tuna have fallen dramatically: there has been an 80 percent decline in the eastern (European) stock over the past twenty years, and a 50 percent drop in the western Atlantic population.

The North Atlantic breeding populations are estimated to have gone down about 90 percent in the last twenty years. Exact counts are impossible, so there are vast gaps between the high estimates made—to no one's surprise—by the fishermen, and the low estimates made by those who would protect the tuna from overfishing. From dock to cabinet ministry, there have been endless discussions about solving the problem at every level, but few protective measures have been taken, because to do so would require unprecedented domestic and international cooperation. In response to dwindling catches, ICCAT's membership divided the North Atlantic into eastern and western sectors, each with its own quota. In 1991, when Sweden submitted a proposal to ICCAT that the bluefin be listed as "endangered," it was immediately voted down by the United States and Japan, countries with a strong economic interest in catching tuna. Conservationists, fishermen, and bureaucrats continued to draft position papers and proposals, while the tuna populations plummeted and the prices rose.

ICCAT is not based in Washington (its headquarters are in Madrid), but it might as well be. The commission is made up of twenty-two member countries on both sides of the Atlantic, plus Japan. Under pressure from powerful commercial fishing interests, ICCAT has consistently supported the fishers at the expense of the fish, while paying only the faintest lip service to the recreational fishers. Even as the Atlantic tuna populations declined, and the fish were listed as endangered, ICCAT allowed the harvest of *Thunnus thynnus* at levels that endangered the species even further.

The commission's charter explicitly specifies that tuna stocks be managed for MSY, an approach that seeks to maximize the annual harvest by holding the population at approximately 50 percent of the predetermined environmental carrying capacity. At this level, the population ought to reproduce at its maximum rate. As long as the population is not reduced below the MSY level, the theory goes, fishermen can harvest the annual surplus indefinitely. But bluefin tuna populations rise and fall for a variety of environmental reasons, and trying to factor in the MSY only contributes to the chaotic nature of the population assessments. Because the MSY may never be determined with any sort of accuracy, quotas can swing wildly from year to year, and impact the population in ways that biologists do not yet understand. ICCAT began keeping records in 1976, but, as Myers and Worm pointed out in 2003, over the past half century, Japanese and other longline fleets around the world have reduced the populations of large predatory fishes (sharks, tuna, billfishes) by 90 percent, leaving the remaining 10 percent for today's fishers. Add to that ICCAT's dedication to maximizing the total allowable catch for all commercial tuna fishers, no matter where they are based or where they fish, and you can begin to understand why North Atlantic tuna populations are at an all-time low. Low population added to intensified fishing effort leads only to disaster, as is happening in the Mediterranean.

Obviously, a robust population requires the addition of new animals, which means that the spawning stocks of the tuna must be protected—or at least not killed off. ICCAT failed to protect the spawning stocks of Atlantic bluefin for many years, and population levels have fallen far below the level required to sustain MSY. Between 1970 and 1993, the recruitment of young fish into the adult population fell from over 300,000 per year to only 50,000 or fewer (Safina 1993). In 1975 stock size estimates for spawning fishes (those over 320 lbs., or 145 kg) were at only 25 percent of levels estimated in 1960; by 1990 spawning stock was at just 7 percent of the 1960 level. Furthermore, ICCAT still operates under the assumption that there are two distinct stocks of northern Atlantic bluefin: one that spawns in the Gulf of Mexico and migrates north along the coast of North America, and another that spawns in the Mediterranean and migrates along coastal Europe and northern Africa. The western stock has declined precipitously over recent decades, and the catch from it is highly regulated by ICCAT and the U.S. National Marine Fisheries Service.

Although Frank Mather had known it in the 1970s, it was not until the late 1990s that NMFS scientists acknowledged that western-stock fish were actually crossing the Atlantic into the eastern ICAAT regulatory zone, where they were being exploited by fishers not under those controls implemented to protect them. Tagging studies confirmed that some subadults or adults hatched in the Gulf of Mexico found their way to the Mediterranean, but tag returns were very limited because eastern fishermen tend not to report the capture of tagged western fish in the Mediterranean—they fear it would cause the imposition of restrictive quotas. Even today, European ICCAT representatives are reluctant to redefine bluefin stock structure, in order to avoid having to consider restrictive quotas. This reticence, coupled with the introduction of tuna "ranches" in the Mediterranean, has resulted in such a massive crash of the eastern Atlantic and Mediterranean bluefin tuna populations that the World Wildlife Fund has called for an immediate closure of all Mediterranean tuna fisheries (Bregazzi 2006).

In 1983 the Greenpeace ship *Rainbow Warrior* was sent to the Bering Sea to document the killing of marine mammals in Japanese drift nets, but with more than five hundred boats operating in the central Pacific, it was a futile gesture. By 1987 the Japanese squid fleet had expanded to 1,200 boats, each deploying up to thirty miles of nets every night during a seven-month fishing season. Japanese and

Taiwanese boats drift netting for albacore kill not only the tuna, but tens of thousands of other animals as well. In 1989 conservationist guerrilla Sam La Budde made a film called *Stripmining the Seas,* which showed drift nets being hauled aboard and drowned dolphins being kicked overboard. The film was seen by many people, including members of the U.S. Congress. Objections to drift netting kept piling up; in 1989 the U.N. adopted a resolution to reduce fleets and ban drift netting in the South Pacific. Japan and Taiwan were the major culprits; a 1989 report showed that the Japanese squid fleet alone was responsible for the death of a million blue sharks, 240,000 seabirds, and 22,000 dolphins. Over worldwide objections, the Taiwanese continued to deploy their "curtains of death," and they are still being used by pirate drift netters. A net that has been cut loose or has broken loose does not shut down; it keeps killing fishes, dolphins, and sea turtles even when there is no one to haul in the net to harvest or release the prisoners. Known as "ghost nets," these gigantic net walls keep on fishing long after the mother ship has departed.

The Mediterranean has been the scene of intense drift-net fishing, but it has supposedly been outlawed. Prohibitions have had little effect, however, and driftnets are still being set by the fishers of various countries, particularly Greece and Italy. In the summer of 2006, two conservation organizations, Greenpeace and Oceana, deployed vessels to document and film the illegal driftnetters at work. The *Rainbow Warrior* spent three weeks confronting (and filming) rogue fishing boats in the Mediterranean, and Sofia Tsenikili of Greenpeace Greece said, "If people are horrified by the images of whales being harpooned in the Southern Ocean, they'd be equally repulsed by the thousands of dolphins and other creatures that are being entangled and killed by fishermen using huge illegal driftnets each season in the Mediterranean." Oceana, a network of marine conservationists, has also documented the use of illegal driftnets by Italian fishers in the Mediterranean. The *Oceana Ranger* sailed through 1,500 miles of the Ligurian and Tyrrhenian Seas, and observed numerous vessels "capturing species like swordfish or albacore tuna with driftnets that are totally banned. We have also witnessed how these catches are unloaded in ports without any type of control . . . and after being loaded into trucks, they are 'laundered' through irregular supply chains." The illegal fishers were reported to the Italian Coast Guard, but it is unlikely that anyone can stop driftnetting—it is just too lucrative.

On May 3, 2005, Andrew Revkin wrote in the *New York Times,* "In just the past 35 years, exploding markets for sushi-grade tuna, combined with intensifying industrial scale hunts aided by satellites and spotters in airplanes, have devastated not only fish but also many fisheries." After discussing her tagging program, as detailed in *Nature* for April 28, 2005, Revkin quotes Barbara Block: "it's hard to believe that a fish of this size and beauty, an animal that has captured the hearts of fishermen and scientists alike for millennia, is slipping off the earth." On the other hand, it's not hard to believe that overfishing is the cause of the decline in fish populations. (What *is* hard to believe is that there is anybody who would disagree.)

Peter Ward and Ransom Myers (2005) compared "the data collected by observers on longline fishing vessels with data collected from a 1950s scientific survey when industrial fishing commenced" in the tropical Pacific. They found that the largest and most abundant predators, such as tunas and large sharks, suffered the greatest decline (up to 21 percent), and that overall, "the biomass of large predators fell by a factor of 10 between the periods." Longliners specifically target the large predators, particularly billfish and tunas, but sharks, often taken "incidentally" in substantial numbers, have not been the object of the fishery. (In recent years, however, because of an exponential increase in the consumption of shark's fin soup in Asian countries, sharks have been specifically targeted, and in some areas have become seriously endangered.) The tunas caught in the tropical Pacific longline fishery are albacore, bigeye, skipjack, and yellowfin, with the yellowfin being by far the predominant species. In the same region, yellowfins are also the object of an intense purse-seine fishery, so their numbers are even further reduced.

With Boris Worm, Ransom Myers was the author of the 2003 report "Rapid Worldwide Depletion of Predatory Fish Communities," in which they put forth the startling statistic that 90 percent of the world's large predatory fishes were gone, the victims of decades of Japanese longlining. (Except for net fisheries, we catch and eat predatory fishes because we can exploit their eating habits; if they didn't bite at baited hooks or other representations of their prey, we wouldn't be able to catch them. Incidentally, *marine* predators are the only predators that humans eat with any regularity.) In a more recent study, Myers and Ward arrived at the same conclusion for the tropical Pacific fish communities: "The index of community biomass is about 10% of its former level and the community is composed of smaller fish and fewer large predators." If the predators are reduced in a particular ecosystem, the prey species on which they feed will increase, and as the prey species—small fishes and cephalopods, mostly—do not usually take longline hooks, the effect of longlining on their populations is an *increase.* Fisheries biologist Daniel Pauly is the author of the phrase (and the concept) of "fishing down the food chain," which means taking the apex predators first (large species like cod, tuna, and swordfish), because they are the most desirable species, and after they are gone, going down a trophic level and taking their prey species (plankton eaters such as anchovies) and then taking what's left. ("Trophic" is from the Greek *trophe,* meaning "food" or "nourishment"; a "trophic level" refers to the ranking of prey species eaten by a particular group of fishes.)

Because it is the top predators that are usually sought first, "it stands to reason that prey populations and their effects on marine communities will increase after release from predator control," said Robert Steneck, a marine ecologist at the University of Maine. Accordingly, fishing alters the organization and structure of entire marine communities via "cascading" trophic chain reactions. Because the top predators are the least numerous, as one moves down a food web biomass increases, but nowadays fish catches have stagnated as fishers have moved from top predators to species at lower trophic levels. Once a top predator has been depleted or exterminated by fishing, alternative predators, which are of no commercial value, thrive in the absence of competition and thus deplete the biomass of prey species at lower trophic levels.

When the stocks of western North Atlantic codfish crashed ten years ago, fishing for cod was banned to allow them to recover, but, reported Bundy et al. in 2000, the population dynamics have been so drastically altered that cod may never reappear in the former numbers. It is not only cod, but virtually all high-quality table fish, such as tuna, haddock, and flounder, that have fallen to about 16 percent of what their numbers were in 1990. In a 2002 article in *New Scientist,* Kurt Kleiner quoted Daniel Pauly: "Jellyfish is already being exported. In the Gulf of Maine people were catching cod a few decades ago. Now they're catching sea cucumber." Pauly and Watson, introducing their 2003 *Scientific American* article, said, "Overfishing has slashed stocks—especially of large predator species—to an all-time low worldwide . . . If we don't manage this resource, we will be left with a diet of jellyfish and plankton stew."

Six

TSUKIJI

Tsukiji Fish Market, Tokyo, 5:00 a.m.

TOKYO. FIVE A.M. The rain-washed streets are quiet. I am in a taxi heading for the largest wholesale fish market in Asia—probably the largest in the world. More than two thousand tons of seafood are auctioned off here every day at 1,667 stalls, and fifty thousand people participate in the action. They say the market covers fifty-six acres, but as booths, trucks, buyers, sellers, and tourists ooze out along the edges, its perimeters seem more than a little fluid. In the aisles, rapidly trundled two-wheeled handcarts, called *neko* (cats) for their agility, threaten to roll over anyone who doesn't understand the traffic patterns of the market—and many of those who do. Compounding the traffic snarl in the market are little tractors *(tāretto)* and miniforklifts that move boxes and fish around the aisles. The selling (and buying) areas—which cover practically all of the acreage—are filled with booths, scales, signs, tables, and every kind of seafood that anyone could imagine—and some that defy the imagination. The market has its own docks on the Sumida River, which opens into Tokyo Bay, so some of the produce has come straight from the sea. Refrigerated trucks have been arriving from all over Japan since midnight, and by five a.m. the pallets have been forklifted from the trucks and delivered to the stalls. Almost everything has been arranged in displays that visibly demonstrate the Japanese penchant for design and packaging. It is a busy, fifty-six-acre bento box.

Tsukiji was originally just marshland along the edge of Tokyo Bay. It was reclaimed in the seventeenth century to satisfy the city's need for more space. When foreigners began to arrive in the latter half of the nineteenth century, it was declared the foreign residents' quarters and all foreigners were required to live there. This arrangement did not sit too well with the foreigners themselves (apparently they didn't like the numerous prostitutes ensconced there for their pleasure). By the 1920s the area had been pretty much abandoned. The great Kantō earthquake of September 1, 1923, burned 700,000 buildings in Tokyo, killed perhaps 100,000 people, and completely destroyed the Nihonbashi fish market. The new market, known as Tsukiji, was started as soon as the rubble could be cleared, but it was not officially opened until 1935. According to Theodore Bestor's history of Tsukiji *(The Fish Market at the Center of the World),* in its first full year of operation the market handled 183,000 metric tons of seafood, valued at 45 million yen, more than $20 million in 1936 dollars.

The Tsukiji market often appears on the itinerary of tourists, who want to get there before everything is sold, and some of the boxes (there are a lot of white Styrofoam boxes here) contain species that even amateurs would recognize: there are fish that look like red snappers; little skinny ones that are probably some kind of sardines (or anchovies); if it looks like an eel, it probably is an eel (I didn't know people ate moray eels); and many more or less recognizable varieties of salmon, bass, groupers, jacks, carp, perch, mackerel, and triggerfish (do people really eat triggerfish?). Even an ichthyologist would be hard-pressed to identify all the kinds of fishes; Japanese ichthyologists often come to Tsukiji to look for new species. There are elongated, silvery fish with tooth-filled mouths; purple fish with big lips; tiny little silver things that look as if they belong in a home aquarium; black fish, yellow fish, red fish, blue fish, striped fish, spotted fish, dried fish, smoked fish, salted fish, long fish, short fish, fat fish, skinny fish. There are flatfishes with both eyes on the same side of the head: halibut, flounder, sole, plaice, turbot. (I know the common names of some of these fishes in English, but these creatures in the boxes are not necessarily the species I know, and of course the names are in Japanese here.) A veritable zoo of fishes: sea horses, squirrel fish, cowfish, roosterfish, goatfish, goosefish, rabbitfish, wolffish. And dolphinfish, sailfish, spearfish, swordfish, cutlassfish, scabbardfish; and the deadly fugu, a pufferfish that can kill you if it's not prepared by a master chef. Mako sharks that look as if they could kill you, but not when they are dead with their black eyes glazed and their fiercely fanged mouths frozen open. Not all the fish are dead; there are many tanks with healthy, live fish swimming in them. These will be selected by wholesalers to be delivered to restaurants as quickly as possible—freshness being the keynote here. There are trays of carefully packed salmon roe and tuna roe, fishes for sale before they would have been born; and here is the roe of sea urchins *(uni),* a delicacy that commands a particularly high price.

At the fish market we also have the not-fish: oysters, clams, mussels, periwinkles, and mollusks in shells that look as if they belong in a shell collector's cabinet. Tiny squid, little squid, big squid (but not giants; the true giant squid is inedible because it reeks of ammonia); cooked octopuses, uncooked octopuses, bright red octopuses, little brown octopuses; cuttlefishes of all sizes and colors; tiny crabs, little crabs, big crabs, giant crabs (the so-called spider crabs of the Bering Sea), jellyfish, lobster, shrimp, prawns, crayfish, sea urchins, sea cucumbers, sea anemones, and many, many varieties of seaweed. If it ever lived in the ocean, it can be found at Tsukiji. And even if it *never* lived in the ocean—even if it never lived *anywhere*—it can also be found here. There are boxes of surimi, a food product made from white-fleshed fish, such as pollock or hake, that has been pulverized to a paste and cooked to a rubbery consistency. Surimi is available in many shapes, forms, and textures, but the most common surimi product is artificial crab legs. (It is often sold as *sea legs* in America, or *seafood sticks* in the U.K.) I pass a counter with big hunks of red meat on it. *Kujira?* I ask the vendor. He nods yes: whalemeat.

All this seafood does not originate in local waters. Every kind of commercial fishery sends its products to Japan. Tsukiji is the final destination for Scottish and Norwegian salmon; Alaskan pollock and king crab; aquacultured shrimp from the Gulf of Mexico and Thailand; octopus from West Africa and the Mediterranean; squid from the Japanese distant-water fisheries in the Falkland Islands and Baja California; minke whales that were killed in the Antarctic; and tuna from New England and also from ranches in Tunisia, Morocco, Spain, Italy, Malta, Greece, Croatia, Turkey, Mexico, and South Australia.

In the stalls there are tuna of every kind, ranging in size from the little bullet tunas through the larger bonitos, albacore, and skipjack. Most of the fishes are displayed in booths, but the star of the show, the main reason the tourists come, is not to be found in a common booth with *neko* bumping by. The bluefin tuna, known as *maguro,* will be found only in special rooms, large open spaces the size of a high school gymnasium without the bleachers. Harsh sodium lights glare overhead; the fish here will be examined very, very carefully. The temperature is kept at a level where the frozen fish will not thaw—everybody

Some of the "not-fish" at the Tsukiji fish market: these are octopuses.

wears a coat and rubber boots, and most of the workers have hats and gloves. You can see their breath, and you can see a mist rising off the ranks of frozen tuna. (Most of this tuna, which will be marketed as *fresh,* starts its journey to a restaurant aboard a fishing boat, where it was flash-frozen soon after it was caught.) Potential buyers, identifiable by the yellow plastic government licenses clipped to their baseball caps, walk carefully among the arrangement of frosted, tailless tuna. The removal of the tail fin, which takes place immediately after the fish is caught, makes the fish much easier to handle and ship. Each fish also has a half-moon-shaped hole where the gills used to be; the breathing apparatus decomposes during shipment. The fish are grouped according to size and ocean of origin: the two-hundred-pounders from the Indian Ocean are together; the five- to six-hundred-pounders from New England or the fish farms of Australia lie side by side, and so on. Each fish has a number on it, and as the buyer examines the fish, he makes notes on a little pad. The auctioneer rings a handbell to signal that his auction is about to begin. He watches the buyers, who bid on each fish with subtle hand signals; his singsong calls continue until each fish in the lot is sold.

In 1992, TriCoastal Seafood Cooperative in Newburyport, Massachusetts, was selling its tuna to Tsukiji Toichi Uoichiba K.K., the fourth-largest auction trading company in Japan. One morning in early August, nine fish, which ranged in size from 275 to 650 pounds, were examined by a Japanese tuna grader named Ryozo Morikawa: "freshness was determined by sight and by touch. The first grade was based on the whole fish. The technician looked at the skin, to see how the colors had lasted. He touched the skin to feel for its resilience, the quality of fat underneath, and the condition of muscle." In *Giant Bluefin,* a book about New England tuna fishermen, but also about their dependence on the *maguro* market, Douglas Whynott quotes Morikawa: "When Japanese people eat, the first thing they do is look. They eat with their eyes. And in Japan red-and-white is an auspicious color combination," partially accounting for the appeal of *maguro,* red-meat tuna. The fish were forklifted into ice-packed "coffins," a rice-paper blanket was placed over them to preserve the color, and they were loaded on a truck destined for Boston's Logan Airport. Next stop: Japan.

Frozen tuna from all over the world are delivered every day to Tsukiji.

Originally, the Cape Cod fishermen would bring their harpooned tuna to the dock at Sandwich, Yarmouth, or Barnstable, where a local dealer would pay them by the pound. They got anywhere from two to thirty dollars a pound, but a four-hundred-pound tuna at two dollars a pound is worth eight hundred dollars, and you can calculate the price at thirty dollars a pound. Payment varied from fish to fish and from season to season, but these giant tuna were worth so much money at the dock that the fishermen could afford to pay a spotter pilot a hundred dollars for each fish that they harpooned. The dealers on the dock would then sell the tuna to Japanese buyers, who paid much more per pound. By the 1992 season, they were selling eighteen-thousand-dollar fish to the Japanese. It didn't take the wily New Englanders long to realize that they could circumvent part of this multi-level system, and they banded together to form a cooperative, called Cape Quality Bluefin (CQB), and negotiated their own deals with Japan. For the 1990 season, CQB handled 100,000 pounds of tuna and grossed $2,153,000. At the docks in Massachusetts and Maine, technicians from Tsukiji instruct the fishermen on the proper techniques for handling and packing tuna for export. As Ted Bestor

(2000) wrote, "A bluefin tuna must approximate the appropriate *kata,* or 'ideal form,' of color, texture, fat content, body shape, and so forth, all prescribed by Japanese specifications . . . Despite high shipping costs and the fact that 50 percent of the gross weight of the tuna is unusable, tuna is sent to Japan whole, not sliced into salable portions."* When some of the CQB fishermen went to Japan to observe the Tsukiji auctions, they saw a 350-pound fish sell for $42,000, or $120 a pound.

Years of experience enable the buyers at Tsukiji to examine the fish with an eye (and nose) toward identifying the richest, reddest meat with the highest fat content. Near the base of what would have been the tail, a slice of flesh has been half-mooned back, revealing the bright red color of the meat. The buyer can cut a sliver off, hold it up to the light, smell it, and taste it—after all, it is precisely the taste and "mouth feel" that determines the quality and price of the *maguro.* He will also rub the sliver between his thumb and forefinger to test the fat content. The tuna are auctioned one at a time—no job lots here. The auction may appear casual to an outsider, but every hand gesture is recorded, and the auctioneer's assistant writes everything down. As each fish is sold, it is marked in red with the name of the buyer. Many of the tuna are sold to local merchants, and the carcass will be delivered, often by handcart, to a booth in the market, where it will be butchered by a man wielding something that looks like a samurai sword. The frozen carcasses are sectioned by a man working a table saw. By noon, most everything has been sold. The trucks belonging to wholesalers, fish companies, and restaurants have taken to the streets of Tokyo to begin the final distribution process.

A giant bluefin, hatched from a tiny egg in the Gulf of Mexico, grew to full size in the North Atlantic. It may have crossed the Atlantic several times. Its life was ended by a harpooner out of Barnstable, Massachusetts, and the fish was gutted on the dock, sold to a

*Bestor also said: "As their long tuna knives reduce the fish into smaller and smaller pieces, and the tub of bones, skin, and other unusable portions fills up, they carefully point to the reality that roughly 50 percent of the weight of a tuna sold at auction ends up as waste before sashimi reaches the consumers' plates. The retail price per kilogram, therefore, has to be at least 100 percent above the auction price just to meet the cost of the fish, even before figuring in any margins for overhead for wholesalers or retailers."

Backbone and muscles of a bluefin tuna before it is carved into manageable pieces

Japanese buyer, trucked to Logan Airport, and airlifted to Narita in Tokyo. From there it traveled by refrigerated truck to Tsukiji, and with many other frozen tuna carcasses it was unloaded and taken to the auction sheds so buyers could ascertain its quality and bid on it. Along the way, it passed through the hands of several distributors, all of whom took a small profit. Cut into small pieces, it was sold to a Tokyo restaurateur, who had it prepared as sashimi. A two-ounce portion will sell for about seventy-five dollars.*

The lessons learned (or not) from the catastrophic demise of the American and Canadian wild-capture codfisheries, however, sent alarm signals to tuna fishermen around the world. It might be possible, some believed, to stave off a similar disaster, by finding a way to farm tuna. But aren't bluefin tuna bigger and wilder than any codfish, and even less susceptible to domestication?

*On February 22, 2008, the Tokyo Metropolitan Government announced that sightseers would be banned from the Tsukiji fish market. They claimed that the trucks and carts made it dangerous for tourists in the aisles, that flash cameras interfered with the auctions, and that all that exposed fish created hygiene issues (Norrie 2008). It is also conceivable that the Tokyo government was reacting to the current furor about the plight of the bluefin tuna and saw no need to contribute to it by allowing tourists to photograph ranks of frozen tuna carcasses, soon to be transformed into expensive sashimi.

FISHING FOR *MAGURO*

Carving the *maguro* with what looks remarkably like a samurai sword

Tokyo, January 5, 2001 (AP). An enormous bluefin tuna—a fish prized as sushi—sold for a record $173,600 Friday in the first auction of the year at Tokyo's main fish market. At $391 a pound, the 444-pound fish was the most expensive auctioned off at the Tsukiji Central Fish Market in years. In 1996, a 250-pound bluefin fetched $44,100. Called *honmaguro* in Japanese, bluefin tuna is popularly served raw as sashimi or sushi in restaurants where a plate of slices can command a bill of more than $100. Both fish were caught in the Pacific Ocean off Aomori Prefecture in northern Japan, an area known for the quality of its tuna. "It's kind of like a brand name," market official Takashi Yoshida said.

ONCE, IN A JAPANESE INN, I was partaking of an elaborate meal, and one of the courses was lobster. The live, uncooked lobster had obviously been removed from the shell, cut into pieces, and then replaced in the shell. When the dish was put in front of me, the lobster's antennae were still waving around, and its beadlike eyes were looking right at me. On another occasion, I was aboard a Japanese whaling vessel, working about two hundred miles south of Tokyo. We killed a whale and brought it alongside to eviscerate it so it wouldn't spoil on the way back to the whaling station. It was held alongside by ropes, and gutted by men working with very long-handled knives. One of them used a kind of corer to remove a softball-sized chunk of meat from the back of the whale, and then brought it, red and dripping blood, to the galley. It was cut into very thin slices and we ate the *kujira sashimi* with soy sauce. To the Japanese, serving fish of the best quality and utmost freshness is a matter of honor and tradition. Therefore the freshest fish will be one that is the most recently dead. Sometimes the fish is killed just before it is eaten, but it still has to be prepared in a manner pleasing to the eye.

The enterprise can be traced back to the seventh century, says Yoshiaki Matsuda, in his history of the Japanese tuna fisheries, "but it was skipjack that was appreciated more by the people. For a long time, skipjack was caught by harpoons, spears and hooks, and eaten as dried or boiled fish and served as offerings at shrines." By the early twentieth century, the Japanese government was supporting distant-water fisheries for skipjack and other tuna species, but during World War II most of the fishing fleet was destroyed. When the war ended, the American occupying forces encouraged the Japanese to resume fishing, first drawing twelve-mile boundaries around the islands, then regularly extending these boundaries—known as the MacArthur Lines—until Japan regained its independence in 1952, at which time all restrictions were lifted. Ever industrious (and probably hungry as well), the Japanese managed to catch 235,912 tons of skipjack and other tunas in 1951. Their longline fishery in the Pacific began in 1952, but within two years suffered a setback that originated in the same terrifying technology that had devastated Japan in 1945. On March 1, 1954, the longliner *Fukuru-maru V* encountered radioactive fallout from the U.S. H-bomb test over Bikini Atoll, and showed up at the Japanese port of Yaizu with contaminated tuna. Hundreds of other vessels that were fishing in the western Pacific had to be abandoned, and the Japanese tuna market collapsed. This panic was followed in the 1960s by the U.S. FDA announcement that tuna with a mercury content of 0.5 ppm (parts per million) would be rejected. The ruling was followed by similar restrictions from Norway, Sweden, Canada, Italy, and France. The large tunas, like bluefins, yellowfins, and bigeyes, had a higher mercury accumulation than the smaller species, and once again the Japanese could not sell their tuna.

Although tuna are found in all major bodies of water (except the polar seas), most of the supply comes from the Pacific Ocean, which accounts for 2.3 million tons, or about 66 percent of the total world catch. The rest of the commercial tuna sold around the world comes from the Indian Ocean (20.7 percent), the Atlantic Ocean (12.5 percent), and the Mediterranean and Black Seas (0.8 percent). Yuichiro Harada, a senior managing director of the Organization for the Promotion of Responsible Tuna Fisheries (OPRT), set up by Japanese fishermen, reports that the number of tuna-fishing boats is decreasing in Japan, because rising fuel costs are forcing financially weak firms out of business. In the meantime, consumption is increasing in the United States, Europe, Taiwan, and China. Americans now eat 100,000 tons of tuna as sashimi a year, says Harada. "Demand [in the United States] will grow further against the backdrop of health-consciousness," he believes. *Toro* is the pinnacle of the expensive sashimi list, but the Japanese consume substantial quantities of other kinds of tuna. The bigeye *(Thunnus obesus)* is one of the most popular varieties. Taiwan is a major supplier of bigeye to Japan, with catches by its boats accounting for some 20 percent of Japan's tuna consumption of about 450,000 tons a year. While skipjack and albacore provide the "light-" or "white-meat" tuna used in such great quantities in salads and sandwiches, three species (bluefin, yellowfin, and bigeye) supply the red meat favored for sashimi. And although restaurants often claim that their *toro* comes from only New England bluefins flown to Japan, red tuna meat is just as likely to originate from the southern bluefins of Australia, or "farmed" bluefins from the Mediterranean or Mexico.

In 1962, Georg Borgstrom, professor of Food Sciences at Michigan State University, published *Japan's World Success in Fishing,* in which he attempted to analyze the phenomenal success of the country's worldwide fishing enterprises and their effects on the rest of the planet. Reading Borgstrom's study, you would have to conclude that it was Japan's intention to fish everywhere, catch everything, and dominate virtually every aspect of the fish-catching industry around the globe. He wrote:

> Japanese overseas fishery enterprises are now in operation or planned in more than 50 countries. They are conducted in most of the oceans of the world, and approximately 200 Japanese fishing vessels are at present based in foreign countries. About 140 of them are tuna vessels, most of them in the longline fishery, and some are fishing for existing plants in foreign countries. Recent information indicates that joint fishery companies are in operation in almost 20 countries, principally in Asia and Latin America, and involve capital participation by foreign interests, private or government, with the Japanese furnishing technical direction, vessels, and gear, and frequently processing machinery.

Because the past has such a strong influence on modern Japanese culture, we tend to believe that many of their customs have been around for a very long time. One of the arguments for the Japanese continuation of whaling, for example, is that they have "always" done it, and it is therefore part of a very long tradition. (They actually began whaling around the middle of the seventeenth century.) Sashimi consumption can also be traced back to the seventeenth century, but eating the belly meat of tuna *(toro)* is not part of any tradition, and goes back only to the mid-1960s, when mechanical refrigeration was introduced into homes, restaurants, and, most important, factory trawlers. Indeed, the red meat of tuna, now the favorite for sushi and sashimi in Tokyo and elsewhere, was originally used for cat food. As Ted Bestor wrote, "The rich flavors of toro fit the new palate and traders could handle, preserve, and ship the fatty toro more easily; in this way, a new market niche was born." The ability to catch and freeze large tuna coincided with the unprecedented rise in popularity (and price) of *maguro,* which has changed the way the world thinks about bluefin tuna.

According to biologists Shingu, Hisada, Kume, and Honma (1975), the Japanese longline fishery began in the Atlantic in 1956, and "made good catch of bluefin tuna" in the Atlantic from 1963 to 1966. The best of those years was 1963, where the catch consisted of sixty thousand fish; only fifteen thousand were caught in 1965. "From 1967 to 1970," they wrote, "the bluefin catch by longline fishery decreased remarkably," but since 1971, "the amount of catch . . . has begun to increase." By 1974 Japan had become the world leader in tuna catches, more than doubling that of the United States, its only serious rival. Joseph and Greenough (1979) reported that Japanese tuna boats had caught 649,000 metric tons, compared to 258,000 for U.S. craft. The Japanese longline fishery first operated off Brazil, but when it collapsed in 1972, they abandoned those grounds and expanded their area of operations to the waters of Spain and Portugal in the eastern Atlantic, and into the Mediterranean, where a host of other fishermen waited to welcome them.

Nowadays, the bluefin tuna fishery is directed almost exclusively toward the insatiable Japanese sashimi market. (Sashimi is raw fish served on a bed of rice; sushi is various items, such as meat, vegetables, and egg, but also including cooked and raw fish, wrapped in vinegared rice or seaweed.) Whether they are caught in the North Atlantic, in the Mediterranean, or off Australia, the large tuna are destined for the Tsukiji, or, to a lesser extent, other Japanese markets. Although scientists have been trying for years, tuna cannot be farmed in the traditional fashion (fish grown from eggs), but they can be "ranched," where juvenile tuna are trapped in nets and towed to pens where they are fattened until they are large enough to be killed for the sashimi market. These tuna "ranches" can be found off almost every country on the Mediterranean, and off Australia, Costa Rica, Malaysia, Japan, Mexico, and Panama as well.

In his 1995 *Giant Bluefin,* Douglas Whynott summarized the early history of commercial tuna fishing, as the fish morphed from the large, undesirable "horse mackerel" into the most valuable fish in the world: "Since 1958 in Cape Cod Bay a small seiner called the *Silver Mink,* captained by Manny Phillips, had been netting 500 to 600 tons of school bluefin per season . . . He was giving fish away. He was selling school bluefin for five cents a pound to a cannery in

Maine . . . Another fisherman, Frank Cyganowski, in his first year as a seiner captain, making his first set off Martha's Vineyard, hauled in sixteen tons of sixty-pound bluefin. The fish were sold to Cape Cod Tuna, the cannery in Eastport, Maine, for five cents a pound, and so Frank Cyganowski's first set was worth $1,600 . . . From 1962 to 1970, a fleet of two dozen Canadian seiners made yearly catches of from 6,000 to 10,000 metric tons of tuna . . . In 1956 Japanese long-line trawlers began to fish in the Atlantic Ocean, first in equatorial waters, later expanding into tropical and temperate waters in both hemispheres. Japanese longliners typically set lines fifty miles long with two thousand hooks per line. By the peak year of 1965, a fleet of several hundred boats was setting 100 million hooks. At first they targeted yellowfin and albacore tuna, and the catches were landed at ports in the Atlantic, but in the 1970s, with increased demand for tuna, target species changed to northern and southern bluefin and to bigeye tuna, which inhabit colder waters."

In his 2001 discussion of tuna conservation, Carl Safina wrote, "Because the prices surrounding them run so high, bluefins are depleted everywhere they swim." When the Japanese learned that giant tunas could be caught off eastern Canada and New England, the bluefin market exploded. Prior to the arrival of the Japanese purse-seiner *Kuroshio Maru 37,* in 1971, the large, red-meat bluefin tuna were worth only a few pennies a pound—if you could find a buyer. When fishermen brought fish to *Kuroshio Maru 37,* they received ten cents a pound, but they didn't know that Cyganowski was receiving another twenty-one cents a pound, thus becoming the first Japanese/New England middleman. The next year, Nichoro Gyogyo K.K. sent *Kuroshio Maru 32,* another purse-seiner with a large cargo capacity, prepared to pay the fishermen a dollar a pound. Fish that had brought forty dollars two years earlier were now worth a thousand. Soon the price of a single "giant" shot up to tens of thousands of dollars. When Atlantic coast fishermen found that they could earn a year's pay in an afternoon, the uncontrolled slaughter of what Barbara Block called "the cocaine of the sea" began.

By 1990 Japan was consuming 800,000 tons (1,600,000,000 pounds) of tuna per year, of which 74 percent was caught by its tuna fleet and the remainder by foreign fishers. As of 2001, wrote Carroll,

Ranks of frozen tuna at the Tsukiji market, Tokyo

Anderson, and Martínez-Garmendia, "More than 45 countries now compete to supply bluefin tuna to Japan. From 1994 to 1997, the main sources of fresh bluefin tuna exports to Japan were Canada, the United States, and the Mediterranean region. In 1997, Japanese bluefin tuna imports of fresh, U.S. product accounted for about 15% of all imports. However, the U.S. share was greater during the summer months, when it comprised more than 25% of total Japanese imports."

Japan now imports some sixty thousand tons of bluefin tuna. The enormous Japanese sushi and sashimi markets are supplied by what is essentially a consortium of tuna fishermen from around the world. Large Atlantic wild bluefins are caught off New England, Maritime Canada, New Jersey, and North Carolina, while their (smaller) Pacific counterparts are fished off California, Japan, and Hawaii. The southern bluefins can be taken off Australia and New Zealand, and nationals from any and every country participate in the international bluefin sweepstakes. With large numbers of people chasing after the remaining bluefins, it is not surprising that these big fish have become scarce. But because greed and gluttony always trump caution and care, a system of creating more giant tuna was developed that appeared at first to solve the problem of scarcity, but the tuna farmers proved to be just as venal and careless as the open-water fishermen, and they managed to overfish the very fishery that they had created. Now the tuna "ranches" are running out of fish too.

What conclusions can we draw from the Japanese consumption of hundreds of thousands of tons of tuna every year? The Japanese will claim that bluefin tuna are plentiful throughout the world's oceans, and their fisheries (and their purchases) are hardly making a dent in the populations—which are on the increase anyway. ICCAT publishes report after report about catch statistics, management, and fishing gear, but appears to be little more than a lobbying organization for tuna fishermen and has no way of reducing fishing pressure. But ask conservation organizations like Greenpeace or WWF, and you will hear that the bluefin tuna is in very serious trouble. No population of animals, no matter how widespread and fecund, can withstand such massive inroads into its numbers. And the practice of tuna ranching, which removes large numbers of potential breeders from the wild population, will surely have unsuspected consequences.

It was always assumed—at least by fishermen—that as a particular fish species grew scarce, they would simply find another place to fish for it. Or if a particular species declined, they would find another species. As Myers and Worm put it in a follow-up to their 2003 report, "Although fishing pressure on large marine predators, such as sharks, tuna, billfish, large groundfish, etc., is high, it has long been assumed that these species are largely extinction-proof. The main reasons for this idea were the seemingly inexhaustible abundance of marine life, the remoteness of many marine habitats, and the extreme high fecundity of marine fish populations."* This rosy attitude has proven to be disastrously flawed. If you take most of the fish of a given species out of the ocean, there may not be enough left to regenerate

*Not everybody agrees with Myers and Worm—about tuna, anyway. In a response to their article, Hampton, Sibert, et al. published "Decline of Pacific Tuna Populations Exaggerated?," in which they questioned the use of the CPUE (catch-per-unit-effort) data for Japanese longliners, saying that the use of such data is invalid because Japanese fishers changed their target species from albacore and yellowfin for the canned tuna market to bigeye and yellowfin for the sashimi market. "Japanese longline CPUE for albacore declined rapidly, not because of declining albacore abundance but because of this change in species targeting." Their results "indicate that biomass decline and fishing impacts are much less severe than is claimed by Myers and Worm."

the species. Gone are the days when codfishermen on the Grand Banks—once the world's richest neighborhood for *Gadus morhua*—could lower a basket on a rope and bring it up filled with wriggling cod.

From 1990 to 1992, Sylvia Earle was NOAA's chief scientist, and as such, she had the (dubious) distinction of participating in many, many meetings. As she recalled (in *Sea Change,* 1995), "I had a front row seat for the debate on numerous unresolved fisheries issues and a chance to seriously study the scientific basis for policies that had led *not* to the widely acclaimed and highly desirable goal of 'sustained use' of natural ocean ecosystems, but rather, to precipitous 'crash and burn' declines." At a meeting to discuss the status of the western Atlantic bluefin tuna, she learned that "the adult breeding population had declined to ten percent of what it had been twenty years ago, when regulation of the taking of these great ocean rangers began. The population as a whole had been reduced by more than half, and most of what remained were immature. Stunned, I blurted out, 'Are we trying to exterminate them? If so, congratulations! We're making great progress.' "

Bluefin tuna have been elevated to the very pinnacle of menu desirability, eaten in various countries, and cooked (or not) in various ways. Always the most fashionable and expensive item on the Japanese menu, *maguro* is the red, fatty belly meat, always served raw. You can order it in expensive Japanese restaurants in New York, Los Angeles, Paris, London, and Rome—and at practically every restaurant in Tokyo that isn't a noodle shop or a McDonald's. Outside of Japan, bluefin tuna is also served grilled, as carpaccio (raw, thin-sliced), or even as tuna filet mignon. But even if the fish was caught off Cape Cod and served in Boston, it is identified as *maguro* because of the pervasive Japanese influence. Fish is healthy food, while for many people, beef has become a potentially unhealthy alternative. Japanese restaurants, where the overwhelming emphasis is on fish, are more popular than ever. (There are a few Japanese restaurants that specialize in beef, but interestingly, "Kobe beef," from specially fed and massaged cows, resembles red-meat tuna.) Thirty years ago, the idea of eating raw fish was repulsive to most Westerners, but nowadays they consume it in restaurants everywhere. Japanese cuisine has revolutionized the way

Westerners think about fish, and Japanese fishermen have revolutionized the way the world thinks about tuna.

In *Tuna and the Japanese,* journalist Takeaki Hori asks tuna auctioneer Hideo Hirahara to define *toro.* Harihara answers: "Well, the part that we call *o-toro* (big *toro*) is only found on the bluefin tuna that visits Japan's coastal areas, on two kinds of tuna that migrate to the Atlantic, and on the southern bluefin tuna (the Indian tuna) that lives in the southern hemisphere. The bluefin and the southern bluefin are very similar types. The ones that we are interested in for *toro* are those that can be caught in the Mediterranean, or the jumbo tuna that can be caught in the waters of New York . . . As a rule *toro* can only be obtained from these tuna." Hirahara goes on to explain that *o-toro* comes from the back of the tuna, but the quality of the *toro* varies from fish to fish; one slice from the back can sell for 4,000 yen ($35), while another, from another fish, can sell for 5,000 yen ($43). "Now, taking this into consideration," says Hirahara, "the *o-toro* from the bluefin that people bid for at Tsukiji goes for anywhere from 10,000 yen [$85] per kilogram up to 20,000 yen [$175]. The first class sushi restaurants and hotels compete for these. They normally charge anything from 3,000 yen [$25] to 4,000 yen [$35] a slice." Hori's book was written in 1996; the prices went up in the next decade.

There are reported to be seven thousand sushi restaurants in Tokyo, ranging from simple stalls at the Tsukiji fish market (where *toro* is too expensive to serve) to elaborate restaurants that appeal to tourists, people on expense accounts, and wealthy Japanese businessmen. One of these establishments is Fukuzushi in Rappongi, opened in 1917 in Tokyo, moved to Sapporo, and then relocated back to Tokyo in 1968. You enter the restaurant through a small courtyard with lighted lanterns and the sound of trickling water, but the interior is slick and modern, done up in panels of black and red lacquer. The set lunches feature sushi, *chirashi-zushi* (assorted sashimi with rice), or eel as the main course. On its website, Fukuzushi explains that *o-toro* (tuna belly) is the most expensive, the fattest of all the tuna—very fatty and meltingly soft. *Chu-toro* is slightly cheaper, darker, and a little less fatty. The restaurant offers "superb Japanese cuisine prepared by expert chefs using the finest fish from Hokkaido."

Until the American occupation changed the diet of the Japanese, they had little taste for fatty foods. As Sasha Issenberg put it his 2007 study *The Sushi Economy,* "after the American occupation, Japanese people got introduced to steaks that were greasy . . . Tokyo palates were acting a lot like those in Paris or Chicago, which associated luxury with fat, whether in foie gras, chocolate truffles, soft cheese, or porterhouses-for-two. In the 1960s restaurants pushed their suppliers for more of what was now known as *toro,* short for *toro-keru,* or 'melting on tongue' . . ." "It was America that raised the price of tuna, if you think about it," says Tsusenori Iida, a Tsukiji tuna buyer quoted in Issenberg's book. *Toro* was headed for America.

In his 2007 *Vanity Fair* article, "If You Knew Sushi," Nick Tosches reviews the introduction of exotic Japanese cuisine into the United States. At first, he says, Americans had no taste for raw fish, and "sushi" was actually sweet rice wrapped in seaweed. In 1923, in the Los Angeles district known as "Little Tokio," a restaurant called Kawafuku Café "may have been the first restaurant in America to serve sushi," but it is not clear if raw fish was incorporated. A restaurant called Kabuki opened in Manhattan in 1961, and *New York Times* restaurant reviewer Craig Claiborne wrote, "Not all of the dishes at Kabuki will appeal to American palates, among these sashimi or raw fish." But a couple of years later, when Nippon opened, Claiborne observed that "New Yorkers seem to take to the raw fish dishes, sashimi and sushi, with almost the same enthusiasm they display for tempura and sukiyaki."

Since then, modest sushi parlors began to appear in New York, followed in recent years by more spectacular Japanese restaurants. The first of the *nouveau japonais* restaurants was Nobu, opened in 1994 by Drew Nieporent, which featured the culinary creations of Nobu Matsuhisa. Trained in Japan, Matsuhisa set up his first establishment (called Matsuhisa) on La Cienega Boulevard in Los Angeles in 1987, but his was not traditional Japanese cuisine. The fusion menu blended Japanese dishes with Argentine and Peruvian ingredients, and eschewed sushi and sashimi. Particularly popular with movie stars, Matsuhisa charges at least $80 per person, and you can easily spend $250 on dinner for two. When chef Nobu came to New York in 2001, he was interviewed by Florence Fabricant of the *New York Times,* who quoted him:

> "Many years ago, before Americans were as adventurous as they are now, a customer in my restaurant Matsuhisa in Los Angeles said I could prepare whatever I wanted for her. But when I served sashimi she said she would not eat raw fish. I took the plate—the freshest fish, so carefully sliced and arranged—back to the kitchen and wondered what to do . . . Then I noticed a pan with very hot olive oil on the stove. I poured some over the sashimi and served it again. She ate one bite, then another. She cleaned the plate. The fish was not really cooked, just about 10 percent, so it was a bit of a joke, but it was enough to make it acceptable to her."

There are now Nobu restaurants in London (opened 1997), Las Vegas (1999), Milan (2000), Miami Beach (2001), and Tokyo (1998). In the upscale Aoyama district, Nobu Tokyo offers the same "fusion" menu as the other restaurants, with untraditional dishes like black cod with miso, or squid in light garlic sauce. There are now three Nobus in New York, and while not one of them serves unadorned raw fish, all three offer farm-raised yellowtail (*hamachi*) and yellowfin tuna (*toro*):

Yellowtail Tartar with Caviar. . . . $21.00
Toro Tartar with Caviar. . . . $30.00
Yellowtail Sashimi with Jalapeño . . . $17.00

Many Japanese restaurants do serve sashimi, however, such as Matsuri on West Sixteenth Street in Manhattan. In a spectacular, soaring space decorated with glowing, giant paper lanterns, the menu is typical Japanese, with tempura, teriyaki, grilled fish and meats, but also a generous selection of sushi and sashimi. "The common denominator of Japanese cuisine," says Matsuri chef Tadashi Ono, "is fresh ingredients"—especially those flown in from Japan.

Although there are other cities that claim the title, the centers of haute cuisine in the United States are Los Angeles and New York. It is no coincidence that the Japanese chef who is currently the most brilliant star in the restaurant firmament opened his first restaurant in L.A., and followed it with one in New York. Masayoshi Takayama operated Ginza Sushiko in Beverly Hills, and in 2004 opened Masa in

Three cuts of bluefin tuna *maguro* prepared by chef Masato Shimizu at the New York restaurant 15 East. The color and fat content of each cut determines the price.

the new Time-Warner Center at Columbus Circle. With a prix fixe of $350 (excluding tax, tips, and beverages), Masa immediately became the most expensive restaurant in a city known for pricey dining. Lunch or dinner for two can easily exceed $1,000. There is no menu; diners get to eat what Takayama chooses to serve that day—a concept known as *omakase.* You might start with an *uni* risotto with white truffles, or tuna tartare and caviar, but the restaurant is really about sashimi. The *New York Times* restaurant critic Frank Bruni gave Masa four stars, a rare accolade, and rhetorically asked himself if the food was worth the price: "The answer depends on your budget and priorities," he wrote. "But in my experience, the silky, melting quality of Masa's uni and toro and sea bream, coupled with the serenity of the ambience, does not exist in New York at a lower price." Of the *toro,* he wrote, "Some of this flesh was so luxurious that it made me feel flushed, giving me a buzz that undulated across a meal and crested with the toro rolls; insanely dense, obscenely intense clumps of fatty red tuna surrounded by rice and seasoned with wasabi and scallions." Adding to the restaurant's cachet (and price) is the suggestion that the fish is flown in daily from Japan; as Bruni wrote, "Mr. Takayama trawls the globe, reeling in bay scallops from as nearby as New England and grouper from as far away as Japan." The *toro,* Masa's pièce de résistance, is likely to have been bought at an auction at Tsukiji, but because many of the prime northern bluefins come from New England, it is possible that Masa's exalted pricing incorporates a toro trip from New England to Japan and back again.

To research his *Vanity Fair* article about sushi, Tosches visited the Tsukiji market (in the company of Harvard anthropologist Ted Bestor and Tomohiro Akasawa, senior commercial specialist for NOAA in Tokyo) and ate a lot of raw fish—at Tsukiji, at restaurants in the Ginza, and also at Masa in New York, which he called "the most expensive restaurant in the country, if not the world." Tosches's dinner for two cost *Vanity Fair* $1,102.74, and consisted of various exotic appetizers, such as icefish, "tiny almost translucent fish with buggy little black eyeballs"; a "hot pot of cherry trout, whose season lasts only a few weeks in the spring"; and then the sushi feast:

> Each piece of sushi is prepared individually and served immediately as Takayama-san slices the fish, reaches into a cloth-covered barrel of rice, applies fresh-made wasabi paste to the side of the sliced fish that will be pressed into the rice, and piece after piece, forms perfect sushi with dexterous rapidity in the palm of one hand with the nimble fingers of the other, placing it before you on a stoneware dish. He tells you to eat it with your hand . . . here, at the most opulent sushi restaurant on earth, the guy is telling me to use my hands.*

Not everybody agrees that the most expensive sushi is the best sushi. In *The Zen of Fish: The Story of Sushi from Samurai to Supermarket*

*Readers might wonder why I relied upon the Masa reviews of others, and didn't just make a reservation there myself. I did go all the way to South Australia to "review" the tuna farms there, and as I live in the city where Masa is located, I could easily have taken a subway there—or a limo, given what I would have spent for dinner—and been able to present my own impressions of the restaurant. But while the *New York Times* and *Vanity Fair* subsidized the visits of Bruni and Tosches, respectively, the editors at Knopf were unlikely to pay for a thousand-dollar dinner for two—even if (well, *especially* if) one of the two was my editor. Besides, the point of writing about ludicrously expensive sashimi is not whether I thought it was worth the price, but whether anybody else did.

(2007), Trevor Corson describes the experiences of Jeffrey Nitta, a Japanese-American restaurant consultant, who first ate sushi with his father in a bowling alley in Los Angeles:

> To Jeff, Masa and his prices were as much an affront to Japanese sushi as were inside-out rolls loaded with mayonnaise and chili—maybe worse. In Japan, the sushi experience was a matter of getting to know the chef at your neighborhood sushi bar, visiting frequently, and letting him choose what he thought you would like from the freshest ingredients of the day . . . In Manhattan, the Chef Masayoshi Takayama wouldn't even consider serving you lunch or dinner unless you handed him $350. Thanks to him, America's most expensive restaurant is, astonishingly, a sushi bar.

Because wild bluefins can be caught off the coast of New England, it seems unnecessary to fly New England tuna to Japan and then back to the United States. Of course, some of the top-quality bluefins taken in the Atlantic are purchased by Japanese buyers and auctioned off at Tsukiji, but many New England bluefins make a much shorter trip: New England waters, Massachusetts dock, New York fish market, New York Japanese restaurant.

Theodore Bestor, professor of Japanese studies at Harvard, points out that sushi has become a global phenomenon: "Japanese culture and the place of tuna within its demanding culinary tradition is constantly shaped and reshaped by the flow of cultural images that now travel around the globe in all directions simultaneously, bumping into each other in airports, fishing ports, bistros and bodegas everywhere. In the newly rewired circuitry of global cultural and economic affairs, Japan is the core, and the Atlantic seaboard, the Adriatic, and the Australian coast are all distant peripheries." Japanese food is in vogue in urban Asia, and demand for sashimi and sushi products is on the rise. This has created opportunities for selling sashimi-grade tuna, particularly in cities such as Shanghai, Guangzhou, Beijing, Hong Kong, Taipei, Singapore, Kuala Lumpur, Bangkok, and Ho Chi Minh City. Many holiday resorts in the Asian Far East have also added sashimi tuna to their buffet lunch and dinner menus. Supermarkets in Singapore and Malaysia have also introduced tuna loins in the seafood section. Fresh sashimi is also an item offered in many upmarket hotels and restaurants in the Asian Far East. China, housing the largest population in the world, is potentially a big market for sashimi as well as canned tuna. Nowadays, Chinese tuna fleets operating in the Pacific Ocean are selling part of their catches on the mainland. In 2005 the Japanese conglomerate Nissho Iwai launched a promotional campaign for sashimi tuna in China.

The rising global popularity of sashimi will surely be the downfall of the bluefin tuna. The all-out onslaught on the species, whether wild-caught or raised in pens, will end badly for everybody, including fishermen, fish dealers, restaurateurs, and consumers. It goes without saying that it will end badly for the tuna. If tuna become scarce the price will go up, which will encourage intensified fishing effort, which will, of course, reduce the populations even more. This was the lesson so painfully learned with codfish. As the fish became scarcer, the codfishermen whose lives and livelihood depended on this fish tried to increase their catches—after all, they had to make payments on their boats, pay off their mortgages, feed their families, and besides, this was the only job they knew. Western Atlantic codfish stocks declined to the point where the U.S. and Canadian governments were forced to close down the fishery permanently, which put the codfishermen out of business whether they agreed with the scientists or not.

The codfishermen of Norway, however, are becoming codfish farmers, and on an increasingly larger scale. According to Paul Greenberg's 2006 article, "In 2002, the Ministry of Fisheries established the Tromso cod breeding facility, at a cost of $18 million. The Tromso-based seafood export council has started a 'Cod TV' cable channel, and a network of government agencies, farmers, and researchers has come together as '*Sats pa Torsk!*' ('Go for Cod!')." Norwegian cod scientists figured out how to keep juvenile cod from eating their pen mates; how to slow their sexual maturation; how to change their diet from uncultivatable zooplankton to cultivatable rotifers; and how to make industrial production economically feasible. They just haven't done it yet. Fishermen of New England and Maritime Canada no longer seek cod in their waters, because there aren't enough to support a fishery (and the governments have shut down the fisheries anyway), but the

Norwegians are banking on their ability to breed cod in sufficient quantities to establish them as viable food fishes once again. The future of the Atlantic cod—and of the Norwegians who would eat *lutefisk* (cod cured in lye)—swims now in pens, not in the open ocean.

One has only to pay a visit to the Tsukiji market to realize that the Japanese approach to marine life is somewhat different from that of most Westerners. As a seagirt island nation with little in the way of natural food resources, Japan is largely dependent upon the sea to provide the major sources of their protein, whether in the form of fish, shellfish, cephalopods, lobster, shrimp, jellyfish, sea cucumbers, seaweed, sharkmeat, or whalemeat. After World War II, with the Japanese economy in tatters, General MacArthur arranged for the delivery of tankers that could be converted to factory ships so the Japanese could hunt whales to feed their starving citizens. From that time onward, the Japanese slaughtered whales in numbers that defied both imagination and logic. Faced with international restrictions, they ignored them. While they paid lip service to the quotas at International Whaling Commission (IWC) meetings, they killed every kind of whale anyway, and then lied about what they had done.* In Matsuda's history of the tuna fishery he says, "The tuna industry is the symbol of Japanese fisheries," and indeed it is, but perhaps not in the way he intended. As with the whales, the Japanese killed as many tuna as they could (and in the case of southern bluefins, lied blatantly about the catch totals), but because they maintain the market for large tunas, they encourage every other fishing nation to participate in this free-for-all. And just as their takeover of the IWC in 2006 changed the rules (and came close to abolishing the moratorium on commercial whaling that had been passed in 1983), their tremendous influence in ICCAT virtually guarantees that bluefin tuna will continue to be caught in numbers that will fulfill the bottomless lust for sashimi, and as an unfortunate side effect, drive the tuna closer and closer to extinction.

*Isao Kondo, a former executive of Nihon Hogei (the Japan Whaling Company) and director of whaling stations at Ayukawa, Taiji, and Wakkanai, published *Rise and Fall of the Japanese Coast Whaling* in 2001, in which he detailed the methods by which Japanese whalers "misreported" the whales they killed: "Cheating on the taking of sperm whales began in 1950. And the degree became more terrible after 1955 . . . Then, the cover up of captured whales escalated. It was said that according to the data of one whaling company, the captured number made public was 30.3% of the actual number—in other words, they captured three times the number made public."

Bluefin meat isn't particularly nutritious—it may even contain dangerous levels of mercury—and its preparation requires little more than the skilled knife work of the carver. No elaborate cooking techniques, no fancy sauces, no exotic accompaniments. Just a piece of red, raw fish on a plate. How can such a menu item command prices higher than traditionally expensive items like beluga caviar or white truffles? Caviar and truffles are rare, and often require esoteric methods of collection (black market sturgeon for caviar, rooting pigs for truffles), but as one can see at Tsukiji every day, bluefin tunas arrive by the hundreds. They may be becoming rare, but they are certainly not in short supply. Bluefin tuna *toro* may be the modern-day equivalent of the Dutch tulipmania of the seventeenth century, when buyers of a particularly rare hybrid would mortgage their houses, sell the family jewels, or otherwise contrive to raise enough money to buy a single tulip bulb. When the bubble burst, many people went bankrupt as the perceived value of the bulbs plummeted to nothing. Will modern consumers ever come to realize that, however succulent, a piece of fish is just a piece of fish, and that the current tunamania may be driving the tuna out of existence?

Once upon a time, the North American passenger pigeon was the most numerous bird on earth. In the eastern United States it numbered in the billions, outnumbering all other species of North American birds combined. Larger and more graceful than the common street pigeon, *Ectopistes migratorius* was a soft gray on its head and back, and rufous pink on the breast, fading to white on the underparts. The bill was black, the feet red, and the eyes orange. Their flocks darkened the skies and extended for miles. Their nesting sites covered hundreds of square miles of forest, with dozens of birds in each tree. As the forests were cleared and converted to farmland, the pigeons' habitat began to disappear, but by far the greatest cause of their decline was the human hunter. Many passenger pigeons were shot for the pot, but untold millions were shot for "sport." In one competition a participant had to kill thirty thousand birds just to be considered for a prize. Thousands of hunters slaughtered millions of birds, which were sold and shipped out by railroad at a price of fifteen to twenty-five cents a dozen (Matthiessen 1959). The last legitimate sighting of a wild passenger pigeon was in 1900 in Ohio. A few indi-

Eight

TUNA FARMING

The offshore tuna pens require constant maintenance.

IN 1937, in a remote bay on the south shore of Nova Scotia, twenty-five miles west of Halifax, some fishermen conceived the idea of trapping big bluefin tuna in mackerel traps and fattening them for sale. But there was no market for "horse mackerel," and the experiment failed (Butler 1977). By the mid-1970s, however, the escalating Japanese sashimi market changed everything for the trap fishermen of St. Margaret's. In 1975, the first impoundment nets (the equivalent of an open-water fish tank with sides of netting) were constructed, about five hundred feet across and fifty feet deep. When they arrived, the tuna had just migrated north from their spawning area in the Gulf of Mexico, and were emaciated and of little commercial value, so they were fattened on two feedings a day of herring and mackerel culls, squid, and whiting. Michael Butler, a Canadian marine biologist, took advantage of the opportunity to observe bluefins as nobody had ever seen them before. He saw how they ate (and how much); he affixed ultrasonic transmitters to monitor their temperature, their swimming depth, and the ambient water temperature; he found out how certain foods affected the fat content of the meat (to make them more desirable for the Japanese market); and he invited underwater photographers to dive with the captive fish and photograph them. Butler turned out many ICCAT reports on St. Margaret's, and in 1982 he wrote "Plight of the Bluefin Tuna" for *National Geographic.*

At St. Margaret's it was Japanese technicians who built the original eight holding pens; even at this early date, the Japanese had recognized the efficacy of shipping bluefin tuna directly to Tokyo. By 1976, the Canadians were sending three hundred frozen bluefins a year to Japan. Business was so good that the following year Janel Fisheries doubled the number of impoundment nets at St. Margaret's, and shipped three-quarters of a million pounds of dressed tuna to Japan at a freight rate of a dollar a pound (Butler 1982). St. Margaret's Bay fishermen have been selling "farmed" bluefin to Japan since the 1970s; because the Canadian tuna are fed on mackerel—a very oily fish—the fat content of these fish makes them particularly desirable at the Tokyo auctions. In 1978, only three years after this lucrative business had begun, the large tuna became scarce, and in 1981 just 116 fish were trapped in St. Margaret's Bay. Janel never closed the operation; the trap net tuna fishery of St. Margaret's continues to this day, a low-profile precursor of the larger, more intensive, and more efficient tuna ranches of Japan, Australia, and the Mediterranean that have come to dominate the news recently.

"Shooting fish in a barrel" is a phrase that denotes an act of consummate facility; the shooter can't miss the fish in the barrel. With the exceptions of fly-fishing, big-game fishing, spearfishing, and surfcasting, fishing is usually not considered a "sport"; for most people it is a hobby or a recreation. In most sports, teams or individuals are supposed to be evenly matched, and the field should be level, so the best player or team might win. But those "sports," such as big-game hunting or big-game fishing, that involve the killing (or the attempted killing) of a living animal, are played on a distinctly unlevel field; the hunter is armed, and the prey doesn't even know it is in the game. The hunters are "smarter," of course; they equip themselves with the latest rods, lures, hooks, rifles, scopes, or ammunition, and then take off after an unsuspecting animal, which is usually doing nothing more than minding its own business. And the hunter "wins" when he kills the animal. Unless the fish is a great white shark and the "hunter" decides to engage it in its own element, armed only with his teeth, man versus fish is a pretty lopsided battle.

No one ever described commercial fishing as a sport. It is difficult, often dangerous, and for most fishermen, a hell of a hard way to make a living. To succeed, fishermen have access to the biggest, fastest boats and the latest fish-finding and fish-trapping technology, but the fish is still doing nothing more than minding its own business. The justification for fishing, or course, is that we have to eat, and somebody has to provide the food to feed us. Even when commercial fishers use Loran, fish finders, or spotter planes, at least the fish are in their own element, and have a "sporting chance" of escaping with their lives. But purse seines, bottom trawlers, longlines, and drift nets have changed the equation completely, entrapping large numbers of fish at once, essentially by scooping up the environment along with the fish. Aquaculture—raising food fishes from eggs to edible sizes—is certainly not a sport, but it is not fishing, either. It is raising "domesticated" animals for food, just as pigs, chickens, and cows are raised. Catching wild fish and fattening them in pens is altogether different. It is not exactly fishing, and it is not exactly mariculture. (It is, as it were, another kettle of fish.) It is changing everything we thought we knew about fishing, and unfortunately, everything we thought we knew about preserving marine resources for the future.

A loophole was discovered in the regulations governing Mediterranean tuna fishing that might signal the total extinction of the sea's bluefin population within a few years. While there are strict quotas on the number of fish that can be caught in nets or by harpoons *(spadare),* there are no regulations whatsoever applied to the practice of "post-harvesting," which means catching wild tuna and keeping them in pens before they are slaughtered. In 2001, there were post-harvesting "farms" in the waters off Spain, Italy, Malta, and Croatia, accounting for some eleven thousand tons of tuna caught, as compared to a total of twenty-four thousand tons landed throughout the Mediterranean by direct fishing. More than 90 percent of the post-harvested tuna

Every country on or in the Mediterranean (except Israel) practices tuna farming, catching half-grown fish and fattening them in pens until they reach the size where they can be killed and shipped to market.

went to Japan. "If nothing is done," says Paolo Guglielmi, of the World Wildlife Fund's Mediterranean Programme Office, "wild bluefin tuna will completely disappear from the Mediterranean, perhaps with no possibility of rebuilding stocks" (Tudela 2000).

The decline in Mediterranean tuna populations was the subject of Elisabeth Rosenthal's *New York Times* story of July 16, 2006. She wrote that "the tuna population in the Mediterranean is nearing extinction, a new World Wildlife Fund report concludes, with catches down 80 percent over the past few years, even for high-tech trawlers that now comb remote corners of the sea in search of hard-to-find fish." Rosenthal visited Sucuraj, Croatia, where she found despondent fishermen, facing ruin because of the disappearance of the tuna. "Today," she wrote, "these majestic predators are rarely if ever caught. 'You have to work harder to catch fish of any kind,' said Lubomir Petricivic, a fisherman who recently opened a restaurant in the harbor here. 'Tuna? Impossible. We don't have any; we can't get it.' "

Post-harvesting—now known as tuna ranching—has completely reshaped fishing in the Mediterranean, and the fish are much the worse for it. Not only are the tuna threatened, but the fish caught to feed them while they are in the pens are also being fished to destruction. Almost all the countries that fish for tuna in the Mediterranean are switching over to this "feedlot" technology. In each of them, the purse-seine catches have declined, while the total catch has increased. The entire catch of the Croatian tuna fleet (growing from nineteen boats in 1999 to thirty in 2000) consists of undersized fish destined for the pens. According to the WWF report, "In the Mediterranean, tuna farming started just a few years ago, but estimated production in 2001 gives an indication of the huge development of this activity in the region. In fact, production in the Mediterranean is likely to make up more than half of the world total and is almost exclusively destined for the Japanese market." Given the eagerness with which Mediterranean nations sell their fish to Japan, it is not a little surprising to learn that Japan maintains a thirty-five-vessel longline fleet in the western Mediterranean, targeting prespawning, large tuna. Perhaps they believe they can avoid the cost of the middleman. In a further attempt to avoid European markups, Japan has now introduced its own tuna farms, with pens in eighteen Mediterranean locations.

In a 2002 article, Sergi Tudela, project coordinator of the World Wildlife Fund's Mediterranean Programme, wrote:

> In sum, all the usual ingredients are there in the case of tuna fattening farms: privatization of a common good (in this case, with the added risk of its probable destruction in the short- to medium-term); concentration of the benefits into a few hands; public aid provided to pillage a natural resource; dispossession of the traditional resource users; social and economic deconstruction of the traditional fishing sector; complete lack of a regulatory framework; connivance of the administration; ineffectiveness of international supra-Statal organizations; and growing demand for the product from a powerful market.

True aquaculture requires that the fish be raised from eggs, not simply moved from one place to another to be fattened. The system of post-harvesting (as now practiced in the Mediterranean) does not qualify as aquaculture under this strict definition, but it still demonstrates all the ills that besiege legitimate aquaculture, such as that practiced with Atlantic salmon. Like salmon, tuna are carnivorous, and must be fed large quantities of small fishes, which themselves may be threatened by overcollecting. This kind of "farming" does not relieve commercial fishing pressure—it increases it. Waste from the pens is another problem, as is their location, close enough to shore and urban centers to disrupt and often pollute the littoral zone. And because tuna farming falls between the definition of a fishery and true aquaculture, it is completely unregulated on a world scale.

For all their vaunted migratory capabilities, no tuna ever swam from New England to Japan, but, as Safina wrote, "probably more bluefins from the east coast of North America cross the Pacific because the next step in the transaction is a one-way air-freight ticket to Tokyo." The same is true of the Mediterranean bluefins. The future of the bluefin tuna, then, is written in Japanese. There is a better than fifty-fifty chance that people enjoying *maguro* in Japan are eating fish that were fattened in pens in the Mediterranean. Compared to almost nothing five years ago, the twelve Mediterranean tuna farms produced eleven thousand tons of tuna in 2001, more than half of the world's total.

The first commercial operation in the Mediterranean was in 1979, at Ceuta, Spanish Morocco, across the Strait of Gibraltar from Euro-

Tuna pens in the Mediterranean are proving to be the downfall of the species.

pean Spain. Lean tuna were caught in *almadrabas*—huge fish traps that consist of miles of nets suspended from buoys and anchored to the sea floor—as they tried to leave the Mediterranean after the June and July spawning season was over. Only two hundred tons were caught, and the product was sold at a premium price to the sashimi market in Japan. With some modifications, what began in Ceuta was the beginning of what would become a major industry throughout the Mediterranean. In the future, tuna fishermen wouldn't wait to trap the tuna in *almadrabas;* they would catch them in purse seines and tow them to the fattening pens.

In 1985, I visited South Australia to research an article on the fish and fisheries of that state for *National Geographic* (it ran in May 1987 as "Australia's Southern Seas"). At Port Lincoln, I interviewed Dinko Lukin, then one of the town's leading tuna fisherman. Beginning in 1977, Lukin's crews caught the fish (southern bluefin tuna) by the pole-and-line method, flash-froze them at sea, and then brought in a hold full of frozen tuna to be sold to Japanese buyers who were waiting on Port Lincoln's docks. The frozen tuna were then transferred to

Workers for Lukin Industries prepare frozen tuna for shipment to Japan, 1985.

a Japanese vessel, which sailed to Japan immediately. Lukin was among the first Australians to sell tuna for the Japanese sashimi market, and the bulk of his catch was shipped to the Tsukiji market in Tokyo. At that time, there were no restaurants in Australia serving raw fish, so almost all of the catch went straight to Japan. (A Google search for "Japanese restaurants Sydney" on July 10, 2006, listed seventy-four in that city and its suburbs, so it is reasonable to assume that some of the tuna now being caught in Australian waters remains there. Indeed, when I returned to Port Lincoln in 2007, I was served grilled bluefin tuna on two occasions.)

In their efforts to supply the expanding Japanese market, Lukin and others overfished the tuna populations on the high seas and in Australian coastal waters, until it became increasingly evident that the stocks were in a steep decline. A trilateral commission, consisting of representatives from Japan, New Zealand, and Australia, developed conservation strategies for SBT that were supposed to protect the stock, but they could not agree on workable programs, and the stock continued to decline. Not long after my visit, South Australian tuna fishing had just about shut down. But Lukin, who had established strong ties with the Japanese, managed almost single-handedly to turn the industry around. Working with a Japanese tuna auctioneer named Hideo Hirahara, he began by catching tuna with baited hooks and trying to transfer them to pens, but this was too harmful to the fish, so they devised a scheme whereby they would purse-seine juvenile tuna in the Great Australian Bight and tow them in floating cages to the waters off Port Lincoln. There they would be fattened for three to six months, killed, frozen, and shipped to Japan. Now a millionaire many times over, the Croatian-born Dinko returned to his home country in 1996 and began tuna ranching there. In a few short years he made tuna Croatia's best-known export product, and tuna farming the most successful investment in that country.

The Australian tuna-ranching industry has prospered since its inception in 1991. After its rise and fall as the heart of South Australia's tuna fishery, Port Lincoln is once again at the epicenter of a thriving fishery, this time without rods. In 2005 the southern bluefin tuna harvest reached nine thousand tons, the biggest to date. The fishery is worth AUS$280 million a year. There are now fifteen tuna ranches in South Australia, located primarily in two areas just east of Port Lincoln: Boston Bay and Rabbit Island. (This locale is famous for great white sharks, and it is not unusual to find one of these "man-eaters" trapped in a tuna net.) Environmental groups have been lobbying for quotas, arguing that the very stock is threatened, but the tuna-ranching industry claims that a reduction in the catch would put people out of work, and besides, the status of the stock is fine. Roberto Mielgo Bregazzi heads Advanced Tuna Ranching Technologies (ATRT), a watchdog organization based in Madrid. "The benefits of ranching have been numerous," he wrote in a 2005 report, "not only to the owners but also to regional employment and the spin-off industries such as tourism, transport and freight. The total export value of the industry has grown from $45 million in 1994 to $252 million in 2001—an increase of 560% in just eight short years."

In an article in the *New York Times* in April 2002, R. W. Apple Jr. wrote about the "new kind of mariculture" that was being practiced off the Pacific coast of Baja California. Mexican fishermen were netting young bluefins and towing them to special enclosures in Puerto Escondido, near Ensenada, where the fish were kept in circular pens and fed live sardines three times a day for six to eight months. When they reached a weight of about 190 pounds, they were killed and frozen. "Despite the lasting slump in the Japanese economy," wrote Apple, "the meat sells for as much as $45 a pound." There are now a dozen offshore fish farms in Mexican waters, and the Norteamericanos are looking to get in on the sashimi bonanza.

When Apple discovered that "new kind of mariculture" in 2002, he could not have predicted the enormous investment that the Japanese were planning to make in Mexican tuna farms. The first farm, off Cedros Island on the Pacific side of the Baja Peninsula, was opened in 1997, and managed to produce only sixty-four tons of tuna in its initial three years of operation. Two more farms were opened in 1998 and 1999, at Salsipuedes and Todos Santos Bay (off Ensenada), and eight more were sited in Bahia La Paz, just north of the city of the same name. As of 2005, there were twenty-two tuna ranches in Baja California, producing more than five thousand tons of tuna. Over the past few years, Japanese fish buyers have paid the tuna farmers more than

U.S.$50 million, securing almost the entire harvest. Bregazzi's report states:

> Donshui, a subsidiary of [the] Japanese conglomerate Mitsubishi, is investing 150 million dollars in building fish farms for tuna in La Paz, Baja California Sur. Mitsubishi, one of the world's most diversified transnationals, is the world's largest tuna trader . . . Mitsubishi is partnering with local entrepreneur Mateo Arjona, in order to invest 18 million dollars building 22 large pens to hold yellowfin tuna ranching near San Juan de la Costa. Waters in the bay of La Paz are considered the best in the world for tuna ranching.

Although the weather and waters in Mexico might be ideal for tuna ranching, no fishing operation is foolproof, and there will always be natural hazards to endanger the fish and the fishermen. Weather anomalies such as El Niño and other storms can wreak havoc with a ranching operation, and there are occasional red tides that can kill all the fish in a pen. In March 2003 a cage full of southern bluefins being towed from the Great Australian Bight to Port Lincoln collapsed, and the netting entangled the fish, preventing them from swimming. Many of them drowned. Although Brian Jeffriess of the Australian Tuna Boat Owners Association said the "deaths were minimal," other observers put the death toll at close to five thousand (Lato 2003). The large quantities of baitfish used to feed the tuna attract sea lions, and sometimes the sea lions get into the pens and attack the tuna. In November 2003 an eighteen-foot-long great white shark made its way into the tuna pen at Coronado Island (Mexico). Although it was shot numerous times, it refused to die. An Australian (Port Lincoln) diver named Deano Stefanek jumped in and spent thirty minutes inside and outside the pen trying to kill the shark with a spear gun. It was finally dispatched, leading to worldwide news items headlined "Diver Attacks Great White."

The great white, the subject of Peter Benchley's best seller and the star of four Hollywood movies, is the most spectacular of the uninvited guests in the tuna pens, but other shark species have also been known to take advantage of the captive buffet. In a 2004 study, Txema Galaz and Alessandro de Maddalena assembled accounts of

"Diver Attacks Great White." Australian diver Deano Stefanek jumped into a tuna net in Mexico and killed an eighteen-foot shark that was menacing the tuna.

blue sharks, makos, and several great whites in the pens, and in the Australian films *Tuna Cowboys* and *Tuna Wranglers,* the divers successfully evict blues, makos, and bronze whaler sharks from the tuna cages. In June 2002, a towing boat on its way from Libya to Spain stopped to check the cage containing sixty tons of bluefins, and watched as a sixteen-foot female white shark tore a hole in the net and joined the tuna in swimming round and round. The shark was never seen to molest the tuna, and after two days of towing, the shark left the cage voluntarily.

In 2004 Rex Dalton wrote that in addition to the tuna farms in the Mediterranean, Mexico, and Australia, "in the United States, legislation is quietly being drawn up to facilitate such fish farming operations in offshore waters—beyond the environmental control of coastal states, in waters difficult to police." A study by Hubbs–Sea World Research Institute of San Diego (funded by Chevron) is to determine the feasibility of establishing a "tuna ranch" in the Santa Barbara Channel, to be affixed to an old Chevron oil-drilling platform. But the same problems attendant upon "traditional" aquaculture are present in tuna farming. Salmon and tuna require smaller fish to eat, so growing these fish in pens compels more, rather than less, fishing in the wild. There are any number of ecological threats that have been shown to derive from penned fish, including concentrated fecal matter and rotting food released into the surrounding waters; genetically altered fish escaping and breeding with wild fish; and probably most frightening of all, diseases spread from captive fish into the wild population.

On June 7, 2005, the Bush administration submitted to Congress for consideration the National Offshore Aquaculture Act, which would grant the U.S. secretary of commerce the authority to issue permits for tuna ranching in federal ocean waters. The bill was introduced by Senators Ted Stevens of Alaska (a state that prohibits aquaculture altogether) and Daniel Inouye of Hawaii. If passed, the measure would allow aquaculture (including fish farming) anywhere from three to two hundred miles off U.S. shores. The Hawaii state legislature has already passed legislation that allows tuna ranching in the waters of the western Hawaiian Islands. (Sometimes referred to as the Leeward Islands, the western islands are a chain stretching some 1,300 miles to the west of Kauai, including Nihoa, Necker, French Frigate Shoals, Gardner Pinnacles, Maro Reef, Laysan, Lisianski, Pearl and Hermes Reef, Midway, and Kure Atoll.) But instead of 1,300 miles west of Kauai, the first proposed tuna ranch called for the Ahi Nui Tuna Farming Company to catch juvenile bigeye and yellowfin and fatten them in eighteen pens to be developed 2,200 *feet* from the western shore of the Big Island. (*Ahi Nui* means "big yellowfin tuna" in Hawaiian.) Community opposition resulted in the tuna farm's being moved twenty miles offshore. Obviously, the market for fattened tuna in Hawaii is limited, and the fish will be "sold to sushi and sashimi markets primarily in Japan" (Bregazzi 2005). With the legislation pending, there are plans to test a tuna ranch off Santa Barbara, California, at the site of the Chevron oil company's obsolete Platform Grace.* There just might be enough sushi restaurants in the Los

*This region has already seen its share of ecological disasters. On January 28, 1969, at Unocal's Platform A, three miles off the coast of Santa Barbara, a drilling operation resulted in an uncontrolled flow of oil from a deep reservoir through oil-bearing sands. Some 3.2 million gallons (79,000 barrels) were released into the Pacific Ocean. Three days after the spill began, winds and currents drove the oil ashore, fouling more than one hundred miles of shoreline; by the fourth day, the oil had spread to

Angeles area so that the ranchers might not have to sell their fish to Japan.

Now the Japanese have entered the tuna-ranching sweepstakes. Japanese ranchers favor smaller tuna because they can compete with fishermen, who get a better price for a smaller fish if they sell directly at the market rather than to a tuna farmer. Still, the Japanese Fisheries Agency estimates that of the 21,500 tons of tuna imported in 2005, 90 percent of the total (19,500 tons) was *chikuyuō*—ranch-fattened fish. Originally, the idea was that sashimi-quality tuna was supposed to be *fresh,* but there is no way to ship unfrozen fish from Greece to Japan, so fresh has been replaced by frozen. The fish are transferred from fishing boats or tuna ranches to large freezer ships, known as reefers, the largest of which has a cargo capacity of 4,500 tons. (A proportion of the vast number of tuna carcasses seen every day at the Tsukiji market has arrived in reefers from Malta, Vietnam, Panama, or other distant ports.) There are also illegal tuna fisheries around the world, illegal, unreported, and unregulated (IUU), that kill an uncounted number of the target species, plus countless sharks, seabirds, marine mammals, and other fishes.

To avoid detection, illegal fisheries fish without a license, fish under a "flag of convenience," fish out of season, harvest prohibited species, use banned fishing gear (such as drift nets), and sometimes do not report catches at all. From Bregazzi's 2005 report: "Most of the IUU blue fin tuna caught in the Mediterranean enters Japan through 'inspection friendly' Chinese and Southeast Asian countries where tuna meat is processed, packed, and shipped under a different denomination product such as ready-to-consume frozen sashimi tuna." As Arata Izawa said (in the 2005 WWF report), "Japanese consumers have no clues to know whether or not the tuna they are eating is illegally caught." "Japan outclasses its rivals by far when it comes to [the] transport of frozen tuna," wrote Bregazzi. "An enormous fleet of Pana-

---

the Channel Islands of Santa Rosa, Catalina, and Anacapa. Eventually the spill covered more than eight hundred square miles and reached all the way to the Mexican border. Marine and terrestrial plants were destroyed; marine mammals, seabirds, fishes, and invertebrates were oil-soaked and killed. Bills were introduced in the California and federal legislatures to create oil-well-free zones, and in 1972 the National Marine Sanctuaries Act was passed, leading to the establishment of the Channel Islands National Marine Sanctuary in 1980.

manian or Japanese flagged reefers insure that the fish is delivered both to Japan and/or processing plants serving the Japanese market with sashimi and sushi." There follows a list of thirty-two reefer-vessel operators—not thirty-two reefers, but thirty-two *companies* that operate reefers. In 2005, more than seventeen thousand tons of ranched Mediterranean bluefin tuna reached Japan on board these ships.

The increase in tuna ranches in Mexico and the Mediterranean has put a crimp in the onetime dominance of the Australians. If a surplus of "product" is supplied to Japan from the Mediterranean tuna farms, the market will crash, and the income that supports the heavily capitalized Port Lincoln ranches will disappear. (Attempts are being made to stir up interest in bluefin tuna in the huge—and untapped—Chinese market.) But the greatest threat to the Australian tuna-farming industry is the declining stocks of the tuna themselves. In 1984, quotas were introduced into the fishery because the catches were dropping dramatically, and by 1988 the Australian tuna quota was cut from 14,500 tons to 6,250. When it was discovered that the southern bluefins migrated from Indonesia and Western Australia to South Australia, the Western Australian fishery was completely curtailed to allow the fish to get to South Australia. But still the S.A. stock continued to decline, and the tuna fishery would have collapsed completely if it hadn't been for the innovations of tuna farming.

Tuna ranching has now become a worldwide industry—and a worldwide threat to the tuna populations. In the Mediterranean, where giant bluefins used to be herded into pens to be killed for local consumption in the *mattanza,* juvenile tuna are now purse-seined and towed to pens for fattening. Harvesting tuna before they are old enough to breed is a guaranteed path toward population collapse. In June 2006 Greenpeace called for the immediate closure of the Mediterranean bluefin tuna fishery, claiming that the population was on the brink of extinction. Greenpeace activists aboard the *Esperanza* observed tuna farms and fisheries around the Balearic Islands, in the waters north of Egypt, and south of Turkey, and talked to the captains of the fishing vessels. The Greenpeacers concluded that "45,000 tons of bluefin tuna may have been caught each year in 2004 and 2005, despite the fact that only 32,000 tons can be caught legally." They found that the bluefin tuna fishery was completely out of control in Europe, and that ICCAT was completely incapable of enforcing regu-

Twice a day, every day of the year, the baitboats pull alongside the pens and the tuna are fed sardines ("pilchards") in Boston Bay, South Australia.

lations on the fishery. A report ("The Plunder of Bluefin Tuna in the Mediterranean and East Atlantic in 2004 and 2005: Uncovering the Real Story") was issued by the World Wildlife Fund (WWF) on June 30, 2006, calling for an immediate closure of the eastern Atlantic and Mediterranean tuna fisheries, because fleets particularly from France, Libya, and Turkey "are greatly exceeding their fishing quotas and deliberately failing to report much of their massive catches." (Unreported catches are slaughtered and processed at sea before being shipped to the Japanese market.) According to Roberto Bregazzi, the author of the WWF report, "Atlantic bluefin tuna stocks risk imminent commercial collapse. In the race to catch shrinking tuna stocks, industrial fleets are switching from traditional fishing grounds to the last refuges in the eastern Mediterranean and Libyan waters."*

Bregazzi was born in London in 1960 and lived there until the age

*The European Union (EU) is a voting member of ICCAT, and EU member countries are represented as a bloc at ICCAT meetings. The membership of ICCAT consists of the following countries, not all of which have anything to do with Atlantic tunas—or the Atlantic Ocean, for that matter: Algeria, Angola, Barbados, Brazil, Canada, Cape Verde, China, Croatia, Equatorial Guinea, European Community, France (St. Pierre et Miquelon), Gabon, Ghana, Guatemala, Guinea-Conakry, Honduras, Iceland, Ivory Coast, Japan, Korea (Rep. of), Libya, Mexico, Morocco, Namibia, Nicaragua, Norway, Panama, Philippines, Russia, São Tomé and Principé, South Africa, Trinidad and Tobago, Tunisia, Turkey, United Kingdom (Anguilla, Bermuda, St. Helena, Turks, and Caicos), United States, Uruguay, Vanuatu, and Venezuela.

of five, when his family moved to Spain. (His mother was English, which explains his command of that language.) He has become the gadfly of the tuna industry, concentrating mostly on the Mediterranean, but everywhere else as well. He began his career in 1996 as a diver for the Cartagena company Tuna Farms of Mediterraneo (TFM), and was soon given responsibility for the logistics of TFM bluefin fishing during the summer seasons, including live transfer and towing operations in Maltese, Tunisian, and Libyan waters. He also supervised the establishment of bigeye tuna farms off the Galápagos (for the Ecuadorans), and off Cape Town, Brazil, and French Polynesia. Bregazzi's reports (another was prepared for the ICCAT meeting in Seville in 2005) are meticulously detailed and documented, and although Bregazzi is now the CEO of Advanced Tuna Ranching Technologies (based in Madrid, as is ICCAT), he is also passionately dedicated to the preservation of tuna stocks. According to Bregazzi, the purposes of ATRT are: vertical integration of tuna-ranching activity, eco-friendly and sustainable tuna-ranching projects, reduction of production costs and timing, and reduction of tuna ranching's inherent risk. On the title page of the 2005 report, he wrote:

> ATRT's Tuna Ranching Intelligence Unity (TRIU) 2004 report was indeed intended to be distributed to tuna fishing and ranching operators throughout the Mediterranean Sea. The 2004 TRIU report has now hit almost every news website worldwide and has enjoyed excellent reviews from many tuna and ranching concerns . . . It has also been spinned and criticised by many in the industry. Our purpose has been served. We have launched a much needed debate among us all, about the future of tuna ranching . . .

Inside the harbor at Cartagena, Spain, in September 2006, Greenpeace activists formed a symbolic "tuna graveyard" with mock crosses and a banner saying "R.I.P. Bluefin Tuna 1996–2006," mourning the decade of tuna ranching there. The environmental organization called for the immediate closure of the Mediterranean bluefin fishery until it could be properly managed, and the adoption of urgent measures, including the establishment of a network of marine reserves to protect

40 percent of the Mediterranean Sea and regenerate its fish stocks. "Tuna ranches like this one are the cowboys of an industry that is directly responsible for wiping out the bluefin tuna from the Mediterranean Sea," said Greenpeace Spain's Sebastián Losada aboard the *Rainbow Warrior.* "A few greedy commercial interests, subsidised by the EU, are employing pirate fishing fleets and fattening tuna to fatten their own wallets. They are depriving hundreds of fishermen from trying to make a legitimate living from the bluefin tuna."

Faced with an incipient crash in Mediterranean tuna populations, what did ICCAT do? At their November 2006 meeting in Dubrovnik, Croatia, instead of a recovery plan for eastern Atlantic and Mediterranean tuna, delegates adopted a weak EU program that included a catch quota of 29,500 tons in 2007 compared to the 15,000 tons recommended by ICCAT's own scientists. The plan also allows fishing during the peak spawning season, which is the worst possible decision for a depleted stock. "This is a collapse plan, not a recovery plan," said Dr. Sergi Tudela, head of fisheries at WWF Mediterranean. "Today's decision sounds the death knell for bluefin tuna in the Mediterranean." According to a recent WWF report, actual catches of bluefin tuna in the Mediterranean are more that 50 percent over the quota set by ICCAT. This illegal activity has meant artisanal fishermen are catching 80 percent less tuna compared to the 1990s. Stocks in the oldest fishing grounds of the Balearic Islands have collapsed, and six farms in Spain closed in 2006 due to lack of tuna. EU fishing fleets are responsible for the bulk of illegal catches of bluefin tuna in the Mediterranean. "The EU has betrayed its obligation to sustainably manage fisheries for the sake of the short-term interests of its own bluefin tuna industry," Tudela added. Several countries attending the meeting, including the United States and Norway, had strongly supported scientific recommendations for stock recovery of Mediterranean bluefin tuna, but their proposals were rejected.

As of June 2007, the European Union ratified the full quota of 29,500 tons for the eastern Atlantic and the Mediterranean, an amount twice as large as their scientists had advocated. William Hogarth, assistant administrator for fisheries at the National Marine Fisheries Service (NMFS) and chairman of ICCAT since 2005, commented that "things are getting worse for the bluefin tuna worldwide," and therefore proposed that the United States should cut its share from 3,000 metric tons to 2,100. This, of course, angered American tuna fishermen, who, like almost all commercial fishermen, believed that quotas were the work of scientists who didn't understand fishing and who wanted to ensure that the fish survived while the fishermen did not. Yankee tuna fishermen might have learned a lesson from the 1990s collapse of the western North Atlantic codfish populations, but they never understood that they were directly responsible, and blamed the government, the scientists, the Japanese—and even the fish themselves—for the permanent closure of the Canadian and American codfisheries. American tuna fishermen were after the big fish that they could sell to Japanese buyers, but the Europeans wanted to catch the smaller fish that they could fatten up and *then* sell to the Japanese.

The bluefin fishery starts in April, when the fish swim into the Mediterranean to spawn. In May and June, once the fish are in the Med, large-scale fishing methods, such as longlines and purse seines, are employed to catch the fish, which are then transferred to tuna farms or shipped directly to Japan. In August 2007, WWF issued a report on its Web site entitled, "Bluefin Witness," in which the global conservation organization wrote that

> longliners, harpooners, and purse seiners all targeted the giants, driven by high prices paid in Japan, which consumes 40% of global bluefin landings and where a single bluefin has sold for over $US150,000! More recently, these fleets have used ever-more sophisticated means to find the tuna, including spotter planes and sonar equipment. . . . Current fishing of the larger eastern population—the last stronghold of the species—is a massive three times higher than the population can sustain. The population is classified as overfished and endangered, and has already disappeared from the North Sea and Black Sea. . . . Few believe that the Med bluefin stock will survive for much longer. Many fear that there is no future in the fishery. In 2007 the fleets aimed to take as much as they could, and in some cases, by any means necessary.

The following month, WWF proposed "the immediate establishment of a sanctuary for the imperiled bluefin tuna around the Balearic

Islands in the western Mediterranean." Situated offshore between Tarragona and Cartagena are Spain's tuna farms, the largest in the Mediterranean. According to the 2006 WWF report "The Plunder of Bluefin Tuna in the Mediterranean and East Atlantic," the Spanish ranching capacity is 11,582 metric tons, almost twice that of Croatia, the country with the next highest capacity. The area around the Balearics is one of the major breeding grounds of the bluefin, and heavy fishing has catastrophically reduced the western Mediterranean population. According to a WWF newsletter dated September 12, 2006, "In 1995 some 14,699 tons were caught there, mainly by French and Spanish fleets—while just 2,270 tons have been fished in the same waters this year." The farming activity has migrated eastward because of overfishing the waters around the Balearics, so a sanctuary sounds like a necessary proposal, but it does not seem likely to pass at the November 2007 ICCAT meeting in Turkey.

According to Sergi Tudela, the European Commission closed the Mediterranean bluefin fishery as of September 19, 2007, because "the European fleets had already well overfished their quotas for 2007." The EU is obliged to close the fishery when the quota has been reached, which actually occurred in June. "The only way to guarantee a future for Mediterranean bluefin is to completely close the whole fishery in June, the key month for the spawning tuna," added Tudela. The quota for Mediterranean bluefins for 2007 was 16,779 tons, of which France was allocated 5,493. The French took some 10,000 tons, exceeding their quota by 40 percent. Even though Cyprus, Greece, Malta, Portugal, and Spain had not filled their assigned quotas, the EU closed the Mediterranean to European tuna fishing for the rest of the year. No penalties were levied against the nations in violation of the quotas, which is just one of the weaknesses of the ICCAT quota system. Even more critical is the insufficiency of real-time reporting and control of activity on the water, all of which leads to massive overfishing. Those fishing countries on the North African coast of the Mediterranean are not affected by the EU closure, and continue to fish at will on this seriously endangered population. France does not have tuna farms, but otherwise, most of the fish caught in the Mediterranean are headed for the fattening pens.

The heretofore obscured news about the imminent collapse of the

Mediterranean bluefin stock is finally coming out. "Sushi Craze Threatens Mediterranean's Giant Tuna" was the headline on an October 2, 2007, article by Reuters reporter Ben Harding. He continued:

> Japanese demand for the fatty flesh to make sushi has sparked a fishing frenzy for the Atlantic bluefin tuna—a torpedo-shaped brute weighing over half a tonne that can accelerate faster than a Porsche 911. . . . Tuna has become a big business throughout the Mediterranean, and the lure of up to US$15,000 for the best and biggest fish attracts dozens of new boats to the industry every year—many controlled by Asian and Italian mafias, sources say. That in turn depresses prices and compels fishermen to break catch limits. Some campaigners say it may already be too late to save the bluefin after high-tech fleets—many guided by illegal spotter planes—this season converged on an area near Libya that had been considered one of its last refuges. "It's over, that's my gut feeling from both a stock point of view and a business point of view," said Roberto Mielgo Bregazzi, a fisheries consultant who set up the first "tuna ranches" ten years ago.

"Tuna Fishery Like the Last Days of Buffalo Hunt," was the headline in the *Edmonton Journal* on October 7, 2007 (and they know about buffalo in Edmonton). The article by Margaret Munro is mostly about the role that Canadians are playing in the decimation of the tuna stocks of the western Atlantic, as they catch what are considered the last of the "giants" in the waters off Prince Edward Island. James Jones of Canada's Department of Fisheries and Oceans maintains that a much heavier toll is taken in the east—and it is, except that the tuna do not restrict themselves to one quadrant of the North Atlantic, and fish from the same population can be caught on either side of the ocean. Compared with the 32,000-ton quota allocated by ICCAT for European waters, 2,100 tons shared by the United States, Canada, Japan, and Mexico seems almost conservationist. "I am very concerned that the western stock has collapsed, and the eastern is, if not collapsed, on the verge of collapse," said William Hogarth, chair of ICCAT—and assistant administrator for fisheries at the National Marine Fisheries Service (NOAA Fisheries). Hogarth supported the cut to the eastern fishery, but was outvoted by European and African countries. "They just decided to blow off the science," said Hogarth. "To me that's inexcusable." In October 2007, Hogarth announced that he intends to ask the autumn meeting of ICCAT to implement a three- to five-year moratorium on all tuna fishing in the Mediterranean. "Given continued blatant violations of catch limits, closed areas, and reporting requirements," he said, "a moratorium is the best hope if we want to avert disaster for eastern Atlantic and Mediterranean bluefin tuna."

Countries currently farming tuna and those considering the activity do so almost exclusively for the Japanese market. The open-water tuna fisheries are in decline (prohibitively high fuel costs are keeping tuna clippers in port, and the fish populations are still falling), but ranching has been a godsend to the fishermen. From the first Australian harvest of tuna and their transfer into the first primitive cages, the ranching industry has spread around the world—wherever young tuna could be corralled and herded into pens. Those countries now engaged in tuna ranching are Australia, Cape Verde, Croatia, Cyprus, Greece, Italy, Indonesia, Japan (of course!), Libya, Malta, Mexico, Oman, Panama, the Philippines, Portugal, Spain, Tunisia, and Turkey. Note that of these eighteen countries, half are on (or in) the Mediterranean. The United States and Malaysia are now contemplating tuna ranching; Costa Rica was considering the establishment of tuna farms, but its Supreme Court ruled against it.

It is interesting—but unsurprising—to note that many tuna-ranching companies are capitalized by Japanese investments or have Japanese citizens as advisers or board members. Remember the visit of Japanese experts Hamano and Koga to Port Lincoln in 1991, with their goal of breeding tuna "so that they could be re-introduced to the sea"? This eleemosynary activity just happened to be a joint venture between the Japanese Overseas Fishery Co-operation Foundation and the South Australian government. The Sicilian tuna ranch operated by New Eurofish of Castellemare del Golfo was incorporated by local businessman Guglielmo Maggio and Mitsui & Co. of Japan. Of the facility of Tuna Fish S.P.A. of Palermo, Bregazzi wrote, "It is still unclear who is behind this operation and if Japanese or Korean concerns are involved in some sort of joint adventure with Sicilian owners." On the north coast of Bali, southeastern Asia's largest tuna-ranching center is a joint venture between Japanese and Indonesian investors, and Malta Fishfarming Ltd. and Melitta Tuna Ltd. were subsidized by Takayama Seafood, which invested U.S.$157,000 in the project in the form of tuna futures.

In Costa Rica, in July 2006, the company Granjas Atuneras de Golfito S.A., backed by Spain, Venezuela, and Peru, proposed a yellowfin tuna farm off the Pacific coast at Golfo Dulce. The operators claimed that in addition to raising wild-caught yellowfins, they would also be breeding them. The plan was opposed by a consortium of diverse groups, including fishermen, chambers of tourism, the indigenous Guayami community, development associations, businesses, scientists, senators, and conservation organizations. Once the groups began investigating, they found that multiple national laws would be violated, and the proposed project would put the entire ecosystem and economy of the Golfo Dulce at risk. One of the key dangers was that the fifteen thousand baby sea turtles that hatch on the beaches right in front of the proposed site would swim through the mesh of the nets, which would be located less than a mile offshore. In May 2007, the Supreme Court, arguing that the tuna farm would violate many Costa Rican environmental strictures, ruled that Granjas Atuneras could not construct such a facility in Golfo Dulce.

Tuna farming creates not only ecological problems, but problems that have to do with the very nature of the fish itself. Until the early 1990s, there was a market in Japan for the prime, very expensive, sashimi-quality meat that came from the wild-caught northern and southern bluefins, and a secondary market for the lower-quality meat served in nonsashimi restaurants and sold in supermarkets. "However," wrote Miyake et al. in their 2003 review of Mediterranean tuna farming, "[northern] bluefin and southern bluefin of smaller size that had been accepted only at the lower quality markets before, now fattened by farming and available in abundance . . . started to constitute a middle category, filling the gap between the two extremes." Farm-fattened fish are now a source of lower-priced *toro,* which can be sold in sushi bars and supermarkets. A glut of lower-priced tuna meat made open-ocean bluefins even more desirable, which encourages heavier fishing pressure on the pelagic populations. Tuna have the misfortune to be on or near the top of the list of "most popular food fishes," so it is more than a little painful to realize that this most beautiful fish in the world is literally being eaten out of existence. At the dock and in restaurants, prices for these fish rise as their numbers diminish. This sounds like nothing more than a traditional "supply and demand" equation, but the difference between fishing and manufacturing is that once the fish are gone, you cannot make any more.

As with fine wines, connoisseurs of fine tuna are evidently able to identify the "vintage." A diner at a sushi bar in Nagoya was served a piece of tuna from a fish supposedly caught in the Tsugaru Strait, between the islands of Honshu and Hokkaido. "This is too fatty," said the diner. "I doubt that this is meat from a fish caught in the wild." According to a December 2004 article in *Yomiuri Shimbun* (an English-language newspaper published in Tokyo), the man was right: "subsequent investigation revealed that the meat was actually from a fish produced on a farm in Turkey. The operator of the sushi bar—which advertises itself as specializing in Hokkaido-sourced fish—served the imported fish because it was unable to secure enough wild bluefin tuna." The Japanese Fisheries Agency reported that of 21,500 tons of [bluefin] tuna imported in 2003, 19,500 tons, or about 90 percent, originated on fish farms. Because the number of wild-caught tuna is declining and the number of ranch-fattened tuna is limited by the fecundity of wild fish (and how many are caught before they can reproduce), it would appear that the only way to continue to appease the insatiable (and discriminating) sashimi god is to figure out a way to breed bluefin tuna in captivity.

Young bluefins have lighter-colored flesh and are less strongly flavored, but as they grow into adulthood, their flesh turns dark red and their flavor becomes more pronounced. The bluefin is generally the variety of choice for fresh tuna connoisseurs. It has more fat, and thus more flavor, than the other varieties. (Even though they are genetically distinct, there is no visible difference between the northern and southern bluefin tuna to sashimi lovers.) At maturity, the flesh is dark red, sometimes even wine-colored, with an appearance very similar to raw beef. (Fresh tuna is usually sold already skinned, as the skin is extremely tough.) As it is served raw, only the freshest and highest-

Nine

THE BLUEFIN'S POPULAR LITTLE COUSINS

Skipjack at the Tsukiji fish market, Tokyo

ALL TUNA ARE SCOMBRIDS, but not all scombrids are tuna. The family Scombridae includes several species of smaller, bullet-shaped fishes that are (correctly) grouped with the tunas, and also the mackerels, which are essentially smallish tunas without the heft and without many of the advanced features that characterize the genus *Thunnus.* The blackfin tuna *(Thunnus atlanticus),* for example, is a small, typically shaped tuna, found only in the western Atlantic, from Martha's Vineyard south through the Caribbean, and along the coast of northeastern South America as far south as Rio de Janeiro. The maximum length is around forty inches, and the all-tackle record (taken off Key West, Florida, in 1996) is 45.5 pounds. A pelagic, schooling fish that feeds near the surface, the blackfin is blue-black on the back, with a golden yellow band that runs from eye to tail but fades out soon after death. Despite its small size, the blackfin is considered a world-class game fish.

The Spanish mackerels, kingfish, seerfish, and the wahoo, a large, elongated game fish, famed for its speed and unwillingness to be landed, are also classified as scombrids. All scombrids are more or less pointed at both ends, with a crescent-shaped tail and a series of finlets on the dorsal and ventral surfaces of the hind end, aft of the second dorsal fin and just before the insertion of the tail fin. The function of these finlets is unknown, but because scombrids are fast swimmers, they are believed to be somehow connected with speed. (The marlins, swordfish, and sailfishes are as fast as or faster than tunas, but while they have the same lunate tail fin, they lack finlets.) Below the big-tuna designations, there are several species of bonitos, "little tunas," "bullet tunas," "frigate tunas," the kawakawa, and the cero. Most of the tunas are considered big-game fishes, worthy of being hunted by fishermen in big boats, but all of the species that can exceed twenty pounds in weight are popular food fishes, and are the objects of some of the world's most extensive fisheries.

Tuna are fished in over seventy countries worldwide, and marketed fresh, frozen, or canned. Only about 1 percent comes to the market to be sold fresh. The rest goes to the cannery, because canned tuna is America's most popular fish. Tuna has been fished from the warm, temperate regions of the Mediterranean and the Pacific, Atlantic, and Indian Oceans since ancient times. Depending on the species, weights average from ten to six hundred pounds per fish. The majority of the commercial tuna harvest comes from California. The average annual consumption of tuna in America is 3.6 pounds per person, most of which is canned.

YELLOWFIN TUNA (THUNNUS ALBACARES)

Yellowfins are probably prettier than bluefins (although some would argue that the giant bluefin is the most beautiful fish in the world), and they are more widely distributed, thus more of them are caught commercially. With its extremely long, canary yellow second dorsal and anal fins, the yellowfin is easily differentiated from other tunas. The pectoral fins, which also become yellow in adult fish, are very long, reaching to the base of the second dorsal. (Only the albacore has proportionally longer pectorals.) The little finlets between the fins

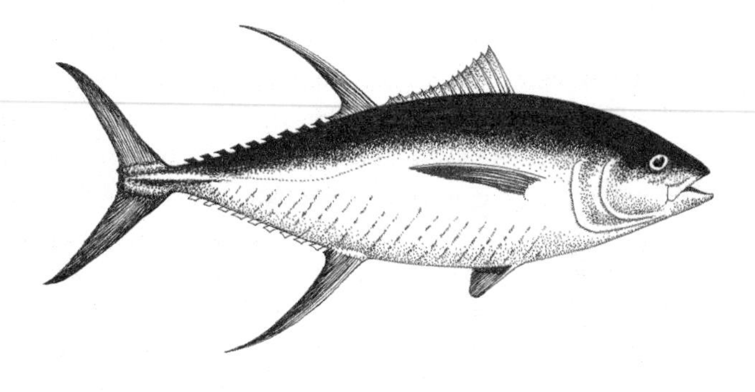

Yellowfin tuna *(Thunnus albacares)*

and the tail are also bright yellow. The most brilliantly colored of the tunas, the yellowfin is metallic blue or greenish black above and pearl white below; in younger fish, the lower flanks are crossed with interrupted, vertical lines. Adult fish have a band of bright gold or iridescent blue (sometimes both) running along the flank. Spawning yellowfins have been observed to "flash" their colors, perhaps as a stimulant to the opposite sex. Conversely, dying tuna begin to lose their bright coloration and soon fade to shades of dull gray.

Like most tunas, the yellowfin is an extremely fast swimmer, but it is one of the few fishes—of any kind—whose actual speed has been measured. It is difficult to calcluate the speed of a fish in the water, but even if this could be done, how would we know that it is swimming at its maximum speed? On a research cruise off the Pacific coast of Costa Rica, Vladimir Walters and Harry Fierstine (1964) designed a device that measured the speed of a line as it was taken out by a recently hooked yellowfin, and found that in the first ten to twenty seconds, it was clocked at around 46 to 47 miles per hour. Zane Grey knew that the yellowfin was fast because in 1925 he wrote, "Tuna of three or four hundred pounds, shooting like a bullet through the water." (Grey admired almost everything about the yellowfin tuna except its intelligence. In *Tales of Fishing Virgin Seas* (1925) he wrote, "The yellow-fins are apparently stupid pigs. They school in myriads and are ravenously hungry. Boats, men, baits, jigs, gaffs, have no significance for them. Nature has not yet warned them.")

Spawning takes place twice a year, and requires a water temperature of at least 79°F. At the Achotines Laboratory in Panama, efforts have been under way for several years to breed yellowfin tuna in captivity. Established by James Joseph in 1985 as part of the IATTC's Tuna-Billfish Program, the Achotines Lab is one of the few research facilities in the world designed explicitly for studies of the early life history of tropical tunas, particularly yellowfins. The laboratory is adjacent to Achotines Bay on the southeastern tip of the Azuero Peninsula on the Pacific side of the Republic of Panama, where the annual range of sea-surface temperature is approximately 21° to 29°C, ideal for spawning tuna. (The Azuero Peninsula is known to big-game fishermen as "the Tuna Coast" because of the yellowfin, but also because it is the home of world-class blue, black, and striped marlins, as well as sailfish, dorado, and wahoo.) The continental shelf off Achotines Bay is quite narrow: the water reaches depths of over six hundred feet less than five miles from shore, affording scientists ready access to oceanic waters where spawning of tunas occurs during every month of the year. Because little is known of the reproductive activities or early life history of the various tunas, the Inter-American Tropical Tuna Commission (IATTC) established a research laboratory to focus on these aspects of tuna biology.

In 1992, scientists at Achotines began a joint project with Japan's Overseas Fisheries Cooperation Foundation to encourage the breeding, spawning, and raising of tuna in captivity. From the inception of the program in 1993 until its conclusion in 2001, several Japanese scientists were based at Achotines, working with IATTC personnel on the complex problems of breeding captive tuna. The tanks at Achotines held yellowfin in virtually every stage of development, from floating eggs and inch-long larvae to 150-pound adults in the large tanks, which are 6 meters (20 feet) deep and 17 meters (56 feet) in diameter. Upon the Japanese departure in 2001, the facilities and equipment (much of which had been provided by Japan) became the property of the Republic of Panama, and remained at the laboratory for use by IATTC scientists.

A captive population of forty-four yellowfins first spawned in October 1996, which involved only two or three pairs of the largest fish. From the Achotines website: "Yellowfin tuna in the main broodstock tank have been spawning almost daily since October 1996; the only successful spawning of yellowfin in land-based tanks anywhere in the world. Spawning generally occurs from early afternoon to late evening." As Kurt Schaefer (2001) described it, "Each spawning event occurred around sunset, and was preceded by courtship behavior during the late afternoon. The courtship behavior included pairing of individuals, chasing, rapid color flashes exhibited by individual fish, and rapid horizontal or vertical swimming. During the following 3 months of 1996 spawning was continuous, with many of the fish exhibiting courtship behavior prior to each spawning event." The numbers of fertilized eggs collected after each spawning in the main broodstock tank range from several hundred to several million. The eggs are gathered by several methods, including siphoning, dipnetting at the surface, and seining with a fine-mesh surface egg-seine. Fertilized eggs are hatched in three-hundred-liter (seventy-eight-gallon) cylindrical incubation tanks.

To learn about the daily lives of yellowfins, Barbara Block and her colleagues tagged several of them off the coast of California, southeast of San Diego. The fish were caught on hook and line, brought close to the boat, and pressure-sensitive ultrasonic transmitters were darted into their back muscles. The results were transmitted to a hydrophone array towed behind the research vessel. Block et al. (1997) found that yellowfins do not spend much time at the surface, but appear to prefer to swim 10 to 20 meters (33 to 66 feet) below, unlike similar-sized bluefins, which pass much more time at the surface. Yellowfins tend to remain at or above the thermocline, except for occasional dives, presumably for feeding, below the interface of warmer surface waters and colder, deeper waters. (Carey and Olson, tagging yellowfins in offshore Panamanian waters, recorded one that descended to 464 meters—1,522 feet.) One "striking" behavior pattern observed by the researchers was that prior to sunrise every day, the tuna rapidly ascended to the surface. The daily habits of yellowfins are consistent with their categorization as "warm-water" tunas, and their depth preference coincides with their inclination to swim beneath herds of spinner and spotter dolphins.

Like many fish species, yellowfins are compulsive schoolers, but they often swim with skipjack, bigeye, and other tunas. Curiously, their association with dolphins has been observed only in the eastern tropical Pacific, and not in the western Pacific, Atlantic, or Indian Oceans. But wherever yellowfins congregate, they are targeted by commercial fishers. Hundreds of thousands of tons are harvested annually, making it one of the most important of all commercially caught tunas. They are taken by longliners and pole-and-line bait boats, but the purse-seiners figured out how to exploit the yellowfins' habit of aggregating under objects at the surface—particularly schools of dolphins. We don't know why they do this, but the habit has proven disastrous for tuna (and dolphins), as tuna fishermen often find their prey by spotting the dolphins at the surface, and when they set their deepwater purse seines around the dolphins, they also catch the tuna, the object of the fishery in the first place.

In addition to its desirability as a commercial species, the yellowfin is one of the world's most popular game fishes. The first person on record to successfully land a giant yellowfin was W. Greer Campbell, of Avalon, California, who caught a 104-pounder in 1896. Two years later, George Frederick Holder pulled in a 183-pounder off Catalina Island, and parlayed his fascination for big tuna into something larger. To attract other gentlemen to the sport of big-game fishing, Holder established the Catalina Tuna Club. (He is better known as the founder of Pasadena's Tournament of Roses.) The members of the club agreed that "the object of this Club is the protection of the game fishes of the State of California . . . and to discourage hand-line fishing, as being unsportsmanlike and against the public interest; that the purposes for which it is formed are to encourage the use of rod and reel fishing and to permit social intercourse among the members of the club" (Matsen 1990). Of the equipment used before Dr. Holder rewrote the rules, Ralph Bandini observed:

> Tackle of those days would make us laugh—and then rub our eyes in amazement: rods pitifully inadequate; lines that broke around 50 pounds; straight-handled reels, with no drag other than a thumb-stall; rowboats to fish from; boards—not swivel chairs—on which to sit! How they took the fish they did only the gods know.

> Certainly they must have been kings among anglers. Men battled great fish for three—five—seven—eight—even fourteen hours. Hands and fingers were smashed by fiercely spinning reel cranks; one corner of the porch of the old Metropole Hotel was dubbed "Tuna Hospital," for there it was that certain angry, discomfited anglers, their hands bandaged, their arms in slings, were wont to sit and nurse their wounds as they grouched about their luck.

In 1899, also out of Catalina, Colonel C. P. Morehouse boated a 251-pounder, an accomplishment so noteworthy that it attracted nationwide press for the Tuna Club of Avalon. Fishing out of the club, Zane Grey bemoaned the shortage of large tuna because the fishing "was spoiled by the Austrian and Jap [anese] net-boats. These round-haul boats have nets a half mile long and several hundred feet deep. When they surround a school of tuna it is seldom that any escape. If the tuna are very large, over one hundred pounds, then a great many of them are destroyed . . . These market fishermen are aliens, and they break the state and federal laws every day during the season."

Despite these alien fishers, Grey managed to catch several "blue-button" fishes (so named because the Tuna Club awarded a snappy little badge to every man who caught a hundred-pounder), and one day in 1919 he hooked a fish so "loggy and rolly" that he decided it must be a yellowfin.* Grey fought the fish (which weighed about 150 pounds) for five hours before it broke the line and escaped. Before he visited Nova Scotia, Grey battled the tuna of California waters, sometimes bluefins, sometimes yellowfins. Both kinds were courageous fighters, but because the yellowfins (which he knew as "Allison tuna") grew larger off California than the bluefins, they were his first "giant" tuna. He caught a 215-pounder off Catalina; and farther south, off Cabo San Lucas (Baja California), he landed a 318-pounder. "The Allison tuna," he wrote, "to my thinking, strike swifter and harder at a

*Oddly, Grey doesn't mention the obvious difference between bluefins and yellowfins, which is, eponymously, the color of the fins. The similarly shaped second dorsal and anal fins of the bluefins are short and grayish blue, but in adult yellowfins these fins are often elongated into graceful, scimitar-shaped blades, and, like the yellowfin's other fins, including the little finlets and the long pectorals (but not the first dorsal), are *yellow.*

moving bait than any of the above mentioned fish [bluefins], barring what the Nova Scotia tuna might do. They make a hole in the sea, and a roaring splash, that would do justice to the plunge of a horse from a high cliff. I never before experienced anything so terrific as the strike of one of the large Allisons."

Ralph Bandini (1884–1964) was, like Zane Grey, an avid fisherman and a prolific writer on the subject of big-game fishing. He employed less purple prose than the author of *Riders of the Purple Sage,* but his enthusiasm for fishing and his love for the quarry shines through his work, particularly *Veiled Horizons* (1939), the title of which is a reference to the fog that enshrouds the waters off Catalina Island in the early morning. For some reason, fishing writers rarely attach dates to the stories of their exploits, so it is left to the reader to try and figure out when something happened. (Even when he includes verbatim entries from his logs, Bandini notes the date, but not the year.) We know he was an early director of the Avalon Tuna Club, which was founded in 1898, so his tuna fishing probably took place before and after World War I. Also, he mentions a time when he was fishing alongside Grey in 1923, and a pod of killer whales chased some "porpoises" right under their boats. Bandini fished for tuna of all sizes, from "small" hundred-pounders to monsters that would have weighed four hundred pounds or more—if he could ever have caught one. One day, out fishing with Captain Roy, he lost two big fish in a row as his line broke. Then, "Not fifty feet from the boat, a tuna came out of the water—and what a tuna! Four times he jumped clear in great twisting jumps that took him fifteen feet into the air. Jumps that showed us his depth, the breadth of his shoulders, his length. If that fish was an inch he was all of fifteen feet long. Remember, he wasn't fifty feet away from us. I had seen some pretty big fish in my time—and I have seen some big ones since. But never before or since have I seen his equal. I couldn't even guess what he weighed, but allowing for excitement, for amazement, I think he could have gone anywhere between fifteen hundred and two thousand pounds." (Some fifty years after Bandini saw the California monster, the current world's record bluefin was caught by Ken Fraser off Auld's Cove, Nova Scotia, in 1979. It weighed 1,496 pounds.) Bandini complains, "Grey, Farrington, Mrs. Farrington, Peel, dozens of others have taken great tuna. How they do it, I don't know—unless there is a difference between our big tuna out here and those of the Atlantic and the North Sea."

Bandini fished off Catalina, where members of the Tuna Club regularly caught yellowfins, some of them of respectable size. But he rarely mentions yellowfins; almost all the fish he described (or caught) were "bluefins." Maybe they were, but it is certain that Bandini thought the yellowfin was an inferior fish. A fisherman known as "Jimmie Jump" had brought in a 145-pound yellowfin that he had caught on light tackle: "Of course he was very proud of it—and justly so. And because he was so proud of it he had to stand a lot of ribbing about it. The fish was a yellowfin, and as such was not regarded as being in the same class as a bluefin . . . Nevertheless the fact remains that, while no handicap in favor of one against the other is set up in the Tuna Club, there has always been a sort of feeling that the yellowfin is not the fish his cousin is." That might explain why Bandini almost exclusively fished for bluefins—or said he did. In fact, the yellowfin is certainly as much of a fighter as his cousin is, and prettier besides.

Around 1924, when Zane Grey was fishing for yellowfins off Cabo San Lucas, Baja California, a certain Captain Heston explained the abundance of fish there:

> "I have not noticed any diminution in the number of yellow-fin tuna. There are as many now as when I first started in this business. And that means millions. Some years there are not so many, other years are good. This year they are thick. Tuna spawn on the bottom, and one tuna will have a thousand eggs where a salmon has one. They multiply tremendously. Probably that fact will save the tuna from extinction."

The same Heston reported a harpooned yellowfin that "was nearly ten feet long and might have weighed seven hundred pounds," but as this is twice as large as the world's record 388-pounder, it might have been a fish story. Grey did not believe that the "yellow-fins" of Baja were the same species as the "Allison tunas" of California: "In my opinion," he wrote, "they were not the yellow-fin we have in Catalina waters in August and September. Also they differ somewhat from the yellow-fin we caught at the Galapagos Islands. Their flesh was remarkably solid and heavy, and that was why, as the scales proved, we guessed their weight so incorrectly. Brilliantly colored, blue, gold, bronze, silver, mother-of-pearl, and so perfectly shaped to combine speed and strength, they were exceedingly beautiful fish." Of course they were the same species; yellowfins are found in every temperate and tropical ocean (except the Mediterranean) in a wide swath between 45° North and 40° South. At least some of the stocks are migratory.

Hunters usually have to see their prey before they shoot at it, but by and large, commercial or recreational fishermen do not see the fish they are after until they catch them. There are some exceptions. In offshore New England waters, the broadbill swordfish is still hunted by harpoon fishermen who spot them from the crow's nest or have someone else do the spotting from an airplane. Longlining is now the preferred method of industrial swordfishing; the longliners are far from the area (and deep below the surface), where the fish take the baited hooks. In New England waters, bluefin tuna are also hunted by harpooners who locate them from the surface, but tuna and swordfish are anomalies. In the eastern tropical Pacific, yellowfin purse-seiners relied on sightings of dolphins to tell them where the tuna might be lurking; they didn't actually see the yellowfins until they were being winched up in the nets. But in other parts of the world, yellowfin fishermen see the huge schools racing along the surface, making as much of a white-water ruckus as any dolphins. Here Zane Grey (1919) describes yellowfins of San Clemente Island (California):

> Then we beheld a spectacle calculated to thrill the most phlegmatic fisherman . . . The dark-blue water, heaving in great, low, lazy swells, showed a roughened spot of perhaps two acres in extent. The sun, shining over our shoulders, caught silvery-green gleams of fish, flashing wide and changing to blue. Long, round, bronze backs deep under the water, caught the sunlight. Blue fins and tails, sharp and curved, like sabers, cleared the water. Here a huge tuna would turn on its side, gleaming broad and bright, and

> there another would roll on the surface, breaking water like a tarpon . . . I saw then that the school, lazy as they seemed, were following the leaders, rolling and riding the swells. These leaders threw up surges and ridges on the surface.

On another occasion, Grey directed his captain to head for a tremendous commotion in the distance: "for a mile across the channel, came the bursting white wall of water, and all over the ocean between us and this wall were white boils and splashes, and hundreds of giant tuna in the air." The tuna were feeding on flying fish, which sometimes crashed into the fishing boat before falling back into the water to be gobbled up by the voracious tuna. In this melee, Grey managed to hook "the largest tuna I had seen . . . I saw his back, dark blue and his thick, wide tail, so instinct with power . . . when my bait alighted he was there. He got it. And so tremendous was the strike that he broke the line."

Because of their size and strength, yellowfins are still highly regarded as game fish—few other species can reach their size and put up such a fight—and therefore they are among the primary objects of sport fisheries around the world. In *Fishing in Bermuda,* James Faiella

Yellowfin tuna photographed in a pen off Puerto Vallarta on the Pacific coast of Mexico. The first dorsal and pelvic fins are tucked in, but the extremely long second dorsal and anal fins, bright yellow in life, are prominently displayed.

wrote, "The yellowfin is considered the most prized of all tuna varieties both around Bermuda and elsewhere for their size, gameness, and the high quality of their flesh." Peter Goadby, one of Australia's most articulate and influential fishermen, wrote *Big Fish and Blue Water,* in which he sings the praises of the yellowfin as a game fish:

> The yellowfin is surely one of the fastest and most tenacious of all fish. The grab of a live bait or lure is followed by a preliminary run then, as if realizing that it is in trouble, the fish really claps on the pace, often sounding as it runs. Then the great circling fight begins with every foot of the line having to be won and held on the reel as the line shortens and the circles narrow and tighten. The time of the tail beats indicates the size of the fish . . . Kip Farrington, one of the most experienced tuna fishermen, rated a 200 lb. yellowfin in deep water as the equivalent of other species, including bluefin, twice the weight.

Goadby was talking about Australia, but charter boats and individual fishermen set out after yellowfins from San Diego, South Africa, Cabo San Lucas, the "Tuna Coast" of Panama, Hawaii, Costa Rica, Mexico, Honduras, Massachusetts, Cape Hatteras (North Carolina), Bermuda, various countries in or on the Caribbean, and throughout the South Pacific, from Thailand and French Polynesia to all Australian coasts.

It is unlikely that the sportfishermen of southern California extracted enough yellowfins to affect the overall numbers. (Of course, Grey believed that the sportfishing would have been that much better if those pesky foreign net boats hadn't been hauling in so many of his precious tuna.) Even so, the numbers quoted by Kip Farrington in *Fishing the Pacific* (1953) are surprising: "Of course there were many marlin and tuna caught by non-club members," he wrote, "but I am quoting the records here to give the reader some idea of the best fishing months. A total of 6,532 tuna were caught by Tuna Club members in 40 years. The best year was 1919 when 911 were taken . . ." Tuna Club members could have caught ten or a hundred times the number they caught over the years, and they wouldn't have come close to the numbers of yellowfins taken by commercial fishers in a given year. In just one part of one ocean, according to a 2002 report in *Pacific* magazine by Robert Keith-Reid,

> the annual yellowfin catch first exceeded 200,000 metric tons in 1980. By 1990 it had almost doubled to 380,000 metric tons. In the past four years it exceeded 420,000 metric tons with a peak of 480,000 metric tons . . . Tagging of yellowfin in the early 1990s, when catches were 10 to 20 percent below present levels, indicated that the stock was not being overfished. But recent research shows that the stock has suffered a "significant" decline of about 35 percent since 1997. The decline is most evident in the western equatorial Pacific where the stock is estimated to be down by more than 50 percent since the mid 1990s. For the whole Western Pacific, the stock is estimated to be 30 percent below what it would have been had it not been fished.

After skipjack, yellowfin is the second-heaviest fished of the tunas. Known as *ahi* in Hawaii, it is less expensive than bluefin, more common, and easier to find in the markets. Its flesh is pale pink, with a flavor somewhat stronger than albacore. It can be served as sashimi, grilled, or canned. When cooked, yellowfin meat is firm and mild-tasting, and tends to have a very light yellowish brown color. In larger fish the meat may become slightly darker and dryer. A report issued by the Japanese Fishing Authority (Commission for Conservation 2005) analyzed recent trends in tuna fishing in the western and central Pacific. In 2004, 1,447 longliners, 176 pole-and-line boats, and 54 purse-seiners accounted for 31,717 tons of bigeye tuna, 41,406 tons of yellowfins, and 303,127 tons of skipjack. By an order of magnitude, skipjack is the predominant tuna species caught in the Pacific (and elsewhere), but yellowfin catches are on the rise. Most skipjack is canned, but more and more yellowfins are destined for Japanese restaurants around the world. Charts with numbers usually cause the MEGO ("My Eyes Glaze Over") syndrome, but the numbers tell a particularly important story about tuna fishing, as in this chart of catches of various tuna species in the Pacific. Figures are in tons, and I have included only a couple of years to show trends. YFT = yellowfin; SKJ = skipjack; BET = bigeye tuna.

	YFT	SKJ	BET	Total
1975	332,098	423,713	108,730	864,451
1985	482,165	651,787	140,923	1,274,875
1995	619,013	1,201,004	190,055	2,010,090
2000	720,311	1,476,809	254,418	2,451,538
2003	866,707	1,549,025	210,647	2,626,379

Recent analyses suggest that yellowfin is exploited to its optimum in the eastern Pacific Ocean, and that there will not be any significant growth in volume for the future in the western Pacific. In the Indian Ocean, exploitation leaves little room for a population increase. Although the yellowfin is not considered endangered, there is a general concern that the increased catches of juvenile yellowfin (especially in the Atlantic, Indian Ocean, and western Pacific) will cause the stocks to suffer in the long term.

For the most part, the behavioral traits of wild animals have evolved in such a way as to be somehow beneficial to each species. Herding, flocking, or schooling, for instance, may be seen as stratagems to protect a given individual by offering too many choices to potential predators. Ultimately, of course, the predators evolved counterstrategies that enabled them to select one animal out of a school or herd, and focus on that one to the exclusion of the others. Lions will pick one zebra out of a herd, and chase only that one, ignoring others galloping by and even those that cut in front of them. Longline fishing, where hundreds of miles of baited hooks are deployed, depends to a certain extent on the gregarious behavior of the target species; if tuna were not schooling fishes, not enough of them would take the baits deployed on the lines, and there would be no point in fishing for a given species in a particular place. The schooling yellowfins that so enraptured Zane Grey have another trait that would ultimately prove to be their undoing: they tend to aggregate near or under floating objects, which can be as insignificant as a floating log, as large as a ship, or as active as a group of dolphins.

For reasons that are still unknown, yellowfins assemble under schools of spotter and spinner dolphins in the eastern tropical Pacific, west of Central America and Mexico. In the 1960s, tuna fishermen out of San Diego learned that they could locate schools of yellowfin by scanning the horizons for disturbances on the surface made by herds of leaping dolphins. In "setting on dolphins," a school is rounded up like cattle by small outboard-motor speedboats, and after the dolphins and tuna are encircled together, the net is closed ("pursed") at the bottom by a cable and rings, and the net with the tuna and dolphins in it is brought aboard by a hydraulic power block. One set could capture anywhere from ten to one hundred tons of tuna, a much more productive method of fishing than the old hook-and-line bait fishing. Production increased dramatically, but so did the dolphin mortality. According to a 1986 study by N. C. H. Lo and Tim Smith, "the annual kill from 1959 to 1972 varied from 55,000 in 1959 to 534,000 in 1961. There were three distinct maxima of 534,000, 460,000 and 467,000, corresponding to peaks in the number of sets made on dolphins in 1961, 1965, and 1970. The total kill from 1959 to 1972 was estimated to be about 4.8 million."

A great environmental outcry resulted in the passage of the Marine

"Tuna clippers" in port in San Diego, California, 1982

Mammal Protection Act (MMPA) in 1972, which made it a violation to harm any cetacean, but the tuna fishermen lobbied for an exemption, and they continued to kill dolphins in staggering numbers—more than 300,000 died in 1972. The fishermen set their nets "on dolphins" until they were sued in federal court by a consortium of conservation groups and forced to suspend their entire fishing operations. They were allowed to commence again only if they would abide by strict quotas imposed by the government, which were to be decreased annually to permit the fishermen to adjust to the new regulations. The first new quota, set in 1976, allowed the fishermen to kill 78,000 dolphins. The number was steadily reduced until it stood at 20,000 by 1981. In 1990, the StarKist Seafood Comapany (a subsidiary of the conglomerate H.J. Heinz) announced it would no longer purchase tuna that had been caught with dolphins, and began to label cans of StarKist tuna "dolphin-safe." Bumble Bee and Chicken of the Sea quickly followed suit, and because these three companies accounted for more than 80 percent of the tuna sold in the United States, the dolphins were spared—for the moment, anyway.

In 1991, Mexico complained under the GATT (General Agreement on Tariffs and Trade) dispute settlement procedure, and the dispute ended with the United States not being allowed to set an embargo on imports of tuna products from Mexico just because Mexican regulations on the way tuna was produced did not satisfy U.S. regulations. The report was never adopted, so Mexico and the United States held their own bilateral consultations aimed at reaching an agreement outside GATT. The outcome of the consultations was the Agreement on the International Dolphin Conservation Program (AIDCP), put in force in 1999. By the end of December 2002, the U.S. Department of Commerce ruled that encircling dolphins with nets a mile wide to catch tuna would not significantly harm them, under the terms set forth in the AIDCP. Should all encircled dolphins be safely rescued from the nets, the tuna could be marketed as dolphin-safe, as opposed to the previous definition of dolphin-safe, introduced by the Earth Island Institute, which denied the dolphin-safe label to any tuna caught by using the practice of encircling.

In 1994, the North American Free Trade Agreement (NAFTA) was signed into law by President Clinton, removing restrictions on trade among the United States, Canada, and Mexico. Under the terms of the MMPA of 1972, American fishermen could not fish in the eastern tropical Pacific without employing devices and procedures that would reduce the dolphin kill, but foreign fishers were not obliged to adhere to American restrictions, and they could therefore fish any way they wanted—including "setting on dolphins." The U.S. government concluded that circling dolphins in nets was harmless and allowed Mexican fishers to sell tuna in the United States under a dolphin-safe label. According to a January 6, 2003, report by the Environment News

The gigantic net being "pursed." The small boat tows the net in a circle until it completely surrounds the school of tuna—and any dolphins that happen to be present.

Service, an Internet watchdog of ecological issues, "On December 31, the National Marine Fisheries Service announced that after new research, it had concluded that the tuna purse-seine industry [practice] of encircling dolphins to catch tuna has 'no significant adverse impact on dolphin populations in the Eastern Tropical Pacific Ocean' " (Lazaroff 2003). Outraged environmentalists planned to contest the new ruling in court. David Phillips of the Earth Island Institute said, "The Bush administration's claim that chasing and netting of dolphins is 'safe' for dolphins is fraudulent and must be overturned."

The IATTC finally came aboard the tuna-protection boat. At a 1992 meeting in La Jolla, California, a schedule of progressively decreasing annual limits on dolphin mortality was set, beginning with 19,500 in 1993 and scaling down to 5,000 in 1999. Dolphin deaths dropped substantially, but there was still the open question of dolphin-safe tuna. Some fishing nations complied with the stricture that only tuna caught under rules established at the La Jolla meeting could be marketed with a dolphin-safe label, but others, released by NAFTA from any such restrictions, could catch tuna any way they wanted and sell it in the United States. In 2005, Jim Joseph of the IATTC and Dale Squires of the National Oceanic and Atmospheric Administration (NOAA) wrote:

> Considering the 99 percent reduction in dolphin mortality and the ecological costs in terms of increasing bycatches and growth of overfishing resulting from a prohibition on fishing tuna in association with dolphins, and recognizing the possibility that the Latin American nations cooperating in the La Jolla Agreement might abandon it unless some relief from the embargoes placed on them by the United States was given, several environmental organizations began discussions with some of the Latin American governments to explore the possibility that an international agreement could be reached that would ensure the protection of dolphins, tunas and the ecosystem to which they belonged, would provide some relief from the embargoes, and would be binding on the participants. These discussions led to the formulation of a binding agreement, the Agreement on the International Dolphin Conserva-

> tion Program (AIDCP), which was adopted in 1998. The agreement was ratified and entered into force in 1999. All nations, and the European Union, that fish for tuna with purse-seine vessels in the EPO have now signed and ratified the agreement, or applied it provisionally pending ratification . . .

Because estimating dolphin populations is an inexact science at best, we were never sure just how many dolphins there were, and therefore, the attempts to determine the effects of the tuna fishery on them were inconclusive. In a study conducted by U.S. government biologists, Frederick Archer and colleagues (2001) reported an insidious side effect of the tuna fishery, in which thousands of dolphin deaths went unreported. They wrote, "although the reported dolphin bycatch has drastically decreased, the populations are not recovering at expected rates." One factor might have been the stress of the high-speed chase and encirclement, where calves would have been unable to keep up with their mothers, and if the mothers were caught in the nets, the calves, having fallen behind, would have escaped, only to starve because their mothers were gone. Thus the counts, showing only the adults caught in the nets, did not include thousands of calves, now calculated to represent an additional 10 to 15 percent spotted dolphin deaths, and 6 to 10 percent more spinner dolphins than were originally estimated.

In 2005, Tim Gerrodette and Jaume Forcada reported on two spotted and spinner dolphin populations in the eastern tropical Pacific Ocean, noting that the dolphins were not recovering as expected, even though the bycatch had been substantially reduced. Visual surveys conducted from 1986 to 2000 over some twenty-six thousand miles (forty-two thousand kilometers) yielded estimates that as of 2000, the population of spotters was about 640,000 animals, and of spinners, 450,000. This was considerably lower than earlier surveys and indicated that even with a lowering of the bycatch by two orders of magnitude, the stocks were not recovering. The reasons for this failure to recover are not evident, but Gerrodette and Forcada suggested several possibilities, including "underreporting of dolphin bycatch; effects of chase and encirclement on dolphin survival and reproduction; long-term changes in the ecosystem; and effects of other species on spotted and spinner dolphin population dynamics." It is probably no surprise to learn that the massive bycatch of spotters and spinners in the tuna purse-seine fishery has had deleterious effects on dolphin populations; it is only surprising that it took forty years to recognize that killing millions of animals in a population might permanently damage it.

Even though the tuna-dolphin problem seemed to be resolved (except for the tuna and the dolphins), the San Diego tuna fishery folded. When the Japanese with their monster tuna seiners entered the fray, the competition outfished the San Diego fleet, and the introduction of Central and South American tuna clippers into the eastern tropical Pacific sounded the death knell for the American tuna industry. The American fleet, once boasting more than one hundred fishing boats, was down to five by 1995, and this fishery, once the pride of San Diego commerce, is essentially over. Foreign competition, overregulation, rising costs, environmental pressure, and a host of other problems caused all of the big corporations to close their canneries on the West Coast, and the great California superseiners were sold to foreign fishers. Now there is very little (if any) tuna processed on the West Coast or anywhere in the United States. Fish is being packed in Mexico, Australia, Samoa, Thailand, Japan, Canada, Spain, and many other countries. The dolphin-safe labeling campaign in 1990 received enthusiastic public approval, yet the requirements for such labeling still leave loopholes for imported tuna. The labeling requires an observer on each fishing vessel to certify that no dolphins are killed or seriously injured during the tuna harvest. Public concern about dolphin mortality prompted the U.S. tuna-canning industry to stop purchasing tuna caught "on dolphins." StarKist, Chicken of the Sea, and Bumble Bee have all pledged to remain dolphin-safe regardless of any future potential changes in the law.

The intervention of the "dolphin huggers" into the tuna-dolphin imbroglio eventually became what is considered one of the greatest triumphs of the environmental movement. Besides the fishermen, it was Bill Perrin, then a graduate student at UCLA, who first noticed the dolphins being killed in the nets, and he published the first popular article on the subject in 1968, called "The Porpoise and the Tuna," and another in 1969, "Using Dolphins to Catch Tuna." Outraged by the tuna fishermen's unwillingness to modify their methods (they claimed it would cost too much to modify their nets), the environmentalists sued under the newly minted Marine Mammal Protection Act—and won. Years of compromise followed, as the dolphin deaths were reduced to around twenty thousand per annum. The Pacific Ocean yellowfin tuna fishery is as robust as ever—but it has moved abroad.

Just as enigmatic as tuna swimming with dolphins in the eastern tropical Pacific is their not doing so in the western tropical Pacific. When the main North American canners decided not to process any more tuna caught through the practice of encircling dolphins, most of the fleets moved from the eastern tropical Pacific to the western sector, where dolphins and yellowfin tuna do not swim together. In this region, the fishermen spot the tuna aggregating at the surface, creating their own commotion as they feed on baitfish (Itano 2000). Just as Zane Grey had spotted his quarry by the "silvery-green gleams of fish that . . . threw up surges and ridges at the surface," the tuna fishers of the western Pacific posted lookouts to scan distant waters for a sign that schools of yellowfins were on the move. Environmental outrage, lawsuits, government regulations, collapsing fisheries, and international treaties finally coalesced to move the yellowfin fishery across the Pacific. It seems beyond stubborn—"stupid" seems a better word—that the tuna fishermen would fight every attempt to get them to change the way they fished, when all they had to do was sail off into the sunset.

We know that certain fish species aggregate under floating (or swimming) objects, but we don't know why. We can't ask the fish, so we have to apply the best theories of animal behavior (or "associative behavior" in this case) in an attempt to understand this unusual—and often maladaptive—activity. It may have something to do with protection (safety in numbers?), breeding, or feeding, but there are many fish species that school or aggregate without benefit of floating (or swimming) objects. It has been suggested that yellowfins rendezvous in numbers to associate with their conspecifics, but this still begs the question of why. One suggestion is the "meeting-point hypothesis," which proposes "that fish make use of animate or inanimate objects to increase the encounter rate between isolated individuals or small schools and other schools in order to constitute bigger schools that are more efficient to the survival of the species" (Fréon and Dagorn 2004).

José Castro, José Santiago, and Ana Santana-Ortego (2002) searched the literature and found records of more than 333 fish species belonging to ninety-six families that aggregated around floating structures such as logs, drifting algae, jellied zooplankton, whales, floats, or anchored fish-aggregating devices (FADs). Castro and his colleagues from the University of Las Palmas, in the Canary Islands, found that "the most widely accepted theory is that fish use floating materials, to some extent, to protect themselves from predators." They wrote that

> aggregation under floats may be the result of behaviour that has evolved to safeguard the survival of eggs, larvae and juvenile stages, during dispersion to other areas. Natural floating structures (e.g., algae, branches of trees) drift in currents that originate in places where the floating objects are frequently found (e.g., river estuaries, coastal areas). These same sea currents also introduce some of the planktonic production generated in these areas into the oligotrophic pelagic environment. Fish associated with drifting floating structures probably feed on invertebrates associated with the structures. However, they may also benefit from the accumulated plankton in the converging waters. . . . In this context, the meeting point hypothesis is only applicable to one specific case, the tuna and tuna-like species.

Further exploiting the inclination of tuna to aggregate, fishermen—particularly those in the South Pacific who fished for yellowfin and bigeye tuna—developed primitive devices that would attract the fish so they could be caught. Indonesian and Philippine fishermen used floating rafts of bamboo or palm fronds, moored to the seafloor and weighted down with baskets of stones. The fish were speared, netted, or caught on handlines. It was only a matter of time before someone realized that attracting fish with one device and catching them with another was redundant, and "vertical longlining" was born. ("Horizontal longlining" is the traditional method of playing out mile after mile of lines, and attracting the fish—and everything else—to the baited hooks.) Combining the best features of FADs and longlines, this system consists of cables or ropes supported at the surface by floats and moored to the bottom with chains and anchors. Each suspended line is rigged with branch lines which are festooned with baited hooks that are designed to attract different-sized fish at different depths. The vertical longlines, therefore, are themselves FADs.

For the benefit of fishermen who want to try this new technique, the Coastal Fisheries Programme of the Secretariat of the Pacific Community (SPC) prepared a detailed manual, explaining everything from rope splicing and knot tying to the selection of bait (and how to catch it), for various targeted species. "Fish aggregating devices," reads the introduction, "are floating rafts or buoys anchored in deep water which, for reasons not yet fully understood, cause tuna and other types of oceanic fish to gather around them . . . This fishing method, which had now [1998] become known as vertical longlining is still evolving, with SPC continuing its work on streamlining the gear and increasing the number of hooks that can be concentrated within a particular area." The manual, titled *Vertical Longlining and Other Methods of Fishing around Fish Aggregating Devices* (Preston, Chapman, and Watt 1998), also details other methods of fishing around FADs, including the Hawaiian night-fishing method named *ika-shibi* (squid-tuna) because it was developed to catch the tuna that were attacking the squid that the Hawaiians were trying to catch. In *ika-shibi,* underwater lights are used to attract small squid, which in turn attract the tuna, now the object of the fishery. The SPC report concludes:

> Vertical longline fishing around FADs can be a productive and potentially lucrative activity. It allows fishermen to target abundant resources of coastal tunas using small boats and simple, relatively inexpensive gear. Where cash markets for fish are well-developed, good-quality fresh tuna can command premium prices. Provided they look after their catch properly, fishermen carrying out vertical longlining around FADs can target this market and make much greater profits than they could from many other styles of fishing.

Fish that are brought out of the water quickly suffocate, and yellowfins caught by South Pacific fishers certainly could not be left on deck to die in the sun. To help fishers solve this problem, the SPC published a little booklet called *Onboard Handling of Sashimi-Grade Tuna,* which succinctly explains how to gaff the tuna ("never gaff the fish in the body, the throat or the heart"); how to kill it ("stun the fish with a sharp blow to the top of the head, then insert a spike into its brain"); how to bleed it ("When the tuna is struggling in the water . . . the blood attains a high organic waste [lactic acid] content and raises the temperature. Bleeding removes the organic waste and helps cool the fish's body"); how to clean it ("carefully rinse the fish inside and outside"); and how to chill and store it ("place it in a slurry of flake ice and sea water . . . protect the fish in a gauze sock or a plastic body bag . . . lower it gently into a refrigerated seawater tank"). The booklet concludes with this admonition: "No matter what methods of handling and presentation are requested by the buyers: always kill, bleed, and chill tuna that weigh over 25 kg as quickly as possible!"

SKIPJACK (KATSUWONUS PELAMIS)

Similar in flesh to the yellowfin, skipjack get their name because they seem to skip out of the water. They are also known as arctic bonito, oceanic bonito, and (in Hawaii) *aku.* Skipjack is the most heavily fished of all tuna species, and one of the world's most important food fishes. It is the most commonly canned fish, with generally the strongest flavor and highest fat content of the small tunas. When canned, it is known as "light-meat" tuna.

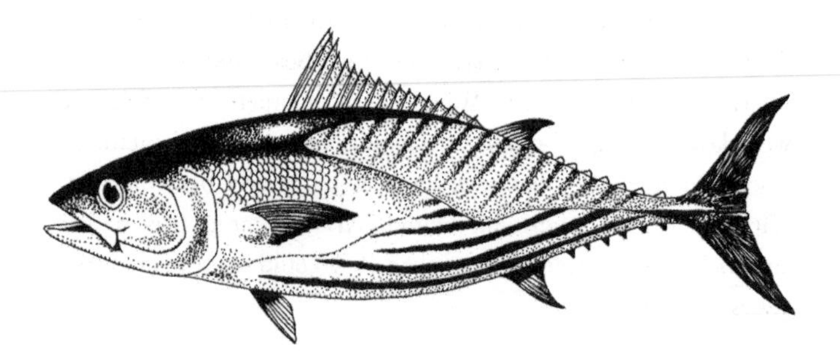

Skipjack *(Katsuwonus pelamis)*

It is of tremendous commercial importance in Japan, Hawaii, and the Caribbean, where it is marketed frozen, salted, and fresh. Skipjacks reach a length of three feet, but most are smaller. The world's record, caught off Mauritius, weighed forty-one pounds. In the Atlantic they frequently associate with blackfin tuna, and in the Pacific and Indian Oceans they school with yellowfins. The skipjack is a popular game fish, and it is the mainstay of the California tuna fishery. Skipjacks are found worldwide in tropical and subtropical waters, often aggregating in schools that may number as many as fifty thousand. In the eastern tropical Pacific, all three species are fished by "setting on dolphins," which means purse-seining spotted and spinner dolphins along with the tuna, resulting in the death of the dolphins as well as the tuna.

Because yellowfin and skipjack aggregate together, they are often caught together. Catches of both species are increasing, and in 2002 the total for skipjack was 2,076,000 tons, amounting to 51 percent of the total catch of all tuna species; yellowfins amounted to another 32 percent, at 1,321,000 tons (de Leiva and Majkowski 2005). In general, the outlook for world production of skipjack and yellowfin tuna is mixed. For yellowfin, most of the fisheries, with the possible exception of the eastern Indian Ocean, are probably fully exploited. For skipjack, catches on the average can possibly be increased in the Pacific, but probably not by much, if any, in the Atlantic and Indian Oceans. In the eastern Atlantic, the catches by the surface fleets targeting yellowfin and skipjack have reached the upper sustainable limit of yellowfin and probably are near that limit for skipjack. This tendency became obvious in the early 1980s, and caused many purse-seine vessels from the Atlantic to transfer their operations to the western Indian Ocean.

The western Pacific supports the world's largest tuna fishery, producing about 60 percent of the world's skipjack and 35 percent of the world's yellowfin. Analyses conducted by scientists of the SPC, based mostly on data from tagging experiments, suggest that the skipjack stocks of the region can support an increase in catch. However, for this increase to become reality the stocks of currently underexploited skipjack must be identified, there must be a demand for the raw material, and they must be vulnerable to fishing gear. A more intense fishing effort may lead to growing catches, but since skipjack are often caught together with small yellowfin and bigeye, in both sectors of the Pacific, the problem lies in ensuring that the increase is in catches of skipjack only (Hampton, Lewis, and Williams 2002).

Almost fifty years ago, the U.S. National Marine Fisheries Service built the world's first research facility specifically designed for maintaining tuna in captivity at Kewalo Basin in Honolulu. Local fishers would bring in adult yellowfin and skipjack, and they would be maintained in large tanks where studies would be conducted on visual acuity, sound sensitivity, olfaction, energetics, thermoregulation, geomagnetic sensitivity, and the spawning and rearing of tuna eggs. High-quality, coral-filtered seawater is used in several twenty-thousand-gallon pools for holding tuna and other pelagic fishes. Skipjack captured at sea were installed in tanks at the Kewalo lab and spawned about eight hours after capture, but the eggs did not survive (Schaefer 2001).

ALBACORE (THUNNUS ALALUNGA)

Identifiable by its winglike pectoral fins that reach beyond the anal fin, the albacore is a medium-sized tuna that inhabits temperate and subtropical waters. It has a maximum life span of about twenty years, by which time it may reach fifty-two inches in length. The present rod-and-reel record is eighty-eight pounds. Unlike some of the larger tunas, which are deepest in the region of the first dorsal fin, the albacore's greatest body depth is just forward of the second dorsal fin.

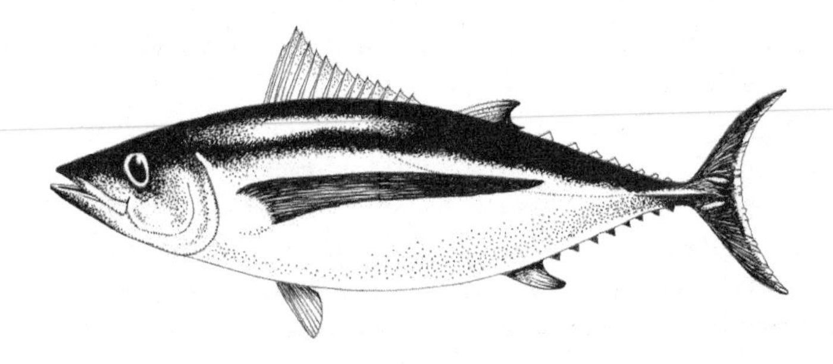

Albacore *(Thunnus alalunga)*

Sport fishermen troll for albacore with feathered jigs, spoons, and lures; or bait their hooks with mullet, sardines, squid, herring, anchovies, or any other available small fishes. Albacore is one of the world's most important food fishes, and is the only tuna that can be labeled "white meat" on the can. It is commercially caught by trolling jigs behind a slow-moving boat, although other nations employ purse seines and longlines.

Albacore is another high-fat variety, rich in omega-3 fatty acids, with the lightest flesh (white with a hint of pink). Its mild flavor and prized white flesh make it the most expensive canned tuna. The albacore fishery is international, with fleets operating in most of the major oceans of the world. Albacore are typically caught between latitudes 10° and 45° both north and south of the equator, with most being taken on the high seas. The South Pacific albacore stock extends from the eastern coast of Australia to the western coast of South America. The longline fleet normally targets adult fish that are usually found near the surface, but they have been caught as deep as 1,500 feet. *Thunnus alalunga* is a highly mobile species that moves throughout a very wide area of the South Pacific. While albacore travel in schools, these schools are generally less dense than those of skipjack and juvenile yellowfin. Consequently the commercial fishery is largely limited to longline or trolling.

According to a 2005 report by John Childers and Scott Aalbers of the NMFS Southwest Fisheries Center at La Jolla, the U.S. troll fishery in the North Pacific accounted for approximately 64,000 tons of albacore. This is approximately 20 percent of the North Pacific catch of albacore; the Japanese account for another 73 percent of the total 320,000 tons, or 233,600 tons per year. The Japanese employ a pole-and-line fishery in the spring, which targets two- to five-year-old fish off the Japanese coast eastward to the Emperor Seamount chain. There are also Japanese, Taiwanese, and South Korean longline fisheries, which harvest albacore in subtropical and temperate waters across much of the Pacific during the winter. Beginning in the early 1980s, Asian high-seas drift gillnet fisheries targeted two- to four-year-old albacore across much of the Pacific, but drift nets have supposedly been outlawed. There is a relatively small Canadian troll fishery for albacore during years when they appear in the waters off British Columbia.

Albacore at the Tsukiji fish market. Note the exceptionally long pectoral fins.

The World Conservation Union has not reassessed albacore in more than ten years, and the last review (1996) was "data deficient," meaning that a population estimate was not possible. Assessments of the stocks of the North and South Atlantic from the same period showed them to be vulnerable and critically endangered, respectively, because of "actual or potential levels of exploitation." The North Pacific population is considered a healthy stock at this time. It is regarded as an "eco-friendly" fishery in that there is very little bycatch and no impact on fishery habitat. Also, unlike some other tuna species, albacore do not swim with dolphins, and therefore no albacore fishery is implicated in the bycatch death of dolphins. The U.S. National Marine Fisheries Service considers the North Atlantic albacore population overfished, but the South Atlantic stock is not considered so.

"Troll-caught" albacore are between three and five years old, harvested by towing artificial lures with unbaited hooks behind a slow-moving boat. Commercial fishermen in North America have used this low-impact, environmentally responsible fishing technique to catch albacore for nearly a century, while albacore fleets from other countries tend to use other methods. Those who advocate the consumption of albacore maintain that younger troll-caught albacore contain more beneficial omega-3 fatty acids than older, larger albacore more commonly available on the market. Processing techniques also affect the omega-3 content of canned albacore. Most canned albacore sold by the big major brands comes from older, larger albacore, which is cooked twice during the canning process. Some fishermen now offer custom-canned, troll-caught albacore, hand-packed and cooked just once to prevent the loss of omega-3s.

Regardless of where (or how) it is caught, most albacore ends up in cans in the United States. In Japan, *shiro maguro* (literally, "white tuna") is not commonly served as sushi or sashimi because the meat is so soft that it is difficult to handle. It also changes color quickly in the sushi case, and although this does not indicate deterioration, it makes the fish look somewhat unappetizing. For these reasons, many sushi chefs choose not to serve albacore. Moreover, albacore accumulates higher levels of mercury than other kinds of tuna, and some groups have urged testing and recall of canned albacore with high levels. Longlined albacore are older fish and have accumulated more mercury than younger, troll-caught fish. The U.S. Food and Drug Administration (FDA) advises women of childbearing age and children to limit their consumption of albacore tuna (chunk white canned tuna) and tuna steaks to six ounces per week or less.

BIGEYE TUNA (THUNNUS OBESUS)

Bigeye tuna are similar in general appearance to yellowfins, but with proportionally bigger eyes and without the yellow fins. The large eyes of this species suggest that it lives and hunts at greater depths than other tunas, and indeed, they are the deepest-occurring of all tuna species, ranging from 150 to 250 fathoms. Bigeyes reach a maximum length of 7.5 feet and can weigh four hundred pounds. (The record is a 435-pounder caught off Peru.) Where the body of a bluefin is deepest around the middle of the first dorsal fin, the bigeye is deepest just forward of the second dorsal. (It is believed that the specific name *obesus* is derived from the depth of the body.) Bigeyes resemble yellowfins—they are both called *ahi* in Hawaii—but *Thunnus albacares* has yellow, scythelike second dorsal and anal fins, and *Thunnus obesus* has longer pectoral fins and a proportionately larger head. Like most of the large tunas, the bigeye is metallic blue-black above and silvery white on its flanks and belly, with pale vertical striping.

In French Polynesian waters, Laurent Dagorn, Pascal Bach, and Erwan Josse (2004) outfitted bigeyes with pressure-sensitive ultrasonic transmitters, and learned that these tuna swam within one hundred meters below the surface during the night and at depths between four and five hundred meters during the day. The fish exhibited a clear relationship with the sound-scattering layer (SSL), a horizontal, mobile zone of living organisms that scatters or reflects sound waves. (It is also known as the deep-scattering layer, or DSL.) The researchers tracked the vertical movements of the tuna at dawn and dusk, and suggested that the fish were probably foraging on the organisms of the SSL. Bigeyes also made regular rapid upward vertical excursions into the warm surface layer, most probably in order to regulate body temperature and perhaps to compensate for an accumulated oxygen debt (i.e., to metabolize lactate). During the daytime, the larger fish made upward excursions about every two and a half hours, whereas smaller fish in Hawaiian waters swam upward approximately every hour. The blood of bigeyes has a "significantly higher affinity for O_2 than blood of other tunas" (Lowe, Brill, and Cousins 2000), which means that they can function better than other species in waters of low ambient oxygen, including some oxygen-poor waters at the depths they inhabit.

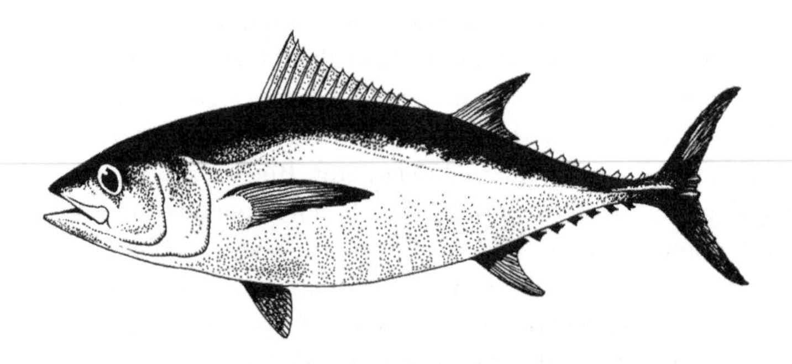

Bigeye tuna *(Thunnus obesus)*

Bigeyes are among the most commercially important tunas, especially in Pacific Rim countries like Japan and Australia. They are usually caught on longlines, not in nets, and are sold fresh, not canned. The bright red meat, higher in fat than that of other large tunas, is prized for sashimi and grilling. Bigeyes are found all over the world in tropical and subtropical waters, and they are popular game fishes on both coasts of North America and off Hawaii and Australia. Among the countries catching bigeye tuna, Japan ranks first, followed by the Republic of Korea with much lower landings. The world catch increased from about 164,000 metric tons in 1974 to 201,000 in 1980, reaching a peak of 214,000 in 1987. In the Indian Ocean, the bigeye tuna fishery was dominated by Japanese fleets up to the end of the 1960s, but as operations of vessels from South Korea became more important, they accounted for more than 60 percent of the catch in the late seventies. The most important fishing gear in the Pacific are longlines, which comprise some four hundred "baskets" (consisting of five branch lines, each with a baited hook) extending up to eighty miles. Day and night operations are common throughout the year, but there are seasonal variations in apparent abundance reflected in changes of fishing effort. In the 1970s, deep longlines employing between ten and fifteen branch lines per basket were introduced. This new type of gear is theoretically capable of fishing down to a depth of 300 meters, as compared to the usual 170 meters reached by traditional longline gear. Catch rates for bigeye increased for about three years and then declined to previous levels again, suggesting that only a portion of the bigeye resources were exploited. Bigeye tuna is exploited in rising quantities as associated catch of the spring and summer pole-and-line fishery in the northwestern Pacific and of the purse-seine fishery in the eastern Pacific, both directed primarily at skipjack and yellowfin tuna. In Japan, its meat is highly priced and processed into sashimi in substitution for bluefin tuna. The catch reported for 1996 to FAO was 328,067 tons, of which 101,591 tons were taken by Japan, 64,498 tons by Taiwan, and 28,418 tons by South Korea.

Bigeye meat is reddish pink in color, and like yellowfin it begins to discolor when exposed to air. For this reason, it is usually not loined or filleted until shortly before use. Larger bigeyes typically have a higher

Ten

THE TUNA INDUSTRY

Loading frozen tuna carcasses aboard a Japanese reefer. Port Lincoln, 1985

THE TUNA "industry" in California began in 1903 by accident. Sardines were the fish of choice then, but the sardine population fluctuated wildly, peaking and crashing, for reasons the fishermen did not understand. Albert P. Halfhill, an enterprising sardine canner, decided to pack the empty cans with albacore, which at that time was considered a "nuisance" fish with no known commercial value. Once consumers began to try canned tuna, they liked the white-meat fish, and more canneries started to pack tuna instead of sardines. By 1913, nine plants were in operation, producing 115,000 cases annually. In 1917, there were thirty-six canners along the California coast. However, getting sufficient supplies of albacore from the Pacific coast proved difficult, so the leading canners approached a local fisherman named M. O. Medina and asked him to fish for tuna south of San Diego. Along with his brothers, Medina headed south: they loaded their boats to the gunwales with yellowfin, skipjack tuna, and bluefin, eventually leading to the birth of San Diego's tuna fleet.

But it was World War I that really put the tuna industry on the map. Troops needed a convenient protein-rich food and canned tuna provided a ready source. Bluefins were harvested off California, while yellowfin and skipjack were the objects of an offshore "high seas" fishery from Southern California through the eastern tropical Pacific, as far south as Ecuador. At first the bluefins—now known as the Pacific species—were sought by sportfishermen, but it was not long before commercial fishermen recognized the possibilities of catching and selling these plentiful fish, and soon a full-fledged fishery was born. It began as a pole-and-line operation, but the introduction of net-fishing put an end to that labor-intensive practice. Prior to 1918, according to S. S. Whitehead in the *California Fish and Game Fish Bulletin,* "the catch was negligible for the fishermen had not found a method of fishing by which they could catch the fish in large enough quantities to pay for the trip," but by 1919, thanks to the newfangled purse seines, the catch reached six million pounds. Schools of fish were spotted from the crow's nest, and a skiff dispatched to surround them in the net. The trapped tuna were brought aboard and delivered directly to the canneries in San Diego. A small amount of bluefin was sold fresh, but the majority of the catch was canned; the most popular size was a half-pound can.

The offshore tuna fishery was rather more adventurous. It involved large boats—the forerunners of the "tuna clippers" of the mid-twentieth century—equipped with live bait wells, and designed and provisioned to remain at sea for weeks at a time. Most of the fishing in those days was done in the eastern tropical Pacific, off the coasts of Mexico, Ecuador, Costa Rica, and the Galápagos Islands, thousands of miles from the home ports of San Diego and San Pedro. The indispensable baitfish, primarily sardines or anchovies, were netted, brought aboard, and kept alive in large tanks. When a school of tuna was spotted, the boat was slowed down, and the bait "chummed" into the water as the fishermen arranged themselves in "racks," along the low-freeboard stern rails of the boat. They would fish with bare hooks, baited hooks, or feather jigs, sometimes with as many as three men to a pole to yank the big yellowfins out of the water. In his 1938 description of this fishery, H. C. Godsil wrote, "the hardest of all is fishing a school of mixed sizes. A ten-pound fish may be followed by one of a hundred pounds or more, and at such times gear is lost or broken wholesale, and not infrequently, men go overboard." Immediately behind the fishermen, the deck tanks filled with flapping tuna, which were cleared away only when the frantic fishing stopped. A tuna boat could carry a hundred tons of fish, most of which was iced in partitioned bins in the hold before being brought back to port. In the San Diego canneries, the fish were cooked, the meat packed by hand, salt and salad oil added, and "the cans pass to a warehouse from which they emerge in attractive labels en route to every part of the United States."

From the early 1930s to the late 1970s, San Diego was known as the Tuna Capital of the World, employing more than forty thousand people directly or indirectly in the industry. Large companies like Van Camp Seafood, StarKist Foods, Westgate California, Bumble Bee Seafood, Pan Pacific, and a host of small canners processed tuna in San Diego and up and down the West Coast. The industry in San Diego was ranked third only to the U.S. Navy and the aircraft industry, bringing in over $30 million a year to the city's economy. Before the outbreak of World War II, most of the boats in San Diego and San Pedro were independent ventures, owned and operated by one or more individuals. During the war, most of the ships in the San Diego tuna fleet were commandeered by the navy and used to shuttle food and supplies to the troops in the South Pacific. By 1954, the United States had become the world's largest producer and consumer of canned tuna products.

The limited capacity of the San Diego boats attracted the Japanese, whose longliners and purse-seiners were already catching thousands of tons of tuna that they were bringing in (duty free) to the California canneries. The Americans responded by building larger and larger superseiners that could travel over three thousand miles without refueling; and switched to nylon nets, power blocks, and other technological advances that would enable them to compete with the foreign fleets. The earlier San Diego "baitboats" had a maximum capacity of about 150 tons, but the new superseiners could carry from 1,200 to 2,000 tons. In *Sea of Slaughter,* Canadian author Farley Mowat's 1984 diatribe on the wholesale destruction of almost every living creature unfortunate enough to have ever come into contact with human beings in the North Atlantic, he describes one of these superseiners:

> The ultimate tuna boat was *Zapata Pathfinder*—a 250-foot superseiner under Panamanian registry that looked more like a Greek shipping magnate's private yacht than a working vessel. Valued at $10–$15 million, she had a satellite navigation system, carried her own helicopter for tuna spotting, and provided her captain with a suite of rooms embellished with a bar and lounge, a kingsize bed, and gold-plated bathroom faucets. She was capable of catching, freezing, and stowing $5 million worth of tuna on a single voyage. She could earn her captain $250,000 a year, but how much she made in overall profits remains unknown, as do the identities of her owners, which are lost to view in overlapping and interlocking companies. Nevertheless, informed observers of the tuna fishery have estimated that *Zapata Pathfinder* probably returned 100 per cent profit on investment every year she operated. The magnitude of the carnage she and her sisters inflicted on the world's tuna population in the process of amassing these obscene profits was so great that, by the end of the 1970s, such superseiners had fished themselves out of business.*

Bigger, faster, more efficient boats looked like the answer to the competition from the Japanese, but a heretofore unsuspected idiosyncrasy of the yellowfin was about to put the California fishers out of business.

Aggregating, with or without floating objects or dolphins, may have served some purpose for yellowfin in the past, but it now appears to be more of a debit than a credit to the populations. Hanging out with dolphins in the eastern tropical Pacific caused the tuna no end of

*Mowat's concern for wildlife is admirable, as are his vituperative attacks on fishermen, whalers, sealers, bird hunters, egg collectors, feather merchants, and anyone else who threatens the existence of living creatures in the North Atlantic. But his books are filled with facts and figures that one cannot check, because he does not bother to tell the reader where his information comes from. *Sea of Slaughter* contains not one footnote and not one bibliographic cross-reference, so there is no way of checking on, say, "the gold-plated bathroom faucets" or "the magnitude of the carnage she and her sisters inflicted on the world's tuna population," except to take Mowat's word for it.

trouble, as it made it that much easier for the fishermen to find them. And for whatever the reason, yellowfins will continue to aggregate around floating objects. It is unlikely that fish can pass along information from one generation to another, and it appears that tuna will never be able to persuade others of its kind to avoid schools of dolphins or FADs. Those that were trapped in the nets were killed, which would have made the transmission of information difficult at best. (A similar situation existed with regard to whaling, where the whales, presumably "smarter" than tuna, could not communicate that those big floating things [ships] could be dangerous, because by the time a whale had learned that, it was already harpooned and en route to the flensers.) FADs, in other words, may be a boon to the fishermen, but they are death to the tuna.

The death of the tuna, of course, is the life of the tuna business, and the life of the tuna business is canned tuna. When the fish are unloaded from the vessel they are thawed in running water or sprays of water. The fish are then quickly gilled, gutted, headed, and finally frozen. After cutting, the tunas are loaded into trays and taken to the precooker. Once they are precooked and cooled, the cleaners remove the skin from the fish and separate the loins from the skeleton. The loins are cut into solid pack or chunks, according to their firmness, and the the last step, canning, is a totally automated process. Canned tuna products are packed in oil, brine, spring water, or sauce. After the cans are sealed, they are cooked a second time ("retort cooking"), for two to four hours. Then the cans are cooled, labeled, and finally packed into cardboard boxes for distribution.

From the earliest days of the California tuna industry, when San Diego had the world's largest tuna fleet, there were canneries close to the docks. Like Monterey, where the fishery was for sardines, San Diego also had a Cannery Row. With the increased fishing activity in the far western Pacific and the closure in the early 1980s of both San Diego tuna canneries, fewer and fewer boats need to come to that port. Instead they are using canneries and getting their repairs and provisions at such places as American Samoa and Puerto Rico. Chicken of the Sea was founded in 1914 when Frank Van Camp and his son bought the California Tuna Canning Company and changed the name to the Van Camp Seafood Company. In 1963, Van Camp Seafood was purchased by Ralston Purina, and then acquired in 1997 by the investment group Tri-Union Seafoods, which changed the name of the tuna-canning business to Chicken of the Sea International. The investment group sold the company in 2000 to Thai Union International, now the sole owner. In 2001, Thai Union closed down the Chicken of the Sea cannery in San Pedro (Los Angeles harbor), the last such facility in the United States. Chicken of the Sea tuna is now processed in American Samoa, where the canneries are the largest private employers in the territory.

High labor costs were responsible for the closure of the mainland American tuna canneries. (In American Samoa, U.S. federal tax laws apply, but minimum wage laws do not.) The operations were often relocated to areas closer to the fisheries, to avoid the additional costs of shipping tuna around the world. For the most part, the canneries employ people at less than what would be considered a minimum wage in developed countries. There are canneries in the Philippines, American Samoa, Puerto Rico, Fiji, Thailand, Japan, Indonesia, Taiwan, and Port Lincoln, South Australia.*

There are many varieties and grades of canned tuna to choose from. Solid or fancy pack contains large pieces of tuna and is usually albacore. Many pay the higher price for white tuna because it has a milder flavor and lighter color. Flaked tuna is broken apart and used in salads where the tuna is mashed and mixed anyway. You can eat tuna raw, cooked, broiled, smoked, grilled, on a roll, as a burger, as a fish cake, as a dip, in a melt, in a salad, in a sandwich, in a pita, in a pot pie, in a casserole, as a sauce (vitello tonnato), in a wrap, with mayonnaise, with noodles, with rice, with olives, with onions, with capers, with relish, with wasabi, with pesto, with soy sauce, with risotto, with fettuccine, with spaghetti, with macaroni, with tarragon, with chick-

*In August 2006, Port Lincoln Tuna Processors announced what was described in the *Adelaide Advertiser* as a "$40m coup for a Port Lincoln company." The cannery had secured a U.S. export order for its G'Day Gourmet brand of canned skipjack and salmon products. The deal followed the signing of the U.S.-Australia Free Trade Agreement in 2005, which removed the 35 percent tariff on canned tuna. Brian Jeffriess of the Australian Tuna Boat Owners Association said that "the penetration into the U.S. market had been quicker than was thought possible. Getting into the U.S. is very difficult as it is the biggest market in the world and everyone is attacking it." Shortly after reading this news item, I went to my local Whole Foods market and bought a little can of G'Day Gourmet "Lemon Pepper Tuna." It contains "light tuna [skipjack], spring water, canola oil, lemon juice, cane sugar, cracked pepper, spices, wheat and soy." (Wheat and soy?)

peas, with cheese, with artichokes, and in combinations that are waiting to be invented.

And your cat can eat it too. One brand advertises that their "Solid Gold Blended Tuna is made with Tuna, Tapioca and Canola Oil with vitamins and minerals. Perfect for adult cats and growing kittens. Ideal as an accompaniment to dry food, to feed alone, or as a treat. Blended Tuna is a highly palatable formula made with 'dolphin friendly,' bite-sized chunks of red and white tuna in a natural gravy." However, if you read the small print that lists the ingredients, you will find that the can contains:

> Tuna | Water Sufficient for Processing | Tapioca | Canola Oil | Cellulose Gum | Calcium Iodate | Biotin | Calcium Sulfate | Tricalcium Phosphate | Potassium Chloride | Choline Chloride | Ferrous Sulfate | Zinc Oxide | Nicotinamide | Vitamin E Supplement | Taurine | Thiamine Mononitrate | Vitamin B12 Supplement | Vitamin A Acetate | Manganous Oxide | Pyridoxine Hydrochloride | Calcium Panthothenate | Folic Acid | Riboflavin | Vitamin D3 Supplement | Copper Sulfate | Menadione Sodium Bisulfate Complex

In 1992, the U.S. Marshall Islands, at the FDA's request, seized thousands of cans of tuna products from Canada that were originally labeled for use as cat food but had been relabeled for human consumption. Although these products were not believed likely to pose a risk to human health, they were fraudulently mislabeled, and the FDA recommended that retail store managers inspect their stocks of tuna and tuna cat food for the suspect products. The tuna could have been labeled as Ocean King Chunk Light Tuna, IGA Chunk Light Tuna, Blue Bay Chunk Light Tuna, National Chunk Light Tuna, Ocean Pride, Shur-Fine, or Ocean Best.

The latest development in producing canned tuna took place in 2000, when vacuum-packed pouches were introduced. Through this new technology, consumers now have access to an easy-to-store, easy-to-open, and easy-to-clean-up food. At the same time, tuna processors have added marinated flavors to canned and pouched tuna, enabling the category to provide more choice to the consumer and better variety. Americans consume over one billion pounds of canned or pouched tuna annually. As long as the tuna populations hold up, it will remain one of America's favorite foods. U.S. processors use either domestic or imported raw (fresh, chilled, or frozen) tuna to produce different varieties of canned and pouched tuna, which are distinguished by the type of meat (white or light), the packing medium (water or oil), and the form. In the United States, most canned tuna is available as either "solid" or "chunk." According to the Food and Drug Administration (FDA), which sets strict definitions for how canned tuna is marketed, "solid" is a portion of the loin cut to fit the can, and "chunk" is cut pieces of varying sizes. Chunk light-meat in water is the most popular light-meat pack, although there remains a demand for oil-packed canned light-meat tuna. The source of most chunk light meat is skipjack, although other species of tuna can be added. Albacore is packed almost exclusively in water in solid form.

The U.S. Tuna Foundation (USTF), a lobbying organization based in Washington, D.C., was established in 1976 to represent the interests of the canned tuna industry, including the distant-water fishing fleets, and the "big three" canned tuna brands, StarKist, Bumble Bee, and Chicken of the Sea International. On their website, the USTF provides this look at the market for canned tuna in this country:

- Japan and the United States are the largest consumers of tuna, using about 36 percent and 31 percent, respectively, of the world's catch.
- Canned tuna is the second most popular seafood product in the U.S. after shrimp.
- In the U.S., Americans eat about one billion pounds of canned or pouched tuna a year. Only coffee and sugar exceed canned tuna in sales per foot of shelf space in the grocery store.
- Of those Americans who eat canned tuna, the vast majority—83 percent—eat it for lunch. In fact, canned tuna is the only regularly consumed seafood at lunch.
- Over one half of canned tuna (52 percent) is used in sandwiches. Another 22 percent is in salads while 15.5 percent is in casseroles/helpers and 7.5 percent is in base dishes.

- Households with children under 18 are about twice as likely to have tuna sandwiches available than households without children.
- Light meat accounts for 75 percent to 80 percent of annual domestic canned tuna consumption; albacore or white meat makes up the balance.

To dispel any fears that anyone might have about eating tuna—or too much tuna—the USTF published this note in the *Tuna Tribune,* an online newsletter dedicated to promoting the consumption of the fish:

> A new study recently published in *The Lancet,* a leading medical journal, reaffirms what the National Academy of Science's Institute of Medicine, Harvard Medical School, and numerous public health organizations have known for years: Women and young children should eat more—not less—seafood, like canned tuna, that is rich in omega-3 fatty acids.*

Japan has a similar organization. The Organization for the Promotion of Responsible Tuna Fisheries (OPRT) is a nongovernmental body, established in Tokyo in December 2000, whose purpose, says its website, is "to link the oceans with the consumers and promote sustainable use of tunas." Its members are longline fishers from Japan, Taiwan, South Korea, the Philippines, Indonesia, China, and Ecuador, as well as traders, distributors, consumers, and public interest organizations in Japan. An organization dedicated to developing tuna fisheries might regard the Myers and Worm report that found Japanese longliners responsible for the demise of large fish populations more than a little objectionable. In the *OPRT Newsletter* for January 2006, Dr. Yuji Uozumi (whose full title is "Director, Research Planning & Coordination Division, National Research Institute of Far Seas Fisheries, Fishery Research Agency, Japan") wrote, "Needless to point out,

*The actual text, from *The Lancet* (Hibbeln, Davis, Steer, et al. 2007), reads: "Maternal seafood consumption of less than 340g per week in pregnancy did not protect children from adverse outcomes; rather, we recorded beneficial effects on child development with maternal seafood intakes of more than 340g per week, suggesting that advice to limit seafood consumption could actually be detrimental. These results show that risks from the loss of nutrients were greater than the risks of harm from exposure to trace contaminants in 340g seafood eaten weekly."

the contents of the [report] entirely differed from the views of scientific committees of the international fisheries commissions responsible for tuna resources. However, there was no way for the general public to know the views of the fisheries commissions." Uozumi agrees with Hampton, Sibert, Kleiber, Maunder, and Harley (2005), that changes in CPUE (number of tuna caught per hundred hooks) did indeed decline since the 1950s, but that was before the Japanese shifted the fishery to target bigeye tuna. "This is a story well-known to all tuna researchers in all countries," wrote Uozumi. "This fact alone shows how the Myers paper is erroneous."

Is there mercury in tuna? Well, yes. But as with so many issues involving these fish, how much—or how much is harmful to people—depends on whom you ask. In a 2005 article in *Discover,* Karen Wright noted that "one particularly common source of low-level mercury exposure is tuna. Because they are large, long-lived predators, tuna accumulate more mercury in their tissue than smaller, short-lived fish. When tested for mercury in parts per million, flesh from albacore tuna, which take five years to mature, was shown to contain about four times as much mercury as chunk light tuna, which is harvested from younger fish." Wright also noted that seafood is one of the most common sources of mercury exposure in adults (the others are coal-fired utilities and industrial boilers which spew mercury into the atmosphere), and that "effects might occur at lower levels than anyone suspected. Some studies demonstrate that children who were exposed to tiny amounts of mercury in utero have slower reflexes, language deficits, and shortened attention spans. In adults, recent studies show a possible link between heart disease and mercury ingested from eating fish. Other groups claim mercury exposure is responsible for Parkinson's disease, multiple sclerosis, Alzheimer's, and the escalating rate of autism."*

*The mercury scare was initiated by the 1956 discovery that families of fishermen near Minamata on the Japanese island of Kyushu were afflicted with a mysterious neurological disease with symptoms that included loss of coordination, tremors, slurred speech, and numbness in the extremities. The symptoms worsened and led to general paralysis, convulsions, brain damage, and death. A Chisso chemical factory that manufactured acetic acid and vinyl chloride and dumped its wastes into Minamata Bay was identified as the source of the mercury that had contaminated the fish and shellfish. Even when the source of the crippling and often fatal disease was identified, the Japanese government did not order the plant closed until the 1970s. By 1997, more than 17,000 people had applied for compensation from the government. Since 1956, a total of 2,262 have died of Minamata disease in Japan.

Most mercury contamination in humans (Americans, anyway) comes from 440-odd coal-fired power plants, which, according to a 2006 essay in *Time* by Jeffrey Kluger, "produce about 48 tons of mercury—40% of the country's total output." Mercury in fish, then, might be only a part of what is becoming an increasingly complex and perilous situation. Kluger noted that "researchers testing birds in the Northeast have found creeping mercury levels in the blood of more than 175 once clean species. Others have found the metal for the first time in polar bears, bats, mink, otters, panthers, and more." It is possible—but not terribly likely—that emissions into the atmosphere can be reduced, but mercury in fish is cumulative, and what is already in the oceans' smaller fish species is going to be consumed by the large predators, no matter what we do up here. The fish with the highest concentration of mercury are swordfish, and some species of sharks, but close behind them are various tunas, not necessarily the larger species.

Wall Street Journal staff reporter Peter Waldman opened his article of August 1, 2005, with "One by one, Matthew Davis's fifth-grade teachers went around the class describing the 10-year-old boy. He wasn't focused in class and often missed assignments. He labored at basic addition. He could barely write a simple sentence." Thinking it was "brain food" and certainly better for him than junk food, Matthew's parents had been feeding him three to six ounces of white albacore tuna every day. In an article two days later in the *San Francisco Chronicle,* his mother described his symptoms as "clouded thinking, bent fingers, acid reflux, lethargy, apathy, lack of coordination, and tingling in his extremities." When his parents took him for tests, they found that Matthew had mercury poisoning; his blood contained more than twelve times the mercury level considered safe by the FDA. His diet was drastically modified, and within a year the symptoms had almost disappeared. In response to Waldman's article, the tuna industry claimed that Matthew's story was "a whole lot abalone" (how clever) and most tuna that was sold didn't have much mercury in it because it came from young fish that had not accumulated mercury to dangerous levels. In her *Chronicle* article, Joan Elan Davis wrote:

> I feel a sense of rage to hear the tuna industry responding to Matthew's and our family's plight with such insensitivity and dis-

> honesty. It is unconscionable that the industry continues to put our children's health at risk by denying that there are significant risks from consuming mercury-contaminated tuna. (The U.S. Tuna Foundation tells consumers that to be at risk, they would have to eat 1,358 cans of tuna per year!) . . . We as consumers need to demand that government, industry and retailers warn the public of the dangers surrounding mercury-contaminated fish, especially tuna, by posting warnings and serving guidelines on cans, in aisles at the store and at the fish counter. Our children should be our top priority. As a society, we need to make absolutely sure that we protect our children's health—especially from easily preventable afflictions such as mercury poisoning. It is inexcusable that any child should have to unnecessarily suffer what my child Matthew has endured.

On a recent plane trip to St. Louis, I met a doctor named Daniel Sulmasy who taught at St. Vincent's Catholic Medical Center in New York City. I told him the story of Matthew Davis, and he said he knew of a similar case, only this time it was a grown man who had decided to eat only tuna: Sulmasy wrote this to me in a letter: "In 2004, a 27-year-old, healthy, unmarried man, working for an Internet-based company in New York City, who had a normal level of cholesterol but was very concerned about his health (his father had died of a heart attack at age 67), told his internist that he was following a strict low-fat diet. Pressed by his doctor about what this meant, he reported that he was eating tuna two to three meals a day: typically canned tuna for lunch and tuna sushi for dinner. Alarmed by this, his internist measured his mercury levels and found elevated blood and urine mercury. The patient, fortunately, had no symptoms of mercury poisoning. Six months after stopping his tuna-only diet, his blood levels had returned to normal."

On December 13, 2005, Sam Roe and Michael Hawthorne, reporters for the *Chicago Tribune,* wrote that "federal regulators and the tuna industry fail to warn consumers about the true health hazards of an American favorite." Their article pointed out that "canned light tuna," which is skipjack and yellowfin tuna, is just as high in mercury as albacore, and the only reason it was still available was that the U.S.

fishing industry had applied pressure to the FDA to keep a mercury warning from the public. The article quoted the U.S. Tuna Foundation: "If consumers were warned about tuna [as they had already been warned about king mackerel, shark, and swordfish] the market for canned tuna would shrink about 20 percent, the U.S. tuna fleet would default on loans, and the seafood industry could face numerous class action lawsuits at substantial cost and adverse publicity." FDA officials were quoted as saying they did not know what is actually put in cans labeled "canned light tuna," and that they stand by their rulings.

Six days later, on December 19, 2005, a full-page ad appeared in the *New York Times,* which read: "Hooked on the Hype; FishScam.com." The website said that FishScam.com was "a project of the Center for Consumer Freedom, a nonprofit coalition dedicated to promoting personal responsibility and protecting consumer choices," and that "a growing cabal of environmental activist groups, public health researchers, and government bureaucrats are using junk science to needlessly frighten Americans about the fish they eat. This scam has very little public opposition, but we're here to set the record straight." According to the website, the "cabal" consisted of:

> Consumers Union, Center for Science in the Public Interest, Earth Share and the Ad Council, Environmental Defense Fund, Environmental Working Group, Friends of the Earth, Greenpeace, Mercury Policy Project, National Environmental Trust, Natural Resources Defense Council, Oceana, People for the Ethical Treatment of Animals, Physicians Committee for Responsible Medicine, Physicians for Social Responsibility, SeaWeb, Sierra Club, and the Turtle Island Restoration Network.

Why would all these organizations want to hype the "mercury scam"? Could it be that some of the "restaurants, food companies, businesses, employees, and consumers" identified in the "Who We Are" page had a connection to commercial fishing? Or were we to assume that the "Center for Consumer Freedom" was honorably guaranteeing our right to eat whatever we please, and not be swayed by those worrywarts at the likes of Consumers Union, Center for Science in the Public Interest, and the Environmental Defense Fund? When

New York Times food reporter Marian Burros investigated the Center for Consumer Freedom, she learned that it was underwritten by tobacco, alcohol, and restaurant interests. She wrote that David Martosko, the director of research, believed that "Americans need to be reminded that the health benefits of eating fish are real, while the risks are imaginary." In the same article (February 15, 2006), Burros quoted Rebecca Goldburg, senior scientist at the Environmental Defense Fund: "I just get disgusted with the view that the public is so dull-headed that they can't understand that fish are generally good for you, but some kinds are to be avoided because they are heavily contaminated."

There seems to be no question that the benefits are real. In a 2006 *New Scientist* article, Bijal Trivedi wrote, "All kinds of foods from Brussels sprouts to peanut butter and potatoes have been touted as brain food. Sadly, these are little more than old wives' tales, but there is one brain food that has solid evidence on its side: fish. No nutrient has garnered as much supporting evidence for promoting mental health as long-chain omega-3 fatty acids, in particular eicosapentaenoic acid (EPA) and docosahexaenoic acid (DHA), which are plentiful in oily fish such as tuna and salmon." It seems that omega-3s are a particularly important component of the membranes that envelop nerve calls in the brain, and when these membranes are flexible, the brain cells are more receptive to incoming signals. In recent experiments, researchers found that rats given a diet deficient in omega-3s lost 80 percent of the DHA in their retinal cells, replacing it with omega-6s, which actually harden the membranes. Omega-3s can even promote neuronal growth to repair brain damage. "Preliminary studies," writes Trivedi, "have also shown encouraging results for schizophrenia, borderline personality disorder, dyslexia, autism, attention deficit hyperactivity disorder, and obsessive compulsive disorder."

When tests showed that a daily dose of fish oil containing omega-3 fatty acids could improve the survival rate of heart attack victims and reduce fatal heart arrhythmia, Europeans embraced this therapy, but Americans did not. In the *New York Times* (October 3, 2006) Elizabeth Rosenthal wrote, "in the United States, heart attack victims are not generally given omega-3 fatty acids, even as they are routinely offered more expensive and invasive treatments, like pills to lower cholesterol or implantable defibrillators. Prescription fish oil, sold under the brand name Omacor, is not even approved by the Food and Drug Administration for use in heart patients." Rosenthal continued: "The American College of Cardiology now advises patients with coronary artery disease to increase their consumption of omega-3 fatty acids to one gram a day, but it does not specify if this should be achieved by eating fish or by taking capsules . . . In Europe meanwhile, research on prescription fish oil, which is now thought to act by stabilizing cell membranes, has gained momentum."

UCLA researchers reported that altering the fatty acid ratio found in the typical Western diet to include more omega-3 fatty acids and decrease the amount of omega-6 fatty acids may reduce prostate cancer tumor growth rates and PSA levels. Published in the August 2006 issue of the journal *Clinical Cancer Research,* this study is one of the first to show the impact of diet on lowering an inflammatory response known to promote the progression of prostate cancer tumors. The omega-6 fatty acids contained in corn oil, safflower oil, and red meats are the predominant polyunsaturated fatty acids in the Western diet. The healthier marine omega-3 fatty acids exist in cold-water fish like salmon, tuna, and sardines. "Corn oil is the backbone of the American diet. We consume up to 20 times more omega-6 fatty acids in our diet compared to omega-3 acids," said principal investigator Dr. William Aronson, a professor in the department of urology at the David Geffen School of Medicine at UCLA. In other words, the omega-3 fatty acids EPA and DHA in fish counteract the effects of the bad omega-6s in cooking oils and red meat.*

In her *New York Times* article, Burros went on to explain that

> it is easy to buy fish that have beneficial Omega-3 fatty acids but do not have a high level of contaminants . . . If you crave a good tuna sandwich, you can eat limited amounts of canned tuna labeled

*Now comes the news (Coghlan 2006) that mercury in fish could put people at risk of developing diabetes because methylmercury kills the cells in the pancreas that make insulin. Experimenters at the Kyungpook National University in Daegu, South Korea, fed mice low levels of methylmercury for a month, and found that the mice produced less insulin, had higher blood glucose levels, and sustained more oxidative damage to their fat cells. Mercury or no, obese people are more likely to develop diabetes, but to be on the safe side, Shing-Hwa Liu, the chief researcher at National Taiwan University, says, "I think people should eat less fish."

> chunk light instead of white (albacore). Even better, simply switch to wild canned salmon from the Pacific . . . For those who do not want to take the bones and skin off the salmon after draining it, some salmon in cans and pouches is sold without them . . . Combining salmon with the same ingredients used in making the best tuna sandwich in the world produces a sandwich just as delicious. And for those who do not like the taste of canned salmon, rinsing it under warm water produces a neutral taste.

Some recent studies have shown that consumption of ocean fish species constitutes no threat at all to either children (born or unborn) or adults. A study published in the *Journal of the American Medical Association (JAMA)* in 1998 indicated that methylmercury in fishes posed no threat at all. Philip Davidson of the University of Rochester, the lead author of the study conducted in the Indian Ocean republic of the Seychelles, noted that mercury did no harm to the more than seven hundred children who were examined, many of whom ate ocean fish several times a week. Indeed, said Thomas Clarkson, another of the authors of the study, the health benefits from fish outweigh the small risk factor associated with mercury residues in some fish. (Some aspects of the study were funded by the Electrical Power Research Institute, an organization dedicated to showing that mercury from coal-powered plants is harmless, so the results should at least be questioned.)

Mercury is certainly bad for you, but some say that the health benefits from eating fish may offset the effects of the small amounts of the substance found in some species. In a follow-up to the *JAMA* article, Clarkson and J. J. Strain said that "ongoing epidemiological studies of heavy fish consumers in the Seychelles Islands . . . do not reveal adverse effects. To the contrary, the results of some developmental tests that were conducted on prenatally exposed children indicate beneficial outcomes that correlate with mercury levels during pregnancy." But the larger the fish, the more mercury it accumulates, so it is probably wise to avoid large species like sharks, tuna, and swordfish. Also whales, but that isn't a problem in the United States. In Japan, however, where the effects of high doses of methylmercury resulted in Minamata disease, they are still eating the meat of whales, and it's very dangerous. In a 2002 article in *New Scientist,* Andy Coghlan wrote:

> Tests on whalemeat on sale in Japan have revealed astonishing levels of mercury. While it has long been known that the animals accumulate heavy metals such as mercury in their tissues, the levels discovered have surprised even the experts. Two of the 26 liver samples examined contained over 1970 micrograms of mercury per gram of liver. That is nearly 5000 times the Japanese government's limit for mercury contamination, 0.4 micrograms per gram. At these concentrations, a 60-kilogram adult eating just 0.15 grams of liver would exceed the weekly mercury intake considered safe by the World Health Organization, say Tetsuya Endo, Koichi Haraguchi and Masakatsu Sakata at the University of Hokkaido, who carried out the research. "Acute intoxication could result from a single ingestion," they warn in a draft paper accepted for publication in *The Science of the Total Environment.*

Unfortunately, it might not actually be whalemeat. In an article published in 2003, Coghlan pointed out: "Much food labeled as whalemeat in Japan is actually from dolphins and porpoises, and exceeds the government's legal limit for mercury . . . Even though it is illegal to mislabel meat, DNA analysis of 17 'whale' products purchased in supermarkets by EIA [Environmental Investigation Agency] investigators in 2002 reveals that 12 were from porpoises and dolphins." The average level of mercury was five times the Japanese government's limit of 0.4 parts per million, but one sample was five thousand times over the limit. In Norway and Iceland, the only other countries in the world still actively whaling, the meat of minke whales—at a maximum length of thirty feet, the smallest of the groove-throated whales—is being served in restaurants. In 2003 Norwegian scientists announced that the meat contained dangerous levels of mercury and PCBs, and warned pregnant women and breast-feeding mothers to avoid it (Doyle 2003).

After the California attorney general tried to get the tuna-canning companies to label the cans with warnings to pregnant and nursing women, Superior Court judge Robert Dondero ruled that such labels were unnecessary because the mercury levels weren't high enough to pose any danger. In her May 13, 2006, article in the *Los Angeles Times,* Marla Cone quoted Deputy Attorney General Susan Fiering: "The people who will be most hurt are women who don't know about the FDA advisory on the Internet and don't have access to good medical care so they won't know about the danger of mercury in this fish." Two months later, the FDA announced that imported canned "chunk light tuna" (mostly skipjack) contained nearly ten times the level of mercury deemed the cutoff level for "low-mercury" fish. The canned tuna, much of which came from foreign sources, particularly Ecuador, was most likely to be consumed by children and pregnant women, because it is a government-subsidized food item for low-income families (Clayton 2006). In other words, while one arm of the U.S. government is issuing stern warnings about mercury, another is distributing cans of mercury-laden tuna.

Fish is now an important source of protein for many people, and in many countries it is becoming more so. It is so important in the United States that in her book *What to Eat,* the nutritionist Marion Nestle, chair of New York University's Department of Nutrition, Food Studies, and Public Health, devotes three chapters to it: one entitled "Fish: Dilemmas and Quandaries"; another on "The Methylmercury Dilemma"; and a third on "The Fish Farming Dilemma." The first of the "dilemmas and quandaries" is the attempt to balance the benefits of eating fish (e.g., omega-3 fats) against the detriments of the toxic chemicals found in fish. You cannot really compare the two; no amount of omega-3 fatty acids is going to neutralize methylmercury (in fact, nothing does), so she quotes the Institute of Medicine (IOM) in Washington: "Everyone should avoid eating much of the fish that are highest in methylmercury—especially shark, swordfish and tuna . . . pregnant women and young children should not eat such fish at all." The first FDA report on methylmercury in fish appeared in 1994, and warned consumers that shark, swordfish, and albacore contained unsafe levels of mercury. They warned consumers about shark and swordfish, but under pressure from the tuna industry, the report said nothing about albacore. The industry believed—probably correctly—that "hardly anyone knows the difference between one kind of tuna and another, the fish companies worried that consumers would interpret advice to avoid albacore tuna as advice to avoid *all* tuna," so no warnings were put on albacore cans at first.

When the Environmental Protection Agency (EPA) entered the picture, it seemed that tuna cans were about to carry some kind of warning, but again the lobbyists prevailed, and while king mackerel and tilefish were added to the FDA's 2001 list of fish high in mercury, there was still no mention of tuna. Now the argument was that if people saw a warning on a can of tuna, they would be reluctant to eat *any* fish, and the entire fishing industry would collapse. By 2004, the FDA and EPA had combined to issue warnings (published on websites and broadsheets) about tuna to pregnant women, nursing mothers, and small children, but there wasn't anything on a can of albacore in 2007 that might even hint at a problem.* Nestle says that the safest tuna is "chunk" or "light" (skipjack), and also, if we have been eating albacore, all we have to do is stop: "Methylmercury, unlike lead, does not stay in the body for long. Its 'half-life' is just two or three months, meaning that if you start now to reduce the amount of methylmercury you eat, half will be gone in a few months, half of what's left will disappear in another few months, and most will be gone within a year."

Selling under the label "American Tuna," a Washington company cans fish that are caught using rods with unbarbed hooks, not vast nets, so there is no unwanted "by-catch," according to an article by

*In fact, tuna cans in 2007 were designed to alleviate your worries. Chicken of the Sea, StarKist, and Bumble Bee "light" and "white" all have a tiny dolphin on them and the words "Dolphin Safe"; and Charlie the Tuna appears on the StarKist albacore cans, saying, "Contains 150mg of Omega-3 Fatty Acids." G'Day Gourmet tuna actually says "low in mercury" on the can, but it's skipjack, which was never a problem.

Florence Fabricant in the *New York Times* (October 17, 2007). The article (which sounds not unlike a press release) continues: "The canning process keeps in flavor and natural fish oils containing omega-3 fatty acids. . . . The fish are flash frozen on the boats, then cooked only once before canning. Most commercial tuna is cooked up to three times then packed with water or soybean oil to replace the natural juices, often with phosphates too. The fish have relatively low levels of mercury . . . as shown in tests by Oregon State University." Except for the one-time cooking and packing the tuna in its own juices, the American Tuna canning process is the same as any other albacore or skipjack cannery. Will people be willing to pay more for less cooking and less oil? At my local market, a six-ounce can of American Tuna's "Firm Pack Pole-Caught Wild Albacore" costs $6.45, compared to a six-ounce can of Bumble Bee "Solid White Albacore in Water," which sells for $2.25.

On January 2, 2008, the alarm bells were rung. Marian Burros's front-page story in the *New York Times* was headlined "High Mercury Levels Are Found in Tuna Sushi Sold in Manhattan." She wrote, "Although the samples were gathered in New York City, experts believe similar results would be observed elsewhere." Dr. Michael Gochfeld and Dr. Joanna Burger analyzed the bluefin sushi from places where it was sold, and identified fifteen restaurants and seven food markets where the mercury content was highest. Five places had mercury levels so high that the FDA could take action to remove the fish from the market. Because restaurants tend to use the meat from larger (for which read: more expensive) bluefins, the mercury content of tuna served in restaurants was higher than that of food stores, which usually sell the cheaper cuts from smaller fishes. Dr. Kate Mahaffey, senior research scientist of the EPA, was quoted as saying, "We have seen exposures occurring now in the United States that have produced blood mercury a lot higher than anything we would have expected to see, and this appears to be related to consumption of larger amounts of fish that are higher in mercury than we had anticipated."

As for the tuna, they have enough problems without worrying about soybean oil, omega-3 fatty acids, or mercury. People are trying to make vegetarians out of captive tuna, a fate worse than, or at least comparable to, death. No studies have so far shown that mercury is actually harmful to born or unborn *tuna,* and it's not people we're concerned about here, it's fish. And throughout the oceans, tuna aren't doing very well at all.

Look at the North Atlantic, for instance. ICCAT recommendations have been published since the mid-1970s, and the results have been a bonanza for the fishermen and an unmitigated disaster for the fish. As long ago as 1975, the director of the National Marine Fisheries Service, in consultation with the U.S. Fish and Wildlife Service, proposed to list the bluefin tuna as threatened under the Endangered Species Act, but the idea was withdrawn when the Atlantic Tunas Convention Act gave the United States authority to implement ICCAT's management recommendations. Big mistake. For decades, ICCAT recognized two distinct populations of North Atlantic bluefins, the western and the eastern. The western population breeds in the Gulf of Mexico, and the eastern in the Mediterranean, and it was assumed that the twain never met. Wrong again. Tagging experiments showed that the stocks intermingled, so harvesting fish from one stock affects the other, and although the quota for the western Atlantic stock (the Gulf of Mexico breeders) was reduced to 2,700 tons, the quota for the "eastern" stock was kept at around 50,000 tons. That's 50,000 tons, not 50,000 tunas. Fifty thousand tons is 100 million pounds, and that's a lot of tuna to take out of the North Atlantic every year.

Mike Leech, former president of the IGFA and now conservation columnist for the sportfishing magazine *Marlin,* bemoans the situation in the March 2007 issue, and offers some suggestions for solving what he calls "A Failure of Management." We might consider listing the bluefin as endangered (again), but its migratory habits preclude protecting it only in U.S. waters; we might list it with the Convention on International Trade in Endangered Species (CITES), which would mean that every international shipment would have to be identified and tracked; or we could pull out of ICCAT as a protest against their failed policies, but none of these measures would have much effect on the fleets of Croatia, Turkey, or Libya, and as long as the Japanese continue to consume all the tuna anyone can catch, the bluefin's future looks grim. Leech concludes his article:

> Scientists tell us the breeding population of western Atlantic bluefin is more than 90 percent depleted and currently stands at an all-time low. So few breeders are left that the population may collapse entirely . . . The bottom line on bluefin tuna: The situation is bad and getting worse . . . The longer we wait, however, the harder it will be to save the bluefin tuna from total collapse. Well-funded conservation organizations need to step up now and get involved in the fight to save the bluefin tuna. Help certainly won't come from ICCAT.

If the bluefin goes down, it will take many fishermen (and wholesalers, retailers, and restaurateurs) with it, but the other tuna species appear to be doing all right, as are those people who fish for them. In a 2007 press release, the Western Pacific Regional Fishery Management Council announced that "approximately 2.189 million metric tons of skipjack, bigeye, yellowfin and albacore tuna were landed in 2006." That's 4,378,000 pounds of fish. If there is anything that will demonstrate what a big business the worldwide tuna industry is, 4 million pounds of fish will probably fill the bill.

CAN WE SAVE THE BLUEFIN?

Mexican tuna ranchers with part of the herd

HUNDREDS OF THOUSANDS of tons of bluefin tuna—which translates to billions of fish—are caught every year, by fishers in the Mediterranean, the North Atlantic, the central and southern Pacific, and the Indian Ocean. They are harpooned, caught on longlines, yanked out of the sea with fishing poles, and drowned in drift nets. The tuna of the western North Atlantic were purse-seined intensively in the 1960s for canning, but by the 1970s commercial fishers targeted larger fish for the Japanese market. In 1981, ICCAT declared the Atlantic bluefin seriously depleted, and tried to set a quota "as near to zero as feasible." Within two years political pressure by the fishermen's lobby rejected this idea, and the quota for 1983 was 2,600 metric tons—a far cry from zero. Continued fishing drove the breeding population to an all-time low, and conservation groups began a futile campaign to save the tuna. In 1992, ICCAT halved the western North Atlantic quota, but when the population was found to have "stabilized," industry lobbyists pounced on this term. ICCAT rescinded the 50 percent cut, and the commission increased the quota again. The spawning population is now at 10 percent of what it was when ICCAT was formed in 1966, and the 2005 IUCN *Red Data Book* lists the western Atlantic bluefin tuna as "Critically Endangered."

Jean-Marc Fromentin of the Institut français de la recherche pour l'exploitation de la mer (IFREMER) has been studying the historic patterns of tuna fishing in the Mediterranean for years. In a 2005 study (with Joseph Powers) Fromentin wrote:

> Our description of Atlantic bluefin tuna fisheries places today's fishing patterns within the two millennium history of exploitation of this species: we discuss trap fisheries that existed between the 17th and the early 20th centuries; Atlantic fisheries during the 1950s and 1960s; and the consequences of the recent development of the sushi-sashimi market . . . While important uncertainties remain, when the fisheries history is confronted with evidence from biological and stock-assessment studies, results indicate that Atlantic bluefin tuna has been undergoing heavy overfishing for a decade. We conclude that the current exploitation of bluefin tuna has many biological and economic traits that have led several fish stocks to extreme depletion in the past.

Subsequently, writing of the eastern Atlantic and Mediterranean bluefin stocks, Fromentin and Christelle Ravier observed:

> there was (and still is) a sharp increase in the efficiency of long-established fisheries together with the rapid emergence of new fisheries. As a result, the eastern Atlantic and Mediterranean bluefin tuna stock is likely to be overexploited and overfished as it currently suffers the highest fishing pressure of its entire history . . . Such a situation is particularly worrisome since their life history traits make the Atlantic bluefin tuna more vulnerable to exploitation than other tuna species.

A WWF report published in September 2006 indicated that Armageddon for the Mediterranean tuna was nearly at hand. The report, written by Roberto Mielgo Bregazzi and presented at the

Brussels meeting of the European Parliament's Fishing Committee on September 13, concluded that there were almost no tuna left in the western Mediterranean, at one time among the most productive of all tuna-fishing grounds. Bregazzi wrote: "Today there is consensus that in a context of sustained increase of fishing and farming overcapacity, all the attempts to achieve a real regional management of this key Mediterranean fish resource have resulted in a complete failure. It is not an overstatement to say that the fishery—itself amongst the oldest in the world—faces a high risk of collapse." In the area around the Balearic Islands, the 1995 bluefin tuna catch was more than 14,000 tons; by 2006 the total catch had fallen to 2,270. The tuna farms, which would have been filling their pens by September, had only managed to achieve a 75 percent capacity, and six Spanish ranches had ceased operations altogether. These latest findings supported the earlier warnings by WWF that huge illegal catches might portend the decimation of the Mediterranean populations, and inevitably, the collapse of the species. At the 2006 ICCAT meeting in Madrid, having learned that the French fleet had illegally exceeded its quota by more than three thousand tons, an international consortium of scientists said that the bluefin tuna would be lost forever if no action was taken to restrain the French and other Mediterranean fishers.

The collapse of the stocks is probably underestimated because of the wholesale quantities of tuna that are not recorded at all, or not included in a particular nation's total, as they are transferred from fishing boats to reefer freezing vessels and shipped directly to Japan. This illegal, unregulated, and/or unreported fishing—primarily by French, Libyan, and Turkish fleets—accounted for some 25,000 tons of tuna in 2004 and 47,965 tons in 2005. Aside from stopping and prosecuting illegal fishers, the 2006 WWF report called upon the commission "to adopt a real long-term recovery plan for the East Atlantic stock of bluefins which should include a set of effective management measures . . . but in the meantime, given the virtual unregulated nature of the fishery and the strong likeliness of a near collapse, *WWF calls for the immediate and complete closure of the fishery*" (emphasis in the original). Given the ineffectuality (or unwillingness) of ICCAT to control its member nations (not to mention IUU fishers), the closure of the Mediterranean fishery seems highly unlikely. The tuna-ranching operations only offered an interim solution and not a particularly good one at that; as many tuna were caught for the fattening pens as were caught in the purse-seine fishery, and besides, a significant proportion died before they reached a saleable size. As long as the Japanese are prepared to pay for sashimi-grade tuna, there will be fishermen to catch them, and the Japanese market shows no sign of abating. If the bluefins of the Mediterranean disappear, where will the sashimi come from?

On April 17, 2007, the WWF website (www.panda.org) announced "Bluefin Tuna in Crisis." The multichapter posting echoed the structure of this book (beginning with "Meet the Bluefin Tuna," celebrating the characteristics that make *Thunnus thynnus* so special); but while it concentrated on the Mediterranean, it also addressed the status of bluefins around the world. The posting coincided with the opening of the fishing season:

> In May 2007, the annual fishing season will begin, amid concerns that the size of official quotas, illegal fishing, and fishing during the spawning season could hasten the collapse of the bluefin stock in the area. Over recent years, fishing fleets, complete with military technology, have hunted down, often illegally, ever declining numbers of these ocean giants. With fishing quotas currently set at a level more than twice as high as science recommends, a devastating collapse of the fishery is likely. This could be the year that the magnificent bluefin tuna, the fish behind the finest sushi in the world, disappears from the Mediterranean.

The website described how the great fish were spotted from airplanes and "scooped up by the tonne"; their populations were in such an alarming decline that WWF was calling for a 50 percent cut in the European ICCAT allocations for 2007. After warning of the imminent collapse of the Mediterranean population, WWF offered possible actions that could be taken by individuals to forestall the decimation. For instance:

- Check with your restaurateur to see if the tuna he serves is from the Mediterranean. If it is, tell him the species is endangered, and not to serve it.

- Do not buy Mediterranean bluefin tuna in a fish-market.
- Join WWF: "Our experts are working in the Mediterranean and around the world by lobbying governments, educating people, and finding sustainable solutions in our efforts to save our waters, but it can't be done without the support of people like you.

Michael Butler's 1982 *National Geographic* article, "Plight of the Bluefin Tuna," brought up the subject of "domesticating" tuna: "Spawning fish captured by French and Italian purse seiners and tonnare can be stripped of eggs and milt, as with salmon and trout. Zooplankton from Mediterranean lagoons can provide food for the larvae and juveniles. An assured supply of fertilized eggs to laboratories in interested countries may turn the bluefin's fortunes." The "interested countries" would turn out to be Japan and Australia, where most of the efforts to raise bluefin tuna are taking place. (At the Tokyo University of Fisheries, Butler was told by Dr. Yutaka Hirasawa that "yellowtail culture currently contributes 150,000 tons to the Japanese market, raised from 75 million fingerlings," but yellowtail are not tuna.) Reflecting on the decline of bluefin populations in 1995, Sylvia Earle wrote: "Some attempts are under way at the New England Aquarium in Boston to discover how to cultivate these ocean giants, to better understand how to protect them in the wild and, perhaps in time, to provide an alternative to wild-caught tuna. In Japan, wild-caught young yellowfin are being raised in special ocean enclosures to market size. These and other efforts to grow top-of-the-food-pyramid predators as food for humans or replacement stock for depleted natural areas are encouraging, but costly, given the volume of food, and time, required to raise these creatures. Meanwhile, fishing continues."

Tuna ranchers in the Mediterranean, South Australia, and Baja California are capturing half-grown bluefins and towing them in nets to offshore pens, where they will be fed and fattened until they are the right size to be killed, frozen, and shipped to Japan. It is becoming clear that tuna farming itself might signal the end of an industry, because the process of scooping up all the tuna leads irrevocably (and obviously) to the end of the fishery. Quoted in Elisabeth Rosenthal's

New York Times article on July 16, 2006, Simon Cripps, director of WWF's global marine program, said (of the collapse of the Mediterranean stock of northern bluefin tuna): "Because of the small fish being harvested for farms, farming is being used to bypass attempts at regulation . . . This is part of the alarm stage; we are seeing a complete collapse of the tuna population. It could disappear and never come back." In an April 2007 article in *National Geographic* called "Global Fish Crisis," Fen Montaigne wrote: "The decimation of giant bluefin is emblematic of everything wrong with global fisheries today: the vastly increased killing power of new fishing technology, the shadowy network of international companies making huge profits from the trade, negligent fisheries management and enforcement, and consumers' indifference to the fate of the fish they choose to buy."

Although it is usually known as tuna *ranching,* the capture of tuna and their transfer to fattening pens is certainly another form of fish *farming,* or aquaculture. But because the procedures differ from most fish-farming operations—most salmon, carp, and tilapia are raised from eggs—tuna farming is usually not included in discussions of aquaculture. In Marguerite Holloway's 2002 article "The Blue Revolution," mostly devoted to salmon farming in Mexico, the word "tuna" appears once, in this sentence: "Carnivorous aquaculture species, such as salmon, shrimp, eel, flounder, halibut, tuna, and sea bass, are fed fish meal and fish oil—essentially ground or pressed anchovies, sardines, capelin, blue whiting, mackerel, Atlantic herring, and other small bony fish." The feeding of captive tuna is perhaps the most serious problem facing the ranchers around the world. In the Monterey Bay Aquarium's Outer Bay tank, the bluefins and yellowfins are fed sardines, smelt, squid, and a "gelatin diet," which consists of Jell-O-like cubes infused with vitamin and mineral supplements. Of course this is not feeding for the market, and it is actually better if the tuna do not overeat (early in the history of the Outer Bay exhibit, tuna in holding tanks were dying of congestive heart failure because they were not getting enough exercise, so they were put on a "lo-fat" diet).

According to Roberto Bregazzi (2005), "about 45,000 tons of bait fish, secured both locally and overseas, were used in 2000–2001. About 60 tons of southern bluefin tuna were also successfully fed a manufactured pellet. An average size southern bluefin tuna increases in weight by 10–20 kilograms [22–44 pounds] during the ranching process," which usually lasts for three to seven months. In the 1999–2000 season, Australian tuna ranches produced 7,780 tons of ranched bluefins, valued at over $2 million. By 2003 the total was 8,308 tons, and the fifteen tuna ranchers in Australia are looking forward to a 9,000-ton year. The tuna ranchers of Australia, however, do not foresee a constantly increasing harvest and a concurrent rise in gross income. In the 2005 ATRT report, Brian Jeffriess of the Australian Tuna Boat Operators Association wrote:

> With the annual tuna ranching industry harvest looming, there are concerns that some of the operators will become unprofitable as a result of a combination of circumstances. In the most difficult year for the industry since tuna ranching built to a formidable producer in the 1990s, farmers are confronting both a downturn in returns owing to a much stronger Australian dollar, and direct competition from industry now located in Mexico, Turkey, Spain, Malta, and Croatia.

As Australia heads for nine thousand tons, European ranches in the Mediterranean will nearly double the Australian production and Mexico is approaching four thousand tons. Europe and Mexico will ship nineteen thousand tons of frozen northern bluefin tuna to Japan, while Australia, the only country farming southern bluefins, will ship nine thousand. Because northern bluefins grow larger than their southern counterparts, the shipments from the Mediterranean and Mexico are favored in Japan, another problem for the Aussies. In 2004, the South Australian government shut down the pilchard fishery, the main source of food for the tuna ranches, because too many dolphins were being killed. According to an article in the *Port Lincoln Times,* the fishery was closed down because nineteen dolphins had been drowned in the pilchard nets in the first six months of 2005.

Back in 1981, when J. H. Ryther of Woods Hole wrote "Mariculture, Ocean Ranching, and Other Culture-Based Fisheries," small-scale culture of marine creatures was considered the only possibility, and it was not even feasible to contemplate farming anything on a large scale. At that time, Pacific salmon fry were raised in hatcheries and released back into the wild to compensate for the loss of wild salmon whose natural spawning beds had been eliminated by dams and other obstructions. According to Ryther, the worldwide mariculture total for 1975 was 3 million tons, including 200,000 tons of milkfish (in the Philippines, Indonesia, and Taiwan), 100,000 tons of salmon raised in nets, and 200,000 tons of ocean-ranched salmon in the United States, Japan, and the USSR. Today, by a substantial margin, China leads the world in aquaculture, and most of the fish farmed in China are carp, used for regional consumption in low-income households. In other parts of the world, farmed tilapia, milkfish, and channel catfish have replaced depleted ocean fish like cod, hake, haddock, and pollock. The worldwide landings for the "capture fisheries" (those in which wild fish are caught at sea) have leveled off at around 85–95 million metric tons per year, with most stocks being recognized as fully fished or overfished. In 1997 the figure for aquaculture was 10 million tons, but by 2000 it had nearly tripled (Naylor et al. 2000), and the FAO reported that the total for 2003 was 42.3 million tons. Global aquaculture now accounts for roughly 40 percent of all fish consumed by humans.

The tremendous increase in fish farming in recent years has been offered as a possible solution to the problems of worldwide overfishing, but aquaculture has its own problems, and in some cases may be contributing to rather than solving overfishing. The species most prominently farmed around the world are carp, salmon, trout, shrimp, tilapia, milkfish, catfish, crayfish, oysters, hybrid striped bass, giant clams, and various shellfish. Of these, shrimp and salmon make up only 5 percent of the farmed fish by weight, but almost one-fifth by value. Farming is the predominant production method for salmon, and aquaculture accounts for 25 percent of world shrimp production—a tenfold increase from the mid-1970s (Naylor et al. 1998).

At Stanford University, Rosamond Naylor's work focuses on the environmental dimensions of intensive food production. She is the director of the Goldman Honors Program in Environmental Science, Technology, and Policy, the director of the Program on Food Security and the Environment, and associate professor of economics. She is perhaps the world's foremost authority on aquaculture. In late 2001 she was flying over the Sonora coast of Mexico looking for shrimp farms, expecting to find, in the words of Marguerite Holloway (2002), "clusters of scattered ponds separated by huge tracts of sere land. Instead, it looked as if the Sea of Cortés had risen and swept across more than 42 square miles of the Sonoran: everywhere were patches of blue, pools of shrimp, one after another, all down the coast." Aquaculture, the "Blue Revolution" of Holloway's article, is by now "a $52 billion-a-year global enterprise involving more than 220 species of fish and shellfish that is growing faster than any other food industry." By 2005, the total for world aquaculture production, including aquatic plants, amounted to 60 billion tons. In 2005, Rebecca Goldburg and Rosamond Naylor wrote:

> Fisheries depletion has created new impetus to expand seafood production through fish farming, often known as aquaculture. Aquaculture is frequently cited as a way to increase seafood supply in a world where greater quantities of fish cannot be obtained from the oceans. It has become an increasingly important source of food; between 1992 and 2002, global production of farmed finfish and shellfish ("fish") almost tripled in weight and nearly doubled in value. Currently, roughly 40% of all fish directly consumed by humans worldwide originate from commercial farms.

Each species of farmed fish (shrimp and shellfish are also known as "fish" in aquaculture-speak) has its own requirements, and it is impossible to generalize about the benefits or detriments of fish farming as a whole. Carnivorous species, such as salmon and shrimp, require food, which is usually provided in the form of fish meal, made from ground-up fish. The cost of providing food for farmed salmon often exceeds the price that the salmon can command; moreover, in this case, farming contributes to overfishing, because the small fish—such as Peruvian anchovies—are harvested almost exclusively for fish meal. (It is not only fishes that eat fish meal, of course; most of the processed fish meal is fed to chickens and pigs.) To feed the carnivores, fishermen are fishing for fish to feed the fish.

James Anderson, a professor in the Department of Environmental

Resource Economics at the University of Rhode Island, introduced a 2002 paper ("Aquaculture and the Future: Why Fisheries Economists Should Care") with this story:

> As they are deliberating, a noisy new gasoline powered Oldsmobile Curved-Dash Runabout interrupts their debate. . . . They have heard and seen the future of transportation and the new problems it will bring, but they continue to discuss the soon-to-be past.
>
> Now, fast forward to the present day and consider a meeting of fisheries experts. . . . When they break to eat dinner, it is likely to consist of salmon and/or shrimp. Yet they seem oblivious to the fact that the seafood they are consuming is farmed. They eat the future of fisheries, but continue to discuss its past.

Anderson's point, of course, is that the future of fisheries is not management, but aquaculture. In fisheries management you are at the mercy of weather, history (as in previous fishers depleting the stock), politics, international treaties (usually a function of politics), international intrigue (as in fishers falsifying records), and any number of other variables, while aquaculture is all about control—of feed, predators, disease, harvest, growth rate, location, technology, exposure to the elements—exactly what farmers need in regard to livestock raised for the market. In his 2006 article about the beginning of cod farming in Norway, Paul Greenberg wrote:

> Taking a long-range view, there is little doubt that we are on the verge of a vast new artificial selection that will determine the characteristics of a future marine ecology. As recently as 20 years ago, aquacultured products were niche items—the bright red slab of lox from Norway, the crawdad from Louisiana. Today, dozens of mainstream fish are being domesticated and will soon appear at supermarket counters everywhere. Yellowtail, halibut, red snapper and even Volkswagen-size bluefin tuna are all coming under some kind of human-controlled production. And whereas animals like sheep and cattle were adapted to fit the farm over thousands of years, many of the ocean species under development today could be tamed in as little as a decade.

The top ten seafoods consumed in the United States, in terms of kilograms per capita, are shrimp, tuna, pollock, salmon, catfish, cod, clams, crab, flatfish, and scallops.* Farmed seafood, particularly salmon, shrimp, and catfish, saw an enormous increase from 1987 to 2000, with salmon rising 265 percent during that period, obviously a function of the exponential increase in salmon farming, and to a lesser extent, the much publicized health benefits derived from eating salmon. Note that tuna, the most popular seafood in the country, is not farmed, at least not in the traditional sense. Bluefins are being ranched, which is not exactly farming, but the fishers of albacore, skipjack, and yellowfin are susceptible to all the ecological variables listed above, with the addition of one more: overfishing.

Wild Atlantic salmon are born in a multitude of rivers in countries with access to the North Atlantic. The far-flung salmon countries include Canada, the United States, Iceland, Norway, and Russia in the north; the United Kingdom, Ireland, and the Baltic countries in the middle of the range; and France and northern Spain on the southern margin. The young salmon leave their freshwater home and migrate thousands of miles to feed in the rich marine environment of the North Atlantic off Greenland and the Faeroes. After one year or more in these feeding grounds, the fish undertake their most impressive return migration to the rivers of their birth, where they spawn and complete the cycle. From time immemorial, the cycle of spawning, ocean feeding, and return migrations went on as if the resource was a permanent feature of the natural world, until the 1950s and 1960s when developments occurred that began to threaten the fish dramatically. The sea feeding grounds, long a mystery, were located, and international exploitation of the fish began at an alarming rate. New types of gear, such as the nylon monofilament net, were introduced with disastrous results, as unregulated ocean fishing fleets started to devastate the stocks of fish while at sea. By the mid-1970s, up to

*In comparison, the leading imports into Japan (and therefore the top seafoods) are shrimp, tuna and swordfish, salmon, crab, cod roe, eel, processed shrimp products, squid, cod (including surimi products), octopus, and sea urchin. In terms of volume, the leading Japanese seafood import is tuna and swordfish at 321.7 million kg, followed by salmon at 276.5 million kg, and shrimp at 256.2 million kg. The high-volume, high-priced products usually arrive by air; 46 percent of the bluefin tuna *(kuromaguro),* 24 percent of the southern bluefin *(minami maguro),* and 21 percent of the bigeye tuna *(mebachi)* come in by plane (Bestor 2004).

2,700 tons of salmon were being taken annually from their ocean feeding grounds, and following this massive loss of stock, their numbers began to fall precipitously.

Large-scale salmon farming is practiced in Norway, Scotland, Finland, British Columbia, Iceland, Alaska, Italy, New Zealand, Australia, Japan, the Philippines, India, France, Bangladesh, Thailand, and Indonesia, and wherever it exists, there is controversy. Fish inevitably escape from open-water pens, especially during storms. Some incidents have resulted in the flight of tens of thousands of farmed fish into surrounding waters. When fish escape from farms and survive in large numbers or establish their own breeding populations, they will compete with wild salmon. Efforts to secure facilities against these accidents may reduce the size and number of releases, but are unlikely to stop them altogether. If they are the same species as the wild salmon (e.g., Atlantic salmon grown in the Atlantic), there is the possibility of interbreeding between farmed and wild fish. When such interbreeding occurs, there is a significant change in the genetics of the population. Genetically engineered salmon in ocean pen farms—a distinct possibility in the near future—adds another layer of concern with respect to interactions with wild populations. If interbreeding were to occur as a result of escapes, such genes could be incorporated into the wild gene pool and possibly diminish the vigor of the wild population. According to an article by Naylor and her colleagues (2001), "up to 40% of Atlantic salmon caught in the North Atlantic and more than 90% caught in the Baltic Sea are of farmed origin. More than a half-million Atlantic salmon escaped on the West Coast of North America between 1987 and 1997; they have been found in 77 British Columbian rivers and are spawning in some locations. In the New Brunswick–Maine region, farmed escapees vastly outnumber wild salmon in some spawning rivers."

The total weight of farmed salmon in 2002 was 1,084,740,000 tons (Naylor and Burke 2005). Farmed salmon are fed meal and oils from wild-caught fish. Each pound of salmon produced requires at least three pounds of wild-caught fish, challenging the presumption that fish farming necessarily reduces commercial fishing pressure. In fact, there is a net loss of protein in the marine ecosystem as a whole when wild catch is converted into meal for aquaculture consumption.

Pens full of salmon produce large amounts of waste—both excrement and unconsumed feed. This may result in water-quality conditions (such as high nutrients and low oxygen) that are unfavorable for both the farmed fish and the natural ecosystem. It is also suspected that nutrients released from salmon farms stimulate microalgal blooms, but proof of this is lacking because little research has been done. The densely packed condition in pens promotes disease, a common problem in most salmon farms. Furthermore, there have been documented disease transfers from farmed salmon to wild populations, and the potential effects are serious. While antibiotics are used to treat some diseases, there are concerns about the effects of antibiotic-resistant bacteria on human health. There has been an emphasis on developing vaccines to prevent specific diseases, which reduces the need for antibiotics.

To consumers, a salmon is a salmon, but there are many different kinds. Atlantic salmon live (mostly) in the Atlantic, but there are six species of endemic Pacific salmon: chinook *(Oncorhynchus tshawytscha),* coho *(O. kisutch),* sockeye *(O. nerka),* pink *(O. gorbuscha),* chum *(O. keta),* and cherry *(O. masu).* All occur on both sides of the Pacific Ocean, except the cherry salmon, which is found only off Japan. Like their Atlantic counterparts, Pacific salmon are anadromous, breeding in fresh water, maturing at sea, and then returning to spawn in the place where they were born. Unlike their Atlantic counterparts, Pacific salmon species spawn only once and then deteriorate and die. In California, Oregon, Washington, and Idaho, the long history of overfishing, the damming of the rivers up which the salmon must swim, and the destruction of the breeding streams by timber interests have driven many populations of Pacific salmon to the edge of extinction. Only the populations that breed in Alaska are considered healthy. (At a recent dinner in Friday Harbor, Washington, I asked if the salmon on the menu was farmed, and I was told that it was fresh—caught yesterday in Alaska.) As a sort of stopgap to extinction, attempts have begun to breed and raise Chinook salmon in Canada, New Zealand, and Chile, and by 2005 some nineteen thousand tons had been successfully raised and brought to market (Naylor and Burke 2005).

While populations of North Atlantic wild salmon fall, more than one hundred local populations in the eastern Pacific have disappeared. Salmon are extinct in 40 percent of the rivers where they once spawned along the Pacific coast. Throughout the world, however, as wild Atlantic and Pacific salmon populations decrease, there is a corresponding escalation in the salmon-farming industry. The potential for interactions between farmed and wild fish and transmission of disease from farmed to wild salmon is especially threatening in the context of these declines. Governments typically encourage aquaculture because it is viewed as economic development, but this often leads to the intensive, large-scale farming methods most often associated with environmental damage. Because the costs of this damage are not borne by the industry, nor are the value of ecosystem services factored into the cost of production, there is no pressure on the industry to operate in environmentally sound ways. The increase in worldwide salmon consumption is coupled not to fishing, but to farming.

Salmon is one of the foods that supplies the "good" type of cholesterol, high-density lipoprotein (HDL). The "bad" cholesterol (LDL or low-density lipoprotein) is believed to be responsible for clogging blood vessels, and HDL is reputed to "exile" the LDL to the liver where LDL is destroyed. In *New Scientist* (August 2002), Meredith Small wrote an article entitled "The Happy Fat," in which she identified some of the ways in which the omega-3 fatty acids found in fishes like salmon and menhaden (and flax seeds, walnuts, and olive oil) are now being investigated as an antidote to depression (Hibbein 1998) and even as a possible inhibitor of prostate cancer (Terry et al. 2001). In a 2002 *Newsweek* article about salmon, Jerry Adler wrote, "Even people that don't like salmon know by now that it contains Omega-3 fatty acids, which are believed to protect against cancer and cardiovascular disease."

There is a concurrent controversy about the "health benefits" to be gained from eating farmed salmon. In January 2004, Hites et al. published a report in *Science* in which they said that the organochlorine concentrations were "significantly higher in farmed salmon than in wild," and additionally, that the levels of PCBs, dioxins, toxaphene, and dieldrin were high enough to incur "the potential for elevation in attendant health risks." They continued: "Although the risk-benefit computation is complicated, consumption of farmed Atlantic salmon may pose risks that detract from the beneficial effects of fish consumption." The Hites article elicited a number of responses, including one from a group of Finnish scientists (Tuomisto et al. 2004) who suggested that it was "nonscientific because the outcome . . . was totally driven by a political variable, whether to ignore the health benefits of fish." Norway is the largest exporter of farmed salmon in the world, so it was no surprise to see that a group of Norwegians (Lund et al. 2004) chimed in with: "Oily fish is an important source of essential fatty acids and lipid soluble vitamins. The positive health effects are well-documented for cardiovascular diseases. Unilateral promotion of very limited health hazards could affect the overall health of populations by urging people to give up a health diet, causing them to substitute fish with less healthy and perhaps less safe foods."

In their response to the Norwegians, Hites et al. wrote: "Although omega-3 fatty acids do protect against sudden cardiac death after a heart attack, young people are not at risk of heart attacks, but their risk of cancer at older ages is increased by exposure to these compounds. In addition, the noncancer effects of exposures, such as reduced cognitive function of children exposed before birth, could overshadow a protective effect against sudden cardiac death in young people and were not factored into the cancer-based risk analysis in our Report."

Salmon are carnivores, and while the first salmon farmers (the Norwegians) fed their captive fish on smaller fish, they soon added pellets enriched with fish oil and vitamins to the mix. This meant that they didn't have to catch so many fish, which reduced the paradoxical situation of catching fish to feed the fish, and therefore decreasing wild fish populations while claiming to protect them. Hites et al. pointed out that most of the contaminants they found in captive salmon could be traced to the feed, so it was theoretically possible to reduce the contaminants by modifying the chemical composition of the feed. Tuna are also carnivores, and also require large quantities of live or frozen fish. As Naylor and Burke point out, "In the case of ranched tuna, which depend largely on live pelagic fish such as sardines, anchovies and mackerel, up to 20 kilograms [44 pounds] of wild fish input are needed to produce each kilogram of ranched fish output." The time required to raise tuna to marketable weight will change dramatically if and when they can be legitimately farmed—that is, raised from eggs.

Real or imagined risks notwithstanding, salmon have become the paradigm for fish farming. "Salmon aquaculture originated in Norway in the 1970s," wrote Goldburg and Naylor in 2005, "and has since bloomed worldwide. Global production of farmed salmon roughly quadrupled in weight from 1992 to 2002, and farmed salmon now constitute 60% of the fresh and frozen salmon sold in international markets. This spectacular increase and the resulting decline in salmon prices have encouraged aquaculturists to begin farming numerous other finfish species, many of them depleted by overfishing." Just as it has set the stage for a boom in worldwide aquaculture, salmon farming has also revealed the extent of the tribulations that are inherent in this extraordinarily complex enterprise. From Marion Nestle's chapter "The Fish-Farming Dilemma" in *What to Eat:*

> Since farm-raised salmon constitute such a large part of the fish-farming industry in industrialized countries and in Chile, it is well worth examining the idea that you are better off eating wild salmon for reasons of safety, health, and environmental protection . . . In part because of what gets fed to wild salmon and in part because farmed fish are less active, these fish have twice the fat and more than twice the saturated fat of their wild counterparts. Their omega-3 content depends entirely on what they are fed, and varies by species and by farm.

Because the fish meal fed to farmed salmon is usually made of ground-up larger fish, and because larger fish accumulate heavier concentrations of PCBs, farmed salmon are higher in PCBs than wild salmon. The Hites report pointed this out (adding dioxins, toxaphene, and dieldrin to the mix), but added that the concentrations vary with geography: salmon from farms in Europe had the highest concentration of PCBs; those from North and South America had the least. Salmon farmers don't take kindly to warnings that their product might be dangerous to health, and they mounted a campaign to demonstrate that *any* kind of salmon was good for you, and you should pay no attention to those scientists who are worried about PCBs. In the end, says Nestle, you should try to find out "where the fish comes from, whether it is farmed or wild, where it is on the food chain, whether it is listed on a state advisory, and how much fat it contains." Armed with this information, one might be able to avoid farm-raised fish from Europe (unless of course, one is *in* Europe), farmed fish that are fed lots of fish meal, and even those that are high in methylmercury.

At one time it was feared that escapees from Atlantic salmon farms would interbreed with wild fish and degrade the species, already in serious trouble in the wild (Schiermeier 2003). Hybridization turns out to be among the least of the problems of salmon farmers. As of October 2006, with the publication of a report in the *Proceedings of the National Academy of Sciences,* an enormous obstacle appeared on the road to sustainable and problem-free salmon farming. Canadian mathematician Martin Krkošek and several colleagues said that sea lice from Pacific salmon farms were killing as much as 95 percent of the young wild salmon that migrate past the pens on their way from inland rivers to the ocean. The pens harbor concentrations of sea lice that are thirty thousand times higher than the open ocean, and young salmon swimming past are infected and killed. Krkošek et al.: "Farm-origin lice induced 9–95% mortality in several sympatric wild juvenile pink and chum salmon populations. The epizootics arise through a mechanism that is new to our understanding of emerging infectious diseases; fish farms undermine a functional role of host migration in protecting juvenile hosts from parasites associated with adult hosts." Ordinarily, adult salmon can withstand the lice infections, but, wrote the authors, "in the areas containing salmon farms, wild juvenile salmon are sympatric with large abundances of domesticated salmon (and their parasites) during their early marine life. Farms provide parasites novel access to these juvenile hosts, resulting in measurable and sometimes severe impacts on salmon survival." Interviewed by Cornelia Dean, who wrote an article on the findings for the *New York Times,* Ransom Myers, one of the authors of the 2003 paper on the worldwide depletion of predatory fishes, said that "the work raises the question of whether we can have native salmon and large-scale aquaculture as it is currently practiced, in the same place."

Tuna and salmon are both fish that people like to eat, but while farmed salmon sales have risen in the United States, there has not been a comparable increase in the sale of "ranched" tuna. Farmed salmon filets are available year-round, but the bluefin tuna supply is uneven, and the price varies with the season. According to the 2006 WWF report, "Costco Wholesale recorded fresh-farmed [salmon] filet sales of over 14 million kilograms in 2002 while fresh tuna sales were 354,545 kg—2.5% of farmed salmon sales." Bluefin and yellowfin tuna are still uncommon menu items, likely to be served in upscale "white tablecloth" restaurants, where signature dishes like "Sesame Seared Tuna" can command hefty prices. Yellowfin "steaks" are often seen in specialty stores, but they turn brown quickly and even though they are still wholesome, they do not sell well. And as much as supermarkets would like to offer some form of fresh tuna, so far it has proven to be more than a little impracticable because of the high prices and the variation in pricing that depends on a daily evaluation of the fish coming in. Nevertheless, after Japan and the European Union, the United States is the third-largest export destination for Mediterranean wild and ranched bluefin tuna, a statistic based on the increase of high-quality sushi and sashimi restaurants in the Northeast, California, and Florida.

In his 2005 article "When Will We Tame the Oceans?," John Marra, a biological oceanographer at Columbia University's Lamont-Doherty Earth Observatory, observed that "fishing in the ocean is no longer sustainable. Worldwide, we have failed to manage the ocean's fisheries—in a few decades, there may be no fisheries left to manage." He acknowledged that "fish farming can harm the environment in many ways; indeed, some mariculture operations have caused whole-scale destruction of coastal ecosystems . . . marine farming can pollute in many ways that are aesthetically, chemically and genetically destructive . . . crowding in aquaculture enclosures or ponds can easily amplify disease and cause it to spread more quickly than it would in the wild . . . the mariculture of carnivorous species puts additional pressure on fisheries to provide ever-larger quantities of wild fish for feed, exacerbating the decline of wild fish populations."

His recommendation? Large-scale domestication of the ocean. Move the mariculture systems farther offshore, he says, to the waters of the outer continental shelves, and deploy much larger fish pens (closed-net structures containing as much as 100,000 cubic meters of water), which could be floated below the surface and towed from one destination to another, feeding the fish along the way. Another suggestion concerns "herding" of tuna at sea, based on the inclination of certain tunas to aggregate under an object that is significantly different from their surroundings. This propensity has already been exploited by tuna fishermen in the design and implementation of fish-aggregating devices (FADs), which can be something as simple as a towed log or platform or as innovative as a firehose directed behind the boat to create a disturbance on the surface. (Research by Dagorn et al. in 2000 demonstrated the affinity of yellowfin tuna for the very vessels that were tracking them.)

Marra says that taming the sea will mean the disappearance of commercial fishermen, and "the bulk of the fish we eat will come from more limited varieties. In such a future we will have to accept an ocean with fish that can be cultured, and we will have to accept less freedom of the seas . . . The common goal should be to maintain the ocean as a sustainable source of food, both economically and ecologically." As on land, sustainability of the ocean's food supply for the world's population means domestication of the seas. In response to Marra's article in *Nature,* Julia Baum, Jana McPherson, and Ransom Myers wrote: "Farming carnivores also results in a net loss of food because of the inefficient energy conversion between trophic levels, as Marra acknowledges. Tuna farming therefore, is not like herding cattle; it is the ecological equivalent of trapping wild shrews and foxes to feed caged wolves." The authors of the response to Marra's article opined that "fisheries and the wild populations that supply them should not be abandoned. Instead, serious effort should be focused on rebuilding depleted fish populations by creating marine reserves and reducing total fishing capacity." (One of the authors of this response was Myers, who, with Boris Worm, published the 2003 paper in which they said that 90 percent of all predatory fish were gone—as a result of overfishing.)

In "Future Seascapes, Fishing, and Fish-Farming," their incisive 2005 analysis of the current state of aquaculture, Goldburg and Naylor wrote:

> The growth in marine aquaculture, and possibly also in hatchery production, will alter not just sources of marine fish and the economics of fishing, but may also transform the character of the oceans from relatively wild, or at least managed for fishing, to something more akin to agriculture. . . . Most marine fish farms will essentially be feedlots for carnivores, particularly if the salmon farming model is copied. Second, although fish farms are unlikely to occupy a large area, the ecological impact on marine resources could be much greater than the geographical extent of fish farms implies. This is because fish farming depends heavily on, and interacts with, wild fisheries.

To that statement Rosamond Naylor (now joined by Marshall Burke) added:

> The oceans are now poised for yet another transformation: the rapid expansion of fish farming, or aquaculture, resulting from the decline in wild fisheries and lucrative business opportunities . . . Although most aquaculture production to date has been of freshwater fish, marine aquaculture has been growing dramatically. Global production of farmed salmon, for example, has roughly quadrupled in volume since the early 1990s. This spectacular increase and the resulting decline in salmon prices have helped prompt aquaculturists to begin farming numerous other marine finfish, including a number of species depleted in the wild. New species farmed in marine net pens include Atlantic cod *(Gadus morhua)*, Atlantic halibut *(Hippoglossus hippoglossus)*, Pacific threadfin *(Polydactylus sexfilis)*, mutton snapper *(Lutjanus analis)*, and bluefin tuna *(Thunnus spp.)*. [Of these, the bluefin is the by far the largest, and to date, the most difficult to raise.]

As Marra pointed out, it is only a matter of time before fisheries around the world crash. Some of the vanishing species—particularly cod, salmon, and tuna—were once exemplars of plenitude. If the wild-capture fisheries are in decline—because the target species are in decline—it seems self-evident that the only recourse available to a fish-eating world is mariculture. Referring to a 2006 FAO report,

"The State of World Aquaculture 2006," Andrew Revkin wrote in the *New York Times,* "Nearly half the fish consumed worldwide are raised on fish farms rather than caught in the wild." The marine and freshwater capture fisheries produce about 95 million tons, of which 66 million is designated for human consumption. (The remainder is used for fish meal, fish oil, and other purposes.) Farmed fish—all of which is destined for human consumption—for 2004 totaled 50 million tons. Most capture fisheries are operating at full or nearly full capacity, which means that the world total will not go up—and as the fisheries decimate the wild populations, it is more likely to go down. It is incumbent upon fisheries managers and governments to solve the problems associated with aquaculture, and to figure out how to replace the decreasing wild populations with farmed fish for a fish-hungry world.

If there aren't enough little fish to feed the big fish, we are simply going to have to find something else for the big fish to eat. In a 2003 article entitled "Eat Your Veg," Kendall Powell wrote, "Over the past few decades, researchers have begun to think that one way to make aquaculture more sustainable is to change the diets of some of our farmed fish—to turn carnivores into vegetarians." Carp, tilapia, and milkfish are herbivores, and can be fed plant food or prepared fishfood, not unlike the food that hobbyists sprinkle into their home aquariums. Could the vast number of fishes and other creatures that are caught incidentally to a particular fishery and usually discarded (the bycatch) be saved and used for fish meal instead of targeting fisheries for fish meal, such as the anchovies? This unfortunately would require that fishers use valuable space aboard their ships for storage of trash fish, which are worth less than the expensive fish they are seeking. In a 2000 article on the effects of aquaculture on world fish supplies, Rosamond Naylor and her colleagues wrote:

> Carps and marine molluscs account for more than three-quarters of current global aquaculture output, and tilapia, milkfish and catfish contribute another 5% of total production. Fed mainly on herbivorous diets, these species provide most of the 19 Mt [megatons] gain in fish supplies from aquaculture . . . But market forces and government policies in many countries favour rapid expansion of

> high-value, carnivorous species, such as salmon and shrimp. Moreover, fish meal and fish oil are already being added to carp and tilapia feeds for weight gain, especially in Asia where farming systems are intensifying as a result of increased scarcity and value of land and freshwater resources. Given the huge volume of farmed carp and tilapia in Asia, significant increases in the fish meal and fish oil content of feed could place even more pressure on pelagic fisheries, resulting in higher feed prices and harm to marine ecosystems.

In "Dollars without Sense," his prescient discussion of big-money tuna ranching, Canadian environmental biologist John Volpe identified one of the major problems with tuna ranching:

> It takes 3 kg of wild fish to produce 1 kg of farmed salmon (i.e., a 3:1 ratio); for farmed cod the ratio is 5:1; but the ratio reaches 20:1 for ranched tuna (in part because tuna are warm-blooded, an energy-intensive physiological state for a cold-water fish). The farms around Port Lincoln alone consume more than 20,000 kg of pilchard, sardine, herring, and anchovy per day. Clearly the consumption of 20 units of edible fish to make one unit of product is no one's idea of a conservation strategy.

But a "conservation strategy" is probably the furthest thing from the minds (or the business plans) of the Port Lincoln ranchers. It is all about the money. It is possible that greed will signal the downfall of the tuna-ranching business: as Mediterranean ranches flood the markets with sashimi-grade tuna, Japanese prices continue to drop, but ranching costs do not. "The industry," writes Volpe, "is showing every sign of becoming a victim of its own success . . . Tuna ranching is viable only as long as the premium price of its product is protected, which in turn demands a scarcity of supply."

Another way to protect bluefin tuna from overexploitation would be to persuade consumers not to eat them. In an op-ed article in the *New York Times* (September 8, 2006) Paul Greenberg suggested that the large, rare fish, like tuna and cod, could be replaced in our diet by smaller, more common fish, like sardines. "Don't eat the big fish," he advised. "Dining on a 500-pound bluefin tuna is the seafood equivalent of driving a Hummer. Ten pounds of little fish are required to produce one pound of bluefin and all the pollutants contained in a tuna's prey 'bioconcentrate' in a tuna's flesh, making it a particularly compromised animal, chemically speaking. And because it takes so many little fish to make a big fish, the sea can sustain only a relatively small amount of large fish. It therefore follows that if we reduce our consumption of the big fish we can reduce our mercury and PCB load and reduce the burden we place on the marine environment. Sardines, mackerel and most fish that are shorter in total length than the diameter of a dinner plate are generally safer to eat." This might work in America and Europe, where most of the tuna are caught, but until the Japanese reduce their passion for *toro,* they will continue to subsidize the hunting of large bluefin.

Japan is an island nation, poor in natural resources and greatly dependent upon the sea for protein, but that does not logically lead to the conclusion that they ought to import every swimming thing and serve it up for sale in the Tsukiji fish market. At one time Japanese whalers hunted various species for food in the nearshore waters of Japan, but this subsistence whaling was replaced by commercial enterprise, and it was not until the conclusion of World War II that Americans reintroduced whaling to Japan as a means of feeding the starving populace. Well into the 1980s, the Japanese continued large-scale whaling, not for food but for commerce, and with an increasing number of catcher boats and factory ships, they included in their hunting grounds great swaths of the eastern North Pacific and the Antarctic. The early Japanese whalers provided food for the people, but by the middle of the twentieth century, hardly anybody ate whalemeat, and although the industry claimed that it had to remain in business to provide a livelihood for its workers, the International Whaling Commission's 1983 implementation of a moratorium on commercial whaling was probably greeted with relief by the thousands of Japanese schoolchildren who had been fed the whalemeat that nobody else wanted to eat.

The Japanese never really stopped whaling. They took shameless advantage of a loophole in IWC regulations that allowed them to pursue a program of "scientific whaling," which meant they could set

Although it looks remarkably like red-meat tuna, right down to the color, this is actually whalemeat for sale in the Tsukiji market.

their own quotas on various whale species, as long as the whales were being killed for "research." After the research, however, the tons and tons of whalemeat not required by science ended up . . . in the Tsukiji fish market. Since the 1983 moratorium, whalemeat has become an expensive delicacy in Japan because it is so scarce. Despite its status and the government's insistence that everybody loves it, few people actually eat whale nowadays. If the Japanese fought so hard—even going so far as to undermine an international treaty—to preserve their "right" to eat whalemeat, think of how they will fight to keep the bluefin tuna flowing into the markets. Sashimi has become a national obsession, and it is almost impossible to imagine anything that would stop them from fishing or subsidizing the fisheries that supply the bluefin to the restaurants of Japan—and elsewhere.

When Rosamond Naylor wrote an editorial ("Offshore Aquaculture Legislation") for the journal *Science* in September 2006, she identified all the problems inherent in aquaculture (escaping farmed fish, interbreeding, parasites, disease, depletion of other species to feed farmed carnivorous fish, effluent pollution, etc.), and argued that "sound environmental standards [should be] adopted in all countries with marine aquaculture, whether near shore or offshore." She was referring to the National Offshore Aquaculture Act of 2005 (S. 1195) introduced by Senators Stevens of Alaska and Inouye of Hawaii, and "crafted by NOAA [which] establishes a permitting process for offshore aquaculture development within the federal waters of the EEZ, and encourages private investment in aquaculture operations, development, and research." She does not, however, identify the specific reason that Stevens and Inouye sponsored this bill in the first place: they were seeking to facilitate the introduction of tuna ranches in the waters of their states. Alaska had prohibited aquaculture altogether, and the pens in Hawaii were originally slated to be set up in the Leeward Islands, but the specifications have now been modified to site them less than twenty miles off the Big Island of Hawaii. It is certainly within their job description for senators to seek income-producing situations for their constituents, but Stevens and Inouye should not have cloaked the bill in the obfuscating rhetoric of environmental protectionism.

The Norwegians estimate that they will be harvesting thirty thousand tons of farmed cod by 2008, and finding enough fish to feed them will present a very serious challenge. At the Aquaculture Protein Center in Ås, Norway, scientists are experimenting with fish meal made of various combinations of soybeans, corn, rapeseed, sunflower seeds, flaxseeds, and wheat gluten. Experimental formulas that have provided the proper nutrients are too expensive, and it has not been satisfactorily demonstrated that a vegetarian fish tastes the same as a carnivorous one. A vegetarian diet tends to reduce the amount of omega-3 fatty acids in the flesh, but this shortcoming can probably be met in humans by supplements.

Experiments are under way in Australia to feed farmed tuna on pellets made of vitamin-fortified grain, and these have met with some success. Researchers have suggested feeding the fish on veggies until about three weeks before they are killed, and then give them a more "natural" diet. "Many researchers," wrote Kendall Powell in "Eat Your Veg," "believe that the final hurdles to making vegetarian feed realistic and affordable will be cleared in the next 5–10 years." It should be possible to solve the environmental problems of the destruction of coastal ecosystems, pollution created by fish farms, and the spread of disease from aquaculture enclosures, by the application of strict controls on the fish-farming community. Anthony Cheshire and John Volkman (2004) of the South Australian Research and Development Institution (SARDI) and CSIRO, wrote that they

> have assessed the health of captive tuna and in all cases the risks ranged from negligible to low under current practices. Other studies have investigated the impact of tuna farms on the composition of sediments and benthic infaunal communities beneath the sea cages . . . research programmes are now aimed at identifying the nature and quantities of farm wastes and understanding regional effects, such as the impact of farming on silver gulls and other key species, including sharks and pinnipeds.

But what can be accomplished in South Australia may not be possible in Croatia, Cyprus, Libya, Tunisia, or Turkey. The Aussies are fishing the main population of southern bluefins, but those Mediterranean countries are in competition for the breeding stock of northern bluefins, and might find that environmental safeguards or studies cut into their profits.*

*Some environmental events are beyond the reach of traditional regulatory agencies. As a result of the Israeli bombing of Lebanon in July 2006, a power station in Jiyyeh, only eighty feet from the sea, was blown up. Tens of thousands of tons of heavy fuel oil gushed into the eastern Mediterranean and uncontrolled fires burned the rest of the oil. Early reports estimated the spill at ten thousand tons, but the eventual total could be thirty-five thousand tons—close to the forty thousand tons that spilled into Prince William Sound in Alaska in 1989, when the 987-foot-long supertanker *Exxon Valdez* ran aground, generally considered the worst oil spill in history. A month after the Lebanese bombing, the oil began to sink, covering the bottom with a four-inch-thick blanket of sludge. As the first large "oil spill" in the Mediterranean, the Jiyyeh event will have massive detrimental effects on the marine life of the region—including the bluefin tuna that breed there.

Most people see tuna in a can or on a plate. Few of us have ever had the opportunity to fish for giant tuna, and because of the decline of the stocks, even if we had the opportunity, we might not see one, let alone catch one. Recently, however, I had an experience that—if possible—made me appreciate the great bluefin even more. At the Monterey Bay Aquarium in California, the Outer Bay tank has some thirty-odd bluefin and yellowfin tuna swimming majestically past the viewing windows. Occasionally, for reasons known only to the tuna, one of them accelerates, and you are forcefully reminded of the amazing speeds attainable by bluefins: fifty miles per hour in the blink of an eye. Admittedly, tuna in tanks are not wild ocean rangers, but they are not going to end up as sushi, either. Some of these giants weigh four hundred pounds, and while they will probably never get the opportunity to dive to one thousand feet or migrate for thousands of miles to their breeding grounds, they can accomplish what none of their pelagic brethren can: they can show us the streamlined quintessence of the most spectacular of all blue-water corsairs. In *Song for the Blue Ocean,* Carl Safina has written this paean to that great ocean ranger:

> In late afternoon the big tuna became very active, and watching them crashing herring thrown behind our boat was so astonishing it kept distracting me from scientific duties on deck. Each handful of fish that hit the surface instantly drew a gleaming, seven-foot-long, wide-bodied rocket from the deep. Often two or three big bluefins raced competitively for the same fish, their fins slicing the surface like drawn sabers of charging horsemen, with the winner sending up geysers and drawing cheers and then blasting the herring like there was no tomorrow. But we were here to help ensure that there was a tomorrow, and seeing all these fish shot us through with great hope and exuberance. These speeding torpedoes were no buffalo, no mere remnants. These were the rushing vitality and power of the sea incarnate.

But spectacular beauty has never been enough to protect a species, and the rolls of extinct or near-extinct creatures include many of the most gorgeous ones. Think of the tiger: in efficiency, grace, power, and beauty, the terrestrial equivalent of the mighty bluefin, now endangered throughout its diminishing range by the needs of Chinese traditional medicine. If the world's tigers are being slaughtered out of existence so that aging Asian men can enhance their virility, what hope is there for a fish, and moreover, a fish that people all over the world love to eat? (When Naylor and Burke wrote about the demise of large carnivorous fish species, they called them "tigers of the sea," an appellation they credited to Rebecca Goldburg.)*

The price commanded by a bluefin tuna makes it possibly the most commercially valuable marine finfish in the world on a per-pound basis. Despite an alphabet soup of regulatory commissions—IATTC, ICCAT, CCSBT, NOAA, NMFS, ATBOA, SARDI, ATRT, UNCLOS—tuna fishing continues unabated, and the wild populations of northern and southern bluefin have been reduced to critical levels. As the tuna populations continue to fall, the Japanese demand for *toro* will increase; fewer tuna will mean higher prices, and higher prices will mean intensified fishing. Intensified fishing will, of course, result in fewer tuna. (Of course, all bets would be off if the Japanese somehow relaxed their demand for *maguro,* but that seems as likely as Americans giving up hamburgers.)

There is another solution, suggested Josué Martínez-Garmendia and James Anderson (2005), but it is difficult to envision its implementation. Their study, published in the journal *Agribusiness,* is entitled "Conservation, Markets, and Fisheries Policy: The North Atlantic Bluefin Tuna and the Japanese Sashimi Market," and examines market considerations with regard to bluefin tuna conservation:

> The integrated process starts by acknowledging that fish populations have limited regenerative capacity. This leads to a yearly harvest cap which means that no more than a certain amount can be harvested this year if we still want to have a viable fishery next year.

*In 2005 I wrote a book called *Tiger Bone & Rhino Horn: The Destruction of Wildlife for Traditional Chinese Medicine.* Like tuna, tigers are being hunted toward extinction, but in a very strange congruence, the Chinese have suggested that they can save the big cats by breeding them on farms (Graham-Rowe 2006). Unlike tuna, tigers breed readily in captivity, but those who object to farming tigers say that if tigers are farmed and then killed for their body parts, it will only increase the pressure on the wild ones by perpetuating the huge, illegal market, and also because wild tiger parts would sell for more than those from farms.

> However, up to this point only a portion of the conservation problem has been solved: we have made sure that a sufficient fraction of the fish population will be preserved to sustain the fishery in the future. The second part of the conservation problem is how to make sure that fish that are not preserved (i.e., the ones harvested) actually attain the greatest social use.

This suggestion, of course, ignores what Garrett Hardin identified as "the tragedy of the commons" in a classic paper published in *Science* in 1968. Hardin spoke of pastures, but the ideas are applicable—and on a considerably larger scale—to the oceans. He wrote:

> The tragedy of the commons develops in this way. Picture a pasture open to all. It is to be expected that each herdsman will try to keep as many cattle as possible on the commons. Such an arrangement may work reasonably satisfactorily for centuries because tribal wars, poaching, and disease keep the numbers of both man and beast well below the carrying capacity of the land. Finally, however, comes the day of reckoning, that is, the day when the long-desired goal of social stability becomes a reality. At this point, the inherent logic of the commons remorselessly generates tragedy.

One herdsman adding one more cow to the pasture encourages others to do the same, and soon the pasture cannot support all the cows, and the farmers fail. Everybody wants there to be more tuna, but nobody is willing to take the necessary steps to preserve them. If you don't take your share (or more than your share), somebody else will take it, and he'll get rich while you don't. Better fish while you can, and let someone else worry about arcane subjects like maximum sustainable yield, migration patterns, or extinction.

The factors that influence the Japanese market price for bluefins are freshness, fat content, color, and shape, but the fisheries pay no attention to these considerations, and are governed by regulations that are based only on population estimates. If the highest-quality fish are selected from North American landings for sale in Japan, the overall profit goes up. The analysis by Martínez-Garmendia and Anderson indicates "that it is more profitable to harvest fewer tuna but of greater size and quality than to focus on a great number of fish captures. This allows for a lower mortality and therefore a greater level of conservation of the resource." A nice idea, but how can we keep the New England fishermen from catching every tuna they can? And more important, how can we convince the Japanese that fewer, better tuna at Tsukiji would be better for them, and better, in the long run, for the tuna?

The tuna fishery is rife with illegal fishers, who ignore quotas, restrictions, boundaries, and any other rules and regulations that might threaten their catch, and the Japanese market is only too eager to absorb thousands of tons of bluefin tuna, regardless of where or how it is caught. (Japanese bluefin fishermen have contrived to circumvent even their own restrictions, bringing in thousands of tons of illegal tuna every year, and then falsifying their records.) It would be good for the tuna, and in the end good for the consumer, if tuna fishing was not practiced in such a remorseless manner, but a modification beneficial to the tuna would entail nothing less than a modification of the fundamentals of human nature, and that is not likely to happen in the foreseeable future.

The ranks of frozen tuna at the Tsukiji market seem endless—and are replenished every day.

In July 2005, a joint research effort funded by France, Germany, Greece, Israel, Italy, Malta, and the Spanish commercial tuna farm Tuna Grasso, S.A., successfully used hormonal induction in captive broodstock of Atlantic bluefins to obtain eggs and sperm. In vitro fertilization was carried out and larvae were produced (Mylonas 2005). According to a report on bluefin tuna research in Spain at the Instituto Español de Oceanografia at Puerto Mazarron, a project has begun

> whose main aim is to develop an aquaculture farming technology for the bluefin tuna, and to improve the knowledge of the reproductive biology of the species in captivity and compare this with populations in the wild. Under study are the hormonal regulation of gonadal development, maturation of gametes, spawning and fecundity, and finally the development of suitable methods to control the reproduction of the bluefin tuna. Work is being done on the development of handling techniques and the completion of wild and caged fish sampling programs for sex and maturity, successful brood stock maintenance, natural spawning and induced spawning success, determination of reproductive hormonal control and successful hatching of eggs. The long term results will determine the global strategy to be followed for the successful and sustainable aquaculture of bluefin tuna.

In the November 1993 edition of *Australian Fisheries,* Roger Nicoll reported that in Boston Bay, "two Japanese fisheries experts will gaze closely into the depths of a Port Lincoln experimental tuna farm for a sign that could spell bonanza for the southern bluefin tuna fishery." Takahiko Hamano and Yoshio Koga were looking for the characteristic swimming display of male and female fish, which would indicate tuna that would spawn in captivity, a heretofore unrealized goal. Only a couple of years after the Australians set up the first "experimental grow-out project" (tuna farm), Hamano and Koga were in South Australia as part of a joint $1 million project involving the Japanese Overseas Fishery Cooperation Foundation, the Australian government, and the Tuna Boat Owners Association of Australia. In Nicoll's article, Hamano defined the objectives of the project: "Our number one objective is to develop fish for the wild fishery. We aim to produce juveniles and release them into the open sea to build up the natural resource . . . It is not for commercial purposes but for the international fishery. Farming is a secondary consideration." The tuna farms were a recent South Australian innovation, and it is possible—but not likely—that the Japanese really intended to raise southern bluefin tuna and "release them into the open sea." As defined, however, the project was almost circular: breed bluefins in captivity, release juveniles into the wild, then catch the juveniles and put them into pens to raise them to adult size so they can be killed and shipped to Japan. Obviously, it would be less complicated to raise the tuna from eggs to adulthood in captivity, but that would require feeding them for eight years or more, so it would be cheaper to let them feed themselves in the open sea, recapture them, and feed them only for a couple of years.

By 1991, when Hamano and Koga first visited Port Lincoln, the Japanese had already been trying to breed (Pacific northern) bluefin in captivity for more than twenty years. Since 1970, fisheries biologists at Kinki University have been growing bluefin tuna from eggs in pens in the Sea of Kushimoto off the Kii Peninsula. In 2002, the Fisheries Laboratory at Kinki claimed to have carried out full-cycle breeding of the fish, and in September 2004 they shipped the world's first fully aquacultured bluefin tuna to market. In Theodore Bestor's 2004 book on the Tsukiji fish market, he noted that the Japanese make a crucial distinction between "wild" or "natural" fish on the one hand and "cultivated" fish on the other. Fish caught in the open water are called *tennen* (literally, "natural" or "spontaneous"); fish raised through aquaculture are *yōshoku* ("cultivated" or "cultured," as in cultured pearls); and those raised in pens are called *chikuyuō*. These distinctions are vanishing, but the Japanese have further complicated the issue by actually raising bluefin tuna from eggs. The fish were 100 percent *yōshoku,* meaning that they were farmed from hatching to adulthood. When the meat of the fifty-pound fish was sold in an Osaka department store for sashimi, it was priced lower than the same meat from farm-raised bluefin, and it sold out almost immediately. Although these fully farmed tunas are relatively small compared with wild-caught juveniles, they are rich in fat, which is the criterion for *toro.* Researchers are now grappling with the challenge of self-sustainable farming: obtaining viable eggs from farmed adults, and raising the fry to become the next generation in a continuous cyclical process.

Headline on page 4 of the *New York Times,* September 26, 2006: "Farming Bluefin Tuna, Through Thick Stocks and Thin." Although the meaning of the headline is not clear, the article concerns Hidemi Kumai, the head of the Fisheries Laboratory at Kinki University, who has "spent more than three decades trying to farm the bluefin tuna—an especially delicate fish, both physically and psychologically, prone to everything from restlessness to cannibalism." It is safe to assume that Kumai knows more than a little about the biology of tuna, so the mistakes in the article are probably attributable to the writer, Norimitsu Onishi. He wrote, "Bluefin bruise easily because of their delicate scales, and their gills take in little oxygen compared to other fish, so they have to swim continuously to breathe—even while asleep." We know that tuna have the largest gills of any fish, and therefore take in *more* oxygen than any other fish, and if the tuna in the pens at Kinki have been seen "sleeping," they are the only tuna anywhere that have demonstrated this behavior. The point is that the Japanese have now succeeded in raising bluefin tuna from eggs, and the meat has been sold in the market. If it isn't as succulent as the meat of wild bluefins caught off Massachusetts, that can probably be fixed.

One problem had been the survival rate of captive bluefins in tanks. Onishi's article said that the little tunas "kept crashing into the nets and dying. We took X-rays and found their necks were broken." An article in the *Journal of the World Aquaculture Society* (Miyashita et al. 2000) also reported that newly hatched bluefins at the Kinki laboratory crashed into the walls of the tank, causing themselves mortal injuries. But larger (and wiser) wild-caught bluefin and yellowfin tuna seem to thrive in the Monterey Bay Aquarium's million-gallon Outer Bay tank, where they form one of the aquarium's most spectacular exhibits. Although a juvenile great white shark was kept in the tank from September 2004 to April 2005, and released when it behaved aggressively toward other residents of the tank, the star attractions of this display are the tuna: yellowfin, skipjack, and Pacific bluefin. Normally inhabitants of deep blue seas, the graceful tunas in constant motion make one of the most impressive of all aquarium exhibits. As Carl Safina put it, "Virtually anyone who sees a living tuna is moved by the beauty of a life so energized and yet so mysteriously cloaked by the sea. Tuna provoke an intensity of interest that few other creatures engender." In his 2001 summary of tunas in captivity, Chuck Farwell (who is responsible for the tunas on exhibit at Monterey, and feels that "it's a privilege to work with an animal so highly evolved") lists several aquariums in Japan that have also built large enough tanks to display tunas: Nagoya, Kaiyukan (in Osaka, where the tuna are dwarfed by a captive whale shark), Kagoshima City, Aburatsubo, Aqua-Marine Fukushima, and Tokyo Sea Life Park. It is obvious that the Japanese *really* love tuna.

Farwell explains the method(s) of capturing tunas for display: yellowfins and skipjack are chummed close to the boat with live anchovies or sardines, and when they take a barbless lure dangled from a liftpole, they are "hooked" and immediately lifted into a vinyl tarp that is suspended over a holding tank on deck. As soon as the tension on the line is released, the barbless hook comes out of the fish's mouth, and the fish ends up in the holding well. Bluefins, on the other hand, will not bite the liftpole lure, and are caught with standard rod-and-reel sportfishing tackle, led onto a sling at the stern of the boat, and then lifted into one of the wells. Before they are introduced to the big tank, each fish is tagged with a positive induced transponder (PIT) tag, which enables researchers to track and monitor individual fishes in the exhibit. The tuna are fed squid, herring, anchovies, and sardines, but sardines appear to be their favorite.

Under the direction of Hamano and Koga, the adult tuna in the pens at Port Lincoln were fed a diet that included the same vitamin preparations that had been used at Kinki, hoping to stimulate spawning. The successful "spawning display" observed in the Japanese labs consisted of males and females swimming together, the female in front and the male following behind. If eggs were actually laid, they would have been collected in a plankton net, and brought to a laboratory where they would have been placed in seawater. "If we get the eggs and hatch them," said Hamano, "then the next step will be to cultivate them to 5–7 cm [2 inches] and then release them into the open sea." Two-inch-long tuna fry would have a very high mortality rate, and it would have been more practical (although less economical) to raise the fry in (increasingly larger) tanks, and then transfer them into pens, so they would not be picked off by every predator that might enjoy a tiny tuna snack. The Japanese plans did not work, and the captive tuna did not spawn. Some fifteen years later, Clean Seas, an adjunct of the Stehr Group's tuna-farming enterprise, undertook much more elaborate projects to breed bluefins in captivity. In July 2006, Jessica Farley, a tuna specialist at CSIRO in Hobart, reported what was happening with the Port Lincoln tuna-raising program.

> I had a chat with the hatchery manager of Clean Seas when I was in Port Lincoln a few months back. They have quite a few 7–9 year-old SBT in sea cages at Arno Bay, which they will transfer to their new land-based aquaculture facility when it's complete later this year—basically 6m deep, temperature- and light-controlled tanks. They don't know if they have males or females at this stage, but are planning to try to get them to spawn. We'll see. Given SBT spawn south of Indonesia, and no one knows what triggers spawning, it might be a difficult task!

Before we can breed tuna in captivity, we have to figure out what inspires spawning in the first place. We know that certain species will spawn only in selected areas, and only in water within very narrow temperature ranges. We have observed the mating dance of yellowfins in captivity at Achotines, where scientists saw "pairing of individuals, chasing, rapid color flashes exhibited by individual fish, and rapid horizontal or vertical swimming." In his 2001 discussion of the reproductive biology of various tuna species, Kurt Schaefer wrote:

> There is a very limited amount of information on size and/or age at maturity in *Thunnus thynnus.* A maturity schedule based on valid histological evidence has not been established . . . Regardless of these inadequate experimental designs and invalid conclusions, subsequent researchers have accepted the age of 100% . . . maturity for the west Atlantic to be 10 years of age (200 cm), and 5 years of age (130 cm) for the eastern Atlantic.

He went on to say that "even less research has been carried out on reproductive biology of *Thunnus orientalis* in the Pacific Ocean, and size and/or age is poorly understood." Schaefer then observed: "The spawning time during 1998 and 1999 of *Thunnus orientalis* broodstock held at the Amami Station, Japan, was between 1900 and 2300 h[ours]. The observed courtship behavior is apparently similar to the captive *T. albacares,* with more than one male commonly chasing after the same female and the same body stripes and/or darker coloration displayed by the males during this period." It may be no more complicated than recognizing that at certain seasons, and under certain conditions, the presence of mature males and ripe females triggers the simultaneous release of eggs and milt, guaranteeing the production of fertilized eggs.

Maybe—just maybe—there is a way to get bluefin tuna to market without catching wild juveniles and fattening them up for slaughter. In the Clean Seas Annual Report for 2006, Hagen Stehr has written the manifesto for his tuna-breeding project:

- Worldwide, wild fish resources are diminishing steadily, in some cases rapidly.
- Worldwide, consumption of seafood is steadily increasing, in many cases quite dramatically.
- Catches of tuna are declining again this year in the Mediterranean due to overfishing, pollutants and health concerns. Scientists believe the resource is doomed and will collapse in the foreseeable future.
- The Mexican tuna harvest is unreliable on account of environmental factors and overfishing. The tonnage harvested is continuing to decline.
- In the Southern Hemisphere, overfishing of the SBT resource is being vigorously addressed by Pacific governments, harvests are being severely restrained and market prices are being driven up.

News item in the *Adelaide* (South Australia) *Advertiser* for October 28, 2006:

> TUNA BROODSTOCK NOW SETTLED
>
> Nine southern bluefin tuna, each weighing around 130 kilos [285 pounds] have been individually transferred by helicopter from a Clean Seas Tuna sea farm to the new onshore breeding facility at Arno Bay. The move took eight hours. Just a few days later the fish were obviously settled in at the hatchery. The broodstock, seven years old, are expected to spawn in February or March 2007.

Clean Seas Aquaculture Growout Pty. Ltd. is a publicly traded company, founded by Hagen Stehr in 1969, and now owned and managed by the Stehr Group in Port Lincoln. The Australian government recently provided a grant of $4.1 million to assist in the commercialization of SBT breeding. Clean Seas is so sure of success in breeding southern bluefins that they have made a public offering of 18 million shares in the company, according to the October 2005 prospectus. They have already raised captive-bred yellowtail kingfish *(Seriola lalandi)* and mulloway *(Argyrosomus hololepidotus),* which are now in significant commercial production. In part, Stehr's "Chairman's Letter" in the prospectus reads:

> Over the past 6 years Clean Seas parent, the Stehr Group, has been actively working towards breeding SBT and has made significant advances in several areas including broodstock management, onshore fish transfer, fish husbandry practices, and the pelletized feeding of juvenile SBT . . . The next step for Clean Seas is the propagation of our own broodstock facility at Arno Bay. Fertilized SBT eggs to be produced in the new facility will be transferred to our existing hatchery facility for the controlled production of SBT fingerlings by adaptation of proven protocols for the other species being propagated. The directors believe that over the next three years this should lead to the establishment of an exciting and long term commercially viable SBT business.

It appears that investing in Clean Seas would have been a good idea. In October 2006 southern bluefin broodstock were airlifted from the pens to a 3-million-liter (790,000-gallon) tank designed to replicate the optimum conditions for spawning. In a 2006 interview in *The*

The Arno Bay (South Australia) facility of the Clean Seas project. It is hoped that southern bluefins will breed here in specially designed tanks.

Australian, Stehr said, "We've got it all on computer, we can make it lighter or darker, we can leave the fish in a state of wellbeing, we've got the sun going up, the sun going down . . . This is a world first, the Japanese won't try it at all, the Americans have tried it and failed and the Europeans have failed too."

During my February 2007 visit to Port Lincoln, Rob Staunton, the farm manager for Stehr Group, drove me to Arno Bay, 150 kilometers north of Port Lincoln, on the western shore of Spencer Gulf. I was granted limited entrée into the holy grail of the tuna business, the giant enclosed tuna tank at the Arno Bay hatchery. I say "limited" because my visit, personally sanctioned by Hagen Stehr, came with severe restrictions, all of which were perfectly understandable. No photography is allowed in the facility itself, because the engineering, water processing, climate control, and every other element in the design of this potential miracle must be carefully safeguarded to prevent corporate theft of the ideas. Along with the grant from the Australian government, Stehr Group has invested millions in the innovative design of this facility, and it would be a disaster if someone borrowed or modified their designs and somehow beat them to the punch. It's hard to imagine anybody replicating this massive facility without all of Australia knowing about it, but of course entrepreneurs in other countries—Japan, for instance—are also very interested in the business of captive-breeding bluefin tuna from eggs. Indeed, the Japanese have already bred bluefin from eggs and raised them to breeding age in the laboratory, but not on the commercial scale being attempted by Clean Seas.* Morten Deichmann, the Danish-born biologist who heads up Clean Seas' hatchery operations, told me that he had visited the breeding facilities at Kinki University in Japan, but for obvious reasons, he could not allow the Japanese scientists behind the scenes at Arno Bay.

To begin our tour, we had to change into special white rubber boots, sterilized to prevent the introduction of alien microbes into the "hydroponic" facilities. (The term "hydroponic" usually refers to growing *plants* in nutrient-enriched water, but the concept is not far from the minds of the tuna breeders.) Staunton and I were chaperoned by Thomas Marguritte, a Frenchman-turned-Australian who allowed us access to these inner sanctums. First we saw the not-so-secret yellowtail kingfish broodstock, swimming in a darkened tank in one of the hatchery buildings. Yellowtail kingfish (more often just "yellowtail") are hefty, baleful fish, with the eponymous tail and a brass-colored band that runs from tail stock to gill covers. They acknowledged our presence by staring directly at us as we lifted a flap door in the side of the tank that enabled us to watch them swimming by. They are large, schooling fishes found throughout the world's temperate and subtropical waters. Also known as Japanese amberjack and called *buri* or *hamachi,* the species is a popular eating fish, grilled or as sushi or sashimi. In the wild, these hard-hitting game fishes can reach a length of six feet and a weight of one hundred pounds, but in captivity, where they do nothing all day but eat, they grow even larger. Historically, yellowtail kingfish farming was based on capturing wild fry

*Spawning bluefins have evidently been observed and even filmed, but I have not been able to find the film. In a 1998 ICCAT paper, Raimondo Sara wrote, "The availability and the considerable technical capacity of Dany Orlando who filmed a group of large bluefin in the Sicilian trap at Favignana has resulted in documented visual proof of the spawning behaviour of bluefin tuna. There is visual documentation of the scientific significance of the term 'school,' the function of grouping by size, and the significance of being the same size to the economy and the strategy of the group. . . . Also examined, for its social significance, is the colouration which components of the group assume at key moments of their life and in particular, during the spawning period. External fertilisation by more than 10 individuals is shown."

in the ocean and raising them to market size, but recent advances have involved controlled spawning and raising juveniles in hatcheries before placing them in pens or cages. As of now, some 40 percent of the Stehr Group's business is farming yellowtail, but unlike tuna, most of the product is sold in Australia. Based on their success with yellowtail, the Stehr Group is hoping to apply its experience and knowledge to the breeding of bluefin tuna.

We exchanged our white boots for blue as we entered the sanctum sanctorum of the Arno Bay hatchery, the tuna-breeding tank. In a cavernous room, illuminated by a battery of fluorescent light fixtures, and with the quiet hum of air-conditioning as the only background noise (the temperature outside was near 100°F), we climbed up on the concrete rim of the 790,000-gallon tank and looked down. (For reference, the Outer Bay tank at the Monterey Bay Aquarium holds a million gallons and is four stories high.) The tank is about eighty feet in diameter and twenty feet deep. Because the light level was fairly low, there wasn't much to see until Marguritte tossed in a couple of small fish. Suddenly the surface broke with an ultramarine and chrome flash as one of the tuna charged at the baitfish. Immediately, the surface was alive with froth, pierced by the sicklelike dorsal and tail fins of the tuna, anticipating a feed, even though, as our docent explained, they had been fed only an hour before. As they circled excitedly beneath us, we could see that these were breeding-size bluefins: seven hundred pounds of sleek, polished torpedo, pointed at both ends, with a dotted line of yellow finlets just before the tail, and the startling parentheses that mark the species' horizontal keels, chrome yellow in the southern bluefin and black in the northern varieties. No one can tell a live male from a live female except another tuna.

Poised on the rim of the tank, we talked about the breeding program. "We can replicate the exact conditions in Indonesian waters where they are known to spawn naturally," said Thomas. "If they usually spawn in the Southern Hemisphere summer when the days are longest and the water temperature is highest, we can make this tank conform to—pick a date, say November 20—and set the length of daylight hours, air temperature, water temperature, and even currents to conform to that moment in the Indian Ocean, south of the Indonesian archipelago." (The seasons are reversed here, of course; even though it falls on December 25 just as it does everywhere else, Christmas comes in the middle of summer for Australians, New Zealanders, southern Africans, and southern South Americans.) The only variable they cannot duplicate is the depth of the water, and they are praying that it is not a critical factor in the breeding of the southern bluefin. Just south of the Indonesian arc of islands—Java, Bali, Flores, Sumba, Komodo, Timor—is the Java Trench, which descends to the deepest point in the Indian Ocean, more than twenty-five thousand feet down. If depth is a factor, the Clean Seas breeding project is doomed.

There were attempts to breed killer whales in captivity in the late 1970s. Of course killer whales are mammals, and do not lay millions of eggs the way tuna do, but up to that time, all attempts to breed orcas, the popular stars of oceanarium shows, had failed. Even those females that had been impregnated in captivity often gave birth successfully, but then didn't nurse the newborn calf, which was destined to starve to death. One time, at Sea World in San Diego, divers entered the tank with Orky and Corky and tried to feed Corky's newborn calf from a tube, but Orky, the thirty-foot-long, ten-ton father, would have none of it, and the divers were forced to feed the calf on the apron of the pool, which technique didn't work and the baby died anyway. It was only when it was realized that orcas spend their entire lives in tight family groups that the breeding of captive killer whales succeeded. Another adult female, introduced into the tank where a baby was about to be born, taught the new mother how to nurse its calf. Newborn tuna, about the size of a grain of rice, do not nurse, and certainly do not need an "auntie," but there may be spawning stimuli that we have not recognized.

At Clean Seas' conference room in their Port Lincoln headquarters, I met with Marcus Stehr, Hagen's forty-two-year-old son, managing director of the company. The day before, Marcus had been aboard one of the tuna purse-seiners a hundred miles off Ceduna, in the Great Australian Bight, as a net cage of perhaps a hundred tons of tuna was started on its journey to the pens of Boston Bay. Like everyone else associated with this venture, Marcus is enthusiastic and optimistic about their potential success and believes it is imminent. When I asked him if that success would completely change the way bluefin tuna are perceived in Australia, he said, "It's not a question of *if,* mate—it's *when.*" Although the Aussies appear to be in the lead, it remains to be seen if they, the Japanese, or the Europeans will win the race to breed the bluefin in captivity. But somehow or other, it has to happen, because the survival of the species (and the tuna industry) depends on it. The fish ranches, once seen as a solution to the problem, are only making it worse.

Harvesting tuna in the Mediterranean is depleting the population at an alarming rate, and the open-ocean tuna fisheries are running out of fish. Roberto Bregazzi called the ICCAT meeting in Dubrovnik in November 2006 "Darwin's Nightmare," and wrote in a personal communication that the organization "has been unsuccessful in the adoption of measures both towards the protection of the bluefin tuna and . . . the economic viability of its Mediterranean and Atlantic fisheries, *even though SCRS (ICCAT's consultative body, formed by biologists from over 48 countries) clearly and conclusively defined the parameters for a sustainable fishery for bluefin tuna.*" The scientists defined a sustainable bluefin tuna fishery, and the fishermen ignored it, voting instead to increase the catch far beyond sustainability. The closure of the Mediterranean fishery proposed by the EU will "do little to restrict the huge market influence exerted by a small group of corporate fishing industries" that are working in Egyptian, Algerian, Cypriot, Turkish, Maltese, and Libyan waters. Because Egypt, Algeria, Turkey, and Libya are not members of the EU, decisions about Mediterranean quotas do not affect them, and they are free to catch anything, anywhere. Moreover, Japan, the country with the highest stake in worldwide tuna fishing, preaches sustainability but encourages free-for-all fishing. "If production prices per kilogram of transferred live bluefin tuna at sea in 2007 remain at current levels," said Bregazzi, "the EEU Mediterranean tuna purse-seining fleets will be unable to even break even in the less likely scenario of scrupulous adherence to shrinking catches yet to be assigned by the EEU for 2007. For this reason, these fleets will be forced once more to over-fish in order to . . . pay for increasing overheads and cope with stringent amortization schedules."

Among the greatest problems facing those—like Bregazzi—who would bring some semblance of order and responsibility to the tuna-fishing industry are the players in the so-called gray market. Tuna fishers, operating outside of any controls, ignore the ICCAT restrictions altogether, and sell sashimi-grade tuna on the gray market to buyers in China, Taiwan, Korea, and Thailand, where 100 million potential customers await the arrival of boatloads of *toro.* Bregazzi followed "Darwin's Nightmare" with another polemic, this one entitled "A Crime of Aggression against Mother Nature." In this essay he angrily condemns the Mediterranean tuna fishers for their utter disregard of the SCRS recommendations, claiming that the fifty thousand tons that the fishers intend to take per year for the next four years (as contrasted with the recommended fifteen thousand tons) will lead to the elimination of "a millenarian extractive activity deeply rooted in the history and culture of many states around the Mediterranean basin," not to mention the extinction of the bluefin tuna.

Even though there were supposed to be some controls governing who could fish where in the Mediterranean, in fact the Mare Nostrum has become a free fishing zone for every nation. Croatian, Greek, and Turkish fleets fish everywhere and anywhere, and the Libyans, probably the most egregious outlaw fishers of all, have staked out grounds all along their 1,300-mile-long coast, have rejected all restrictive quotas, and have unilaterally opened their fishing zones to boats of all nations. More than two hundred "fishing units" will be operating in Libyan waters, accounting for 60 percent of all purse-seine catches in the Mediterranean. These catches, most of which will go unreported, will be towed to the pens of Spain, Italy, Sicily, Algeria, Egypt, Greece, Croatia, and Turkey, and will total upwards of thirty-five thousand tons, already more than twice the SCRS-recommended maximum.

The free-falling tuna populations will have an effect on the Japanese sushi market. A front-page article by Martin Fackler in the *New York Times* for June 25, 2007, entitled "Waiter, There's Deer in My Sushi," reported that because of the shortage of tuna, Japanese chefs are resorting to highly unusual ingredients for their sushi, including raw deer meat and even raw horse meat. (Other sushi chefs are serving smoked duck with mayonnaise or crushed daikon with sea urchin.) Shigekazu Ozoe, the owner of the Tokyo restaurant Fukuzushi, tried to sell raw horse meat sushi: "We tasted it," he said. "It was soft, easy to bite off, had no smell"; but in the country that Fackler calls "seafood-crazy," horse meat isn't going to replace fish.* Another problem facing the Japanese is the rising popularity of bluefin tuna in other countries, including the United States, China, and Russia. The import of some sixty thousand tons of bluefin tuna (southern, northern, and Pacific) per year supports more than the restaurant industry. The importers, market workers, shippers, fishers, and even the giant conglomerates rely on a steady supply of *maguro.* If the bluefin goes down, it will take more than the sushi business with it.

The day after Fackler's article appeared in the *Times,* the paper followed up with an editorial entitled "Japan's Tuna Crisis":

> The real issue here is not the deprivation of Japanese taste buds but the decline of the tuna. Along with sharks and other ocean-dwelling species, tuna have been in free fall for decades, in part from rapacious overfishing by big industrial fleets, of which Japan's is by far the most aggressive . . .
>
> Ellen Pikitch, director of the Pew Institute for Ocean Science, and other experts believe that the new limits may not be tough enough to prevent commercial extinction of the much-prized Atlantic bluefin tuna. Ms. Pikitch also points out that in any case illegal and unreported fishing will push actual catches well above those levels.
>
> In that sense, Japan's sushi crisis may be just the wake-up call both the consuming public and the regulators need.

As of June 2007, attempts to get southern bluefins to breed at the Stehr Group's Arno Bay facility in South Australia have failed. According to operations manager Rob Staunton, "The males were producing milt [sperm cells]; chasing the females, all the signs we needed, but the females didn't release the eggs . . . we even helped them with hormones, but without success." The breeding cycle in southern bluefins occurs only in the spring, so the fish in the Arno Bay

*Although the United States is emphatically not among them, there are several countries where horse meat is considered a delicacy, and even included in everyday menus. In France, butcher shops offer *cheval;* in Austria it is used in stews; Kazakhs eat it salted, dried, and smoked; in various regions of Italy it is served as steaks, sausages, or carpaccio; and the Quebecois import large quantities of horse meat from the United States. Japan has a long tradition of eating horse meat *(sakura),* raw or barbecued.

tank will have to swim around their tank for another year, "relaxing" and getting ready for the next spawning season. Next year, said Staunton, "we may remove a few tuna, 'milk' them of their eggs and sperm, or at least check their stage of maturity without killing them, quite a hard task on a 150kg [330-pound] tuna!" The exclamation point (in Staunton's letter to me) points up the difficult and desperate nature of the breeding project. Repeated failures will not only result in the loss of the millions of dollars invested in this project, it will also set the possible preservation of this endangered species back to the point where we have to depend on the fishermen and the Japanese fishmongers, whose very livelihood depends on more fish being caught and sold, to reduce their consumption of tuna. Some years ago, at a whaling commission meeting, I asked Sylvia Earle, then deputy commissioner of the U.S. delegation, why she thought the whalers appeared so eager to destroy the resource upon which their industry depended. She answered: "Think of oil, timber, coal, and any number of fishes, birds, and mammals. Whenever humans have had the opportunity to exploit a natural resource, they have over-exploited it."

On March 4, 2008, Clean Seas announced that male and female bluefins in the tank at Arno Bay had spawned. In an article in *The Australian* (McGarry, 2008), Chairman Hagen Stehr said, "We have proven what can be done, even with southern bluefin tuna, which is the holy grail of aquaculture. . . . In the future this will be a staggering industry of immense proportions. It depends on us, the state government and the federal government, how big we want this to be. In years to come, this will give us a sustainable bluefin industry, that no one in the world will be able to attack." Stehr's company has already shown that they can raise big fish such as yellowtail kingfish and mulloway from eggs, so raising bluefins should not be an insurmountable problem. Because there are no quotas on fish raised artificially, Clean Seas can essentially set the market for tuna exported to Japan, China, Europe, and the United States. If they can raise the tuna to market size, there is essentially no limit on the amount they can sell. It will be years before the Clean Seas bluefins grow to marketable size, so we must wait to learn if the Australian captive-breeding program will arrive in time to rescue the Mediterranean bluefins from the rapacious overfishing by the Europeans.

EMINENT DOMAIN

Man and fish: Kiwi White of Port Lincoln, South Australia, and friend

IN THE BOOK OF GENESIS, because God saw that "the wickedness of man was great in the earth," he decided to destroy mankind, except for Noah, whom he commanded to build an ark so he could save a few chosen members of the human race and all the animals ("every living thing of all flesh . . . Of fowls after their kind, and of cattle after their kind, of every creeping thing of the earth, after his kind, two of every sort shall come unto thee, to keep them alive"). Noah gathered the animals two by two, it rained for forty days and forty nights, the earth was flooded, "and all the flesh died that moved around the earth, both of fowl, and of cattle, and of beast, and every creeping thing that creepeth upon the earth, and every man." Every living thing that wasn't on the ark died. Then the earth dried out, the human and animal passengers disembarked, and they became fruitful and multiplied. And that is why we have all these people, all these beasts, all these creepy-crawly things.

During the creation, God "made the great whales and every living creature that moveth, which the waters brought forth abundantly, the moving creature that hath life, the fowl that may fly above the earth and the open firmament of heaven," but with the exception of those "great whales," there is no mention of sea creatures. We must assume that the fishes, sharks, dolphins, oysters, squid, octopuses, clams, crabs, starfishes, and jellyfish were not bothered by a little rain, and somehow contrived to follow in the wake of the ark. (As gill-breathers, however, they might have had a problem when "the waters dried up from off the earth.") God refilled the ocean basins (think of Jonah and the whale, or Moses parting the Red Sea), and all that marine life seems to have reestablished itself. And indeed, after the waters receded, God told Noah that "the fear of you and the dread of you shall be upon every beast of the earth, and upon every fowl of the air, upon all that moveth upon the earth, and upon all the fishes of the sea, into your hand they are delivered." Even before the Flood, however, in Genesis 1:26, we learn that God said, "Let us make man in our image, after our likeness; and let them have dominion over the fish of the sea, and over the fowl of the air, and over the cattle, and over all the earth, and over every creeping thing that creepeth upon the earth."

"Dominion" means ownership or authority, so it is obvious that the Bible clearly defines marine and terrestrial animal life as the property of mankind, to do with as he wishes, without fear of repercussion. At first, our "dominion" consisted of beasts of burden, such as camels, horses, and oxen, and maybe some chickens, but this was soon extended to include farmed animals, such as sheep, cattle, and for some, pigs. The slaughter of cattle, sheep, pigs, and chickens for human consumption is largely kept out of sight of the consumer. Most people don't want to know about abattoirs; they are only interested in the appearance of cuts of meat in the market or in a restaurant. Cows and sheep have big brown eyes and placid dispositions, and there are even some species of cattle that are bred to be aggressive—bulls used in the rodeo sport of bull riding and Spanish fighting bulls, for example—but for the most part, cattle are amenable to just about everything, including being herded into an abattoir where they will be slaughtered. Just about every part of the cow, from the outside to the insides, is utilized.

In India, however, the cow is considered sacred, and millions roam freely through the towns and cities. The Hindu religion forbids the eating of meat, so the cows remain unfarmed and unharmed. (Indeed, there are shelters where hungry or sick animals can be taken for treatment and recuperation.) There are also several regions in southern India—the state of Kerala in particular—where beef is eaten regularly. In Hinduism, practiced around the world by nearly 900 million people, respect for animal life is a central theme, and the cow is esteemed above all other animals. "If someone were to ask me what the most important outward manifestation of Hinduism was," said Mahatma Gandhi, "I would suggest that it was the idea of cow protection." So while millions of Indians go hungry, a potentially enormous source of protein wanders untouched through the streets. Whatever the reasoning, however, India's Hindus are as unlikely to kill and eat their sacred cows as Americans are to kill and eat their almost equally sacred horses. Although not mentioned by name in the Bible, we can assume that some of the "domesticated" animals—especially cows and horses—were among the most important animals brought aboard the ark, and perhaps because they are not mentioned at all, the fishes were not among those animals considered worth saving.

Horses too have big brown eyes, and many breeds—ponies, draft horses, show horses, racehorses—are integrated into our society in ways denied cows and sheep. (Pigs make good pets, and are considered "smarter" than cows, sheep, or chickens, but the hog industry in America is enormous, and provides ham, bacon, pork, fatback, and pork rinds.) The animals that supply beef, lamb, and pork are slaughtered around the world. (Sheep are killed for food, but they are also sheared for wool while they are alive, which makes them doubly important.) Human consumption comfortably justifies the killing of these animals, but why do few Americans eat horses? Is it because they serve mankind so nobly, pulling his plows, carts, chariots, and coaches? Because heroes from Alexander the Great to Napoleon rode into battle on noble steeds? Because horses are considered beautiful? Members of royalty and the upper classes often had their portraits painted on horseback, and there are many artists who earned a good living painting pictures of the horses themselves. When the racehorse Barbaro broke his leg during the 2006 Preakness Stakes, people everywhere prayed for his recovery. Several surgeries failed to restore him to health, and when it was determined that he could not be saved, he was euthanized. The tragic story of Barbaro was front-page news throughout America. Although many horses are slaughtered in America, killing them is still considered barbaric. Somehow, horses have achieved a sort of "sacred" status, not unlike the cows in India, which cannot be touched for religious reasons.

For the most part, horses have been replaced by machines for transportation and farming, but there are many people—cowboys and jockeys, for instance—who still depend on horses for their livelihood. We think of horses as graceful, fast, powerful animals, noble servants of man, and perhaps this is why we don't eat them. But the bluefin tuna is also a graceful, fast, powerful animal, and these characteristics do not protect it. Indeed, nothing really protects it—not the Bible, not even the organization nominally designed to do so. Rather than establishing guidelines that would at least ensure the continuation of the species, ICCAT appears to be fulfilling Carl Safina's revised definition, the "International Conspiracy to Catch All Tunas." The recent allocation of 29,500 tons to European fishermen points out the glaring inadequacy of ICCAT and will be another step toward the extinction of the species.

Fishing for food has been practiced for as long as shore- (or riverside-) dwelling humans have existed. As Andres von Brandt says in his history of fish-catching methods, "Fishing, like agriculture is a form of primary production. Its history began when man had to be content with what food nature could provide . . . It mattered not whether it was taken from the water or gathered from dry land." Until the development of modern fishing technology, subsistence fishermen, even if they were fishing for an entire village, could not threaten even a local population. The development of purse seines, longlines, factory ships, and drift nets (not to mention refrigerator ships and floating factories) have elevated commercial fishing to a level of efficiency that threatens not only entire species, but their ecosystems as well. Too long have we ignored the plight of marine creatures mostly out of our sight—until they appear in a fish market or in a restaurant. But why should we even think about protecting the bluefin tuna? Is it not just a fish, like a halibut, a trout, or a sardine? And aren't fish for eating?

Of course they are—the inhabitants of many countries depend on fish for most of their protein—but you can't eat them if there aren't any left. In 2006 fourteen marine biologists published a major study in the journal *Science,* in which they warned that unless things change dramatically, in fifty years there will be nothing left to fish from the oceans. The study, innocuously called "Impacts of Biodiversity Losses on Ocean Ecosystem Services," points out that as of 2003, 29 percent of all ocean fisheries were in a state of collapse; and as Ransom Myers and Boris Worm reported in 2003, 90 percent of the big fishes are already gone, and we are fishing for the remaining 10 percent. Of course, it is not only fishing that has contributed to this sorry state of affairs; worsening water quality, toxic algal blooms, dead zones, invasive exotic species, and the disappearance of animals and plants that filter pollutants from the water have all played a role. Notice that all these calamities are anthropogenic—caused by man. *Homo sapiens* is by far the greatest threat to environmental stability (especially if you throw global warming into the mix), but also its only hope. Solutions to the problems of overfishing are painfully obvious but very difficult to implement. If we can't keep the tuna fishermen from catching all the tuna, then it is our responsibility to somehow ensure that the tuna will survive and prosper far from longlines and nets. We owe it to ourselves (and to the tuna, whose populations we have ravaged) to preserve this wonderful creature for the balance of life on earth. If we can't protect the most celebrated of all fishes, what can we protect?

People have to eat, and what's the difference if they eat tuna, salmon, or codfish? Well, they can't eat much cod anymore, because for all intents and purposes, the codfishery is over. There are simply not enough cod to sustain any sort of a fishery, so the American and Canadian governments wrote *finis* to codfishing. Yes, people have to eat, but they don't have to eat the last of a species. The bluefin has been hunted—and is still being hunted—toward the abyss of extinction. If fishing at the current level of wanton carelessness continues, we will run out of bluefin tuna. The uncooked fatty belly meat of the bluefin has reached unprecedented heights in gastronomy, even though there are those who would argue that it is soft, mushy, and tasteless. It would be more than shameful to lose an entire species (or three) to a food fad. We may not be able to ameliorate the Japanese lust for *maguro,* but at the very least, we have to find a way to keep the bluefin from disappearing altogether.

In his 2002 book *Dominion (The Power of Man, the Suffering of Animals, and the Call to Mercy),* Matthew Scully examines "the many ways our society has turned its gaze away from animals, and countenanced a shameful climate of exploitation and cruelty toward them." Much of the book is devoted to the hunting of terrestrial and marine mammals (whales and dolphins), but Scully is also concerned about those who misinterpret the Bible's injunction for man to exercise dominion over the animals. He writes, "My copy of the Good Book doesn't say, 'Go forth to selleth every creature that moveth.' It doesn't say you can baiteth and slayeth and stuffeth everything in sight, either, let alone deducteth the cost." Although Scully includes only the most limited discussions of fish, his remarks about, say, whales, are equally applicable to tuna, especially as regards the attitude of the Japanease. Scully quotes Masayuki Komatsu, a member of the Japanese delegation to the IWC:

> "Japan believes that difference of view, in particular culture, should be respected. Culture is local, at most national. However [the] other side's view is that their judgement, should prevail all over the globe, and they try to impose their judgement, their views, over us . . . Killing whales is not different from killing fish in Japanese culture, because whale is a kind of fish. So the Japanese fisherman does not care."

The Japanese know perfectly well that a whale is not a fish, but they have demonstrated the same cavalier attitude toward fish as they have toward whales. Fish—of every species—are obviously there for the Japanese to eat, and regardless of how they are killed, how many are killed, or whether or not they are endangered, "the Japanese fisherman does not care." Neither does the Japanese tuna farmer, the sushi chef, the restaurateur, or the consumer. A nationwide insensitivity to the plight of the bluefin is not good news for the fish.

It is admittedly a stretch to include fish in the category of animals that should be protected. Fish are very different from our fellow mammals: they are legless, scaly creatures that live in a totally alien environment. They are uncountably numerous, lay millions of eggs, and suffocate if they are removed from the water. Fish don't make very good pets. People keep goldfish in a bowl or tropicals in a tank, but for the most part, these represent animals as room decoration (or as lessons in the teaching of "responsibility" to children), and not a physical interaction with a living animal. Empathy with a fish is difficult to engender. Captive fishes are largely unresponsive; they swim in endless circles, their facial expressions do not change, and they breathe water. Most familiar fishes are considered food items, so it is more than a little difficult to imagine befriending a salmon or a trout that will soon be served on a plate. (Turkeys and chickens are also largely unresponsive as pets.) The bluefin and yellowfin tuna on display at the Monterey Bay Aquarium were demoted to a position of unimpressive inferiority when a juvenile great white shark was introduced into their tank.

There are massive campaigns to protect endangered species, such as pandas, tigers, polar bears, and wolves, as we do not comfortably countenance the slaughter of our fellow mammals, except those that we are planning to eat. (Whales are also the object of large-scale conservation efforts, but cetaceans are anything but cuddly, and the save-the-whale movement is based more on intellectual considerations—the whales' and ours.) Cattle, sheep, and pigs are also mammals (and relatively cute when young), but because we raise them to be eaten, we have no discernible moral problem with killing them. The passenger pigeons are gone, eaten into oblivion, but other endangered birds, such as whooping cranes and California condors, are now the objects of concerted conservation efforts, and there are even a few reptiles and amphibians that we worry about losing. We have a moral obligation to protect those animals that cannot protect themselves—especially those whose decimation we have engineered, often for reasons that are embarrassingly inadequate. As naturalist William Beebe wrote in 1906: "The beauty and genius of a work of art may be reconceived, though its first material expression be destroyed; a vanished harmony may yet again inspire the composer; but when the last individual of a race of living beings breathes no more, another heaven and another earth must pass before such a one can be again."

Even the most fervent admirer of the tuna recognizes that a fish—

especially a very large fish—cannot assume the iconic status of a panda, an eagle, or a dolphin. Despite its beauty, power, and grace, the lordly bluefin is still "only" a fish, and as such has been acknowledged first as a food item, and at a distant second as a conservation object. But even if they do not fulfill our traditional criteria for species-that-must-be-saved, certain fishes need protection. The bluefin tuna must be put on that list, if for no other reason than that it is being eaten out of existence. (It is sad indeed that we must have a "reason" for not driving a species to extinction.) The bluefin tuna has been hunted commercially for centuries, and because its numbers are now running low, efforts are under way to breed the great fish in captivity. "Breeding," however, is very different from "ranching."

Tuna ranching might be seen as analogous to cattle ranching, but the tuna are not raised from birth, but rather fattened after being caught, so the analogy fails. To date, we have only begun to figure out how to make more tuna after the current crop is exhausted. "We must think hard," writes Scully, "on the idea of breeding and farming wild animals for any commercial purpose, let alone for tawdry and silly purposes like culling ivory or harvesting 'racks' . . . Even if the notion were true—if mankind's only alternative in the century to come were to turn even the remaining elephants and lions of the world into 'exotic livestock,' for ivory and sport hunters—this would be numerical survival only for those species, and their disappearance a morally preferable alternative. We would then have lost all appreciation for the elephant and the lion, and there are fates worse than extinction . . . it would be better just to let them all go." Amen. If it is the destiny of the bluefin to be farmed, slaughtered, auctioned, chopped up, and served in a Japanese sushi bar, maybe we ought to let them all go too.

Of all the species that have ever lived on earth, 99.9 percent are extinct. Since humans have been around only for the last couple of hundred thousand years, we had nothing whatever to do with the disappearance of the trilobites, pterodactyls, ichthyosaurs, velociraptors, and the millions of other creatures that disappeared long before the fateful appearance of *Homo sapiens.* Extinction is inevitable, but we don't get to decide when it happens and to which species. We don't have the right to remove any species from the earth, even those that

we find bothersome, like cockroaches and rats. Every living creature, aquatic or terrestrial, has as much right to existence as we do. As Henry Beston put it in his 1929 book *The Outermost House:*

> We need another and a wiser and perhaps a more mystical concept of animals. Remote from universal nature and living by complicated artifice, man in civilization surveys the creature through the glass of his knowledge and sees thereby a feather magnified and the whole image in distortion. We patronize them for their incompleteness, for their tragic fate for having taken form so far below ourselves. And therein do we err. For the animal shall not be measured by man. In a world older and more complete than ours, they move finished and complete, gifted with the extension of the senses we have lost or never attained, living by voices we shall never hear. They are not brethren, they are not underlings: they are other nations, caught with ourselves in the net of life and time, fellow prisoners of the splendour and travail of the earth.

The big-game fisherman sees the bluefin tuna as a sleek and powerful opponent; to the harpooner it is an iridescent shadow below the surface, flicking its scythelike tail to propel it out of range; the purse-seiner sees a churning maelstrom of silver and blue bodies to be hauled on deck; the longliner sees a dead fish, hauled on deck along with other glistening marine creatures; the tuna rancher sees the bluefin as an anonymous creature to be force-fed until it is time to drive a spike into its brain. The auctioneer at Tsukiji sees row upon row of tailless, ice-rimed, tuna-shaped blocks; Japanese consumers see it as *toro,* a slice of rich red meat, to be eaten with wasabi and soy sauce; to the biologist the tuna is a marvel of hydrodynamic engineering, its body packed with modifications that enable it to outeat, outgrow, outswim, outdive, and outmigrate any other fish in the sea; and to those who would rescue *Thunnus thynnus* from the oblivion of extinction, it has to be seen as a domesticated animal, like a sheep or a cow. For some, such a shift is almost impossible; the bluefin tuna, the quintessential ocean ranger, the wildest, fastest, most powerful fish in the sea, cannot be—and probably *should* not be—tamed. But if it isn't, we will only be able to say that we loved the tuna not wisely, but too well.

EPILOGUE TO THE VINTAGE EDITION

WHEN THIS BOOK was first published in July of 2008, its title was *Tuna: A Love Story*. Because I thought up the title and even designed the jacket, I was very proud of my work until people began telling me that, while it was certainly clever enough, it didn't come close to conveying what the book was about. *Tuna: A Love Story* could have been about anything: recipes, sushi, sandwiches, carpaccio, or even my reverence for the bluefin tuna, the fish whose portrait I painted for the cover. It was in fact about all those things, but upon looking at the title, no one would know that the book was also about biology, overfishing, mercury poisoning, fish farming, the "tuna-dolphin problem," and even the possible extinction of the tuna. For this edition, and for future editions, if there are any, the title has been changed to *Tuna: Love, Death, and Mercury*.

In Marrakech, Morocco, from November 17–24, 2008, the ICCAT held its annual meeting. On the table were recommendations to reduce the total annual catch (TAC) of Mediterranean bluefin tuna by a substantial amount in an effort to prevent the total collapse of the population. ICCAT scientists also urged a seasonal closure during the spawning months of May and June, but both proposals were overwhelmingly defeated. Tasked with preventing a collapse of the Mediterranean bluefin tuna fishery, the commission instead voted for catch quotas higher than its own scientists recommended, leaving industrial fleets free to scoop up tuna at the height of the spawning period. The commission rejected its own description of its management of the bluefin fishery as "an international disgrace" and endorsed a TAC of twenty-two thousand tons for 2009. The decision was driven by the European Union, supported by Morocco, Algeria, Tunisia, Egypt, and Syria, and later joined by Japan. The Japanese, of course, have more than a vested interest in tuna, wild-caught, farmed, or otherwise.

In an article in the *Port Lincoln Times*, published on November 20, 2008, Clean Seas, the South Australian company that is attempting to breed bluefin tuna in captivity, announced that they did not succeed in the spring of 2008 but that they were "confident" that they would have fingerlings in 2009. The most interesting aspect of this article, however, was the acknowledgment that Clean Seas, previously so intent upon keeping the details of their operation secret, was working with Japanese scientists at Kinki University, the only hatchery to date that has successfully "closed the bluefin tuna life cycle." The ICCAT decision to *raise* the Mediterranean quotas was a disaster in the making.

Immediately after the decision, in an article titled, "Bluefin Tuna—Magnificent Fish Too Valuable to Save," Charles Clover wrote, "Tucked away out of sight of the eyes of the world in Marrakech this week, the European Union played some of the dirtiest tricks imaginable to outflank a proposal by the United States to cut catches of bluefin to the upper limit that scientists said was justifiable, fifteen thousand tons. Instead, a greedy mob of countries led by the European Union pushed through a proposal to set quotas at twenty-two thousand tons, 50 percent higher than scientific advice." In a comment posted to Andrew Revkin of the *New York Times* on November 24, 2008, Carl Safina said, "I must say that based on their whole history I would have been astounded if ICCAT had set an Eastern quota that complied with the science. I'm ashamed of what they do, but no longer surprised. It's what we've come to expect. . . . ICCAT is a body resolutely incapable of doing the right thing. It's less than worthless. If that wasn't true, we wouldn't have an ocean full of such disasters. We'd have fish." Referring to the unconscionable performance of ICCAT, Sergi Tudela of WWF Mediterranean said, "Today's outcome is a recipe for economic as well as biological bankruptcy with the European Union squarely to blame. . . . Bluefin consumption in the main consumer market of Japan is expected to drop from eighteen thousand tons due to the economic crisis, with around thirty thousand tons of frozen bluefin already in Hong Kong and Japan—and additional unknown amounts in other Asian countries and in freezer ships. Our industry sources also tell us that there are seven thousand tonnes of illegally fished tuna in fattening cages across the Mediterranean that nobody wants to buy."

Maybe it's because of the mercury.

Mercury does not occur in a liquid form in nature, but is obtained primarily from cinnabar (also known as mercury sulfide or HgS), which has been mined since ancient times. The Romans used mercury mines as penal institutions for criminals, slaves, and other undesirables, because they recognized that there was a high likelihood that the prisoners would become poisoned, thus sparing their captors the need for formal executions. According to Leonard Goldwater's 1972 *Mercury: A History of Quicksilver*, cinnabar was mined (and used) in forty-three countries, including Austria, Brazil, Sweden, Spain, and Turkey. Spain still has the largest cinnabar mines in the world. After cinnabar is mined it is heated in retorts or furnaces until the mercury vaporizes and then condenses into its familiar liquid form. (It was called "quicksilver" because of its resemblance to molten silver.) Because of their bright red color, cinnabar crystals were once used in the manufacture of jewelry and red paint, but the mercury content made these uses too dangerous. Thousands of tons of cinnabar were lost when sixteenth-century Spanish treasure ships sank while transporting it in one direction or the other between South America and Europe, and the cinnabar has been leaking its poisons into the ocean for four hundred years.

In the hardcover edition I wrote that mercury in fish comes mostly from precipitates that are spewed into the air from coal-fired power plants. The mercury is ingested by bacteria, converted to methylmercury, and then passed up the food chain from the smaller fish to the larger ones until it gets to the apex predators such as tuna. I discussed the dangers of mercury in tuna, cited various studies, and concluded that while mercury is in fact poisonous to humans, the small amounts in bluefin tuna (the nominal subject of my love affair) probably wouldn't harm anyone very much. The FDA warned pregnant women and nursing mothers against eating tuna, but everyone else could go right ahead and eat it. In fact, you probably *should* go right ahead and eat it, because fish is so good for you. It is an excellent source of lean protein and the omega-3 fatty acids which are essential for brain and eye development. It's lower in cholesterol than most meats and usually cheaper in the supermarket. The American Heart Association suggests eating at least two servings of oily fish every week to help keep your heart healthy.

Bad idea.

Mercury is so pervasively dangerous that it will probably do you more harm than good to eat certain kinds of fish regularly. Mercury is used in extracting gold from ore; it was used in hat making (remember the Mad Hatter?); it's used as a disinfectant (remember Mercurochrome?); it's in amalgam dental fillings; it's also used in the manufacture of chlorine.

In nature, chlorine is found in a combined state only as sodium chloride (NaCl) or common salt. Breaking the chemical bond in sodium chloride produces chlorine, a greenish-yellow gas that combines with nearly all elements.* Chlorine is widely used in everyday products, most importantly safe drinking water, but also in the production of paper, dyestuffs, textiles, petroleum products, medicines, antiseptics, insecticides, food, solvents, paints, plastics, and many other consumer products. Chlorinated compounds are extensively employed for sanitation, pulp bleaching, disinfectants, and textile processing. (Along with these products, chlorine also appears in chlorofluorocarbons (CFCs), responsible for the destruction of the ozone layer; and polychlorinated biphenyls (PCBs), which wreak havoc with wildlife everywhere.)

The manufacture of chlorine has long been one of the primary sources of mercury release into the environment. According to a recent study by the advocacy group Oceana, there are still factories where chlorine is being produced in a way that "creates numerous tons of mercury wastes with associated disposal and cleanup prob-

*Chlorine gas was used by the German army in April, 1915, against the French Army at Ypres, Belgium. At first, the French officers assumed that the German infantry were advancing behind a smoke screen and orders were given to prepare for an armed attack. When the gas arrived at the Allied trenches, soldiers began to complain about chest pains and a burning sensation in their throats. When they realized they were being gassed, many soldiers ran away, leaving a four-mile gap in the Allied line. But the Germans were also worried about what the chlorine gas would do to them so they hesitated to advance, allowing Canadian and British troops to retake the position before the Germans could burst through the gap that the gas had created. Chlorine is a powerful irritant that can inflict damage to the eyes, nose, throat, and lungs. At high concentrations and prolonged exposure it can cause death by asphyxiation. Ninety soldiers died from gas poisoning before they could be taken to a dressing station; of the 207 brought to the nearest aid station, forty-six died almost immediately and twelve after long suffering.

lems, pumps up corporate electric bills unnecessarily, and in some cases turns neighboring communities against the companies." Oceana identifies five American chlorine plants ("the filthy five") in Ohio, Tennessee, Georgia, West Virginia, and Wisconsin, that have not converted to the mercury-free technology now used in caustic soda plants around the world (including thirty-six in Japan) to reduce mercury contamination of the atmosphere. Caustic soda (sodium hydroxide or NaOH) is an important ingredient in the pulp and paper industries and in the production of textile dyes, soap, detergents, solvents, and herbicides.

For hundreds of years, mercury was prescribed as the cure for syphilis, but there is little evidence that it worked; indeed, it probably hurt more than it helped. In his 1874 *Materia Medica*, which described the uses of various substances in medicine, John B. Biddle, M.D., discussed the preparations of mercury that could be used to cure or ameliorate syphilis and various other diseases:

> While it retains the liquid or metallic state, mercury is inert; but, when taken internally, it sometimes combines with oxygen in the alimentary canal, and thus becomes active. In the state of vapour, it frequently proves injurious—in some instances exciting salivation, ulceration of the mouth, &c.; in others, inducing a peculiar affection of the nervous system, termed *shaking palsy* (*tremor mercurialis*), which is often attended with loss of memory, vertigo, and other evidence of cerebral disturbance, and sometimes terminates fatally.

It was administered various ways, including rubbing it on the skin and taking it orally. One of the more curious methods was fumigation, in which the patient was placed in a closed box with his head sticking out; mercury was placed in the box and a fire was started under the box which caused the mercury to vaporize. It is now known that vaporizing mercury is one of the best ways to poison the patient. Indeed, some of the symptoms of syphilis—tremors, hearing loss, joint pain, forgetfulness, and delirium—are much the same as those of mercury poisoning. In *Mercury*, Goldwater wrote, "The use of mercury in the treatment of syphilis may have been the most colossal hoax ever perpetrated in the history of a profession that has never been free of hoaxes." Thus the application of quicksilver for patients suffering from syphilis gave rise to the saying, "A night in the arms of Venus leads to a lifetime on Mercury."

The following information comes directly from the U.S. Environmental Protection Agency's website:

> Mercury is a naturally occurring element that is found in air, water, and soil. It exists in several forms: elemental or metallic mercury, inorganic mercury compounds, and organic mercury compounds. . . . Pure mercury is a liquid metal, sometimes referred to as quicksilver that volatizes readily. . . . When coal is burned, mercury is released into the environment. Coal-burning power plants are the largest human-caused source of mercury emissions to the air in the United States, accounting for over 40 percent of all domestic human-caused mercury emissions. EPA has estimated that about one quarter of U.S. emissions from coal-burning power plants are deposited within the contiguous United States and the remainder enters the global cycle. Burning hazardous wastes, producing chlorine, breaking mercury products, and spilling mercury, as well as the improper treatment and disposal of products or wastes containing mercury, can also release it into the environment. . . . Mercury in the air eventually settles into water or onto land where it can be washed into water. Once deposited, certain microorganisms can change it into methylmercury, a highly toxic form that builds up in fish, shellfish, and animals that eat fish. Fish and shellfish are the main sources of methylmercury exposure to humans. Methylmercury builds up more in some types of fish and shellfish than others. The levels of methylmercury in fish and shellfish depend on what they eat, how long they live, and how high they are in the food chain.

Mercury begins its journey upward from the moment it lands on the bottom of the sea (or a river or lake), where it is absorbed by bacteria and converted to methylmercury, after which the now toxic bacteria are ingested by small animals, which themselves are eaten by larger and larger animals until we reach the pinnacle of the food chain. Big fish are recognized as the *natural* pinnacle of the food chain, but of course, in the same way that humans provide the mercury that works up the food chain, they have also replaced the big fish as the apex predators. In other words, we are the ultimate reapers of the deadly system we created.

The largest fish—tuna, swordfish, marlins, some sharks—are at the top of the food chain. Yellowfin and bigeye tuna are the red-meat tunas that are popularly served in restaurants as grilled tuna steaks, tuna carpaccio, tuna teriyaki, and of course, tuna sushi and sashimi. As top predators—yellowfin and bigeye tuna can be six feet long and weigh four hundred pounds—these fish have a significant mercury content. The bluefin, the largest tuna of all, will naturally have the most mercury aboard, but because the primary destination for bluefins caught around the world is Japan, Americans don't give much thought to the mercury content of *maguro*. For the Japanese market, however, the bluefin is being so heavily fished in the Mediterranean (a bluefin spawning area) that the World Wildlife Fund has called for a complete shutdown of the tuna fisheries to save the remaining tuna from extinction.

There are fish that are safe to eat. The Monterey Bay Aquarium's "Seafood Watch" lists every seafood item regularly consumed in America and tells you whether it's safe to eat, ecologically or toxicologically. Among the "Best Choices" are Alaskan halibut, anchovies, Arctic char (farmed), bluefish, Pacific cod (the Atlantic cod has been fished to near extinction), sole, herring, mackerel, Atlantic dorado (mahimahi), wild salmon, and sardines. In other words, there are plenty of other fish in the sea (and in restaurants); you shouldn't be eating tuna for health reasons—the tuna's or yours.

Late in 2008, Island Press published Jane Hightower's *Diagnosis: Mercury: Money, Politics & Poison*. Hightower is a San Francisco doctor whose patients included a woman who complained that "her house seemed to be making her sick"; her symptoms included fatigue, headache, trouble concentrating, and hair loss. She felt as if she had a hangover; sometimes she couldn't get out of bed for a couple of days. To Dr. Hightower's questions about her diet, the woman said she was a vegetarian and didn't eat meat, but she ate fish—tuna, swordfish, sushi, sea bass, halibut—at least nine times a week. Testing her, Hightower found that her blood mercury level was 26.0 micrograms per liter, five to six times higher than the EPA guidelines. A couple brought their seven-year-old son to see her because the boy was experiencing stomachaches, headaches, and lethargy, and turned red when he was in a warm bath. The parents told Dr. Hightower that they believed that fish was good for you, so they had been feeding their son canned albacore and yellowfin tuna steadily since he was two. The boy was tested, and his hair found to have a mercury level of about 15 micrograms per gram, fifteen times over the EPA standard. When the boy was taken off this dangerous, all-fish diet, his health improved, but, says Hightower, "He will most likely need special education and help for the rest of his life, as he still has difficulty with schoolwork, language skills, and social skills."

December 19, 2008. Forty-three-year-old actor Jeremy Piven, starring on Broadway in the David Mamet play *Speed-the-Plow*, complained about "excessive exhaustion and fatigue" so his doctor asked him what he had been eating. When Piven (best known for his role in the television series *Entourage*) said that he had been eating sushi twice a day, Dr. Carlton Colker tested him for heavy metals. The actor, found to have a "very, very elevated level of mercury . . . five to six times the upper limit that is usually measured," withdrew from the show and was briefly hospitalized. (David Mamet quipped that "he is leaving show business to pursue a career as a thermometer.")

Few consumers are aware of the mercury content in the tuna they are eating. Even now, the only warnings given to potential consumers of tuna can be found on the EPA website, where pregnant women and nursing mothers are told not to eat tuna because their babies, born and unborn, are susceptible to mercury poisoning. On its "Seafood Watch" handout, based on factors that include human and species endangerment, the Monterey Bay Aquarium says that one should *avoid* bluefin tuna. In her book, Dr. Hightower reports that albacore tuna has three times the mercury content of "light meat" tuna, which is skipjack, a small tuna canned in the billions and found on supermarket shelves around the world. "White meat tuna" sounds somehow better than ordinary "light meat tuna," but in fact, it contains more mercury. From the EPA website:

> Outbreaks of methylmercury poisoning have made it clear that adults, children, and developing fetuses are at risk from dietary exposure to methylmercury. During these poisoning outbreaks some mothers with no symptoms of nervous system damage gave

> birth to infants with severe disabilities; it became clear that the developing nervous system of the fetus may be more vulnerable to methylmercury than is the adult nervous system. Mothers who are exposed to methylmercury and breast-feed their babies may also expose their infant children through their milk.

Yet there are no warnings on cans of tuna and hardly any in restaurants. Of course, the absence of certain fish species on restaurant menus would serve as an implicit warning, but how is the customer to know that? Because of overfishing, Caroline Bennett of London's Moshi Moshi restaurant chain serves no bluefin in her conveyor belt sushi bars—and says so. Thanks to the marine conservation group Oceana, warnings are beginning to appear in supermarkets. Oceana "campaigns to protect and restore the world's oceans by winning specific and concrete policy changes to reduce pollution and to prevent the irreversible collapse of fish populations, marine mammals, and other sea life." Their current projects include halting the slaughter of sea turtles, stopping offshore drilling, banning Mediterranean drift-netting, protecting sharks from finning, saving bluefin tuna, and encouraging supermarkets to post signs warning of the dangers of mercury in fish. According to Oceana's Jacqueline Savitz, "We now have convinced thirty-six percent of major grocery stores in the United States to post signs." Among the chains now posting signs that contain the FDA warnings are Trader Joe's, Whole Foods, Kroger, Harris Teeter, Costco, Albertsons (SuperValu owned), and Safeway, and they are working on Wal-Mart, the world's largest publicly owned corporation. A typical sign, posted adjacent to the canned fish display shelves, reads:

NOTICE!

Pregnant and nursing women, women who may become pregnant, and young children should not eat the following fish:
SWORDFISH—SHARK—KING MACKEREL—TILEFISH

They should also limit their consumption of other fish including:
FRESH OR FROZEN TUNA

There is no mention of mercury, the reason for the warning in the first place. Is this because the supermarkets don't want to frighten their customers? Because the fishing industry doesn't want any of their products to be associated with mercury?

Because everyone knows that mercury is poison, one would assume that there would be some sort of guideline posted somewhere about what a "safe" level might be—assuming there *was* a safe level. But as Dr. Hightower learned, the "guidelines" are often vague, inconclusive, and largely unavailable to the fish-consuming public. When Hightower questioned a research scientist at the California Department of Public Health, she was told that "a blood mercury level of two hundred micrograms per liter was okay in adults." That was four to ten times what she had been seeing in her patients, and *forty* times the ceiling recommended by the EPA. When Dr. Hightower looked up mercury poisoning in a medical textbook, she learned that the symptoms included insomnia, nervousness, mild tremor, impaired judgment and coordination, decreased mental efficiency, emotional liability, headache, fatigue, loss of sex drive, and depression, severe paresthesia (prickling or tingling sensation of the skin), trouble speaking, trouble walking, tunnel vision, hearing loss, blindness, microcephaly (small brain size at birth), spasticity, paralysis, and coma. In the *Cecil Textbook of Medicine*, she found that the "reference range"—what is considered the maximum acceptable to maintain good health—was less than fifty micrograms per liter for whole blood. No further information as to diagnosis, treatment, prognosis, and so forth was included in the textbook.

Diagnosis: Mercury is not only about Dr. Hightower's patients. It is about the grim history and consequences of mercury poisoning, especially in Japan. Originally published in Japanese in 1977, Akio Mishima's *Bitter Sea: The Human Cost of Minamata Disease* was reissued in English in 1992. In the introduction we read:

> As a result of the bay's pollution with toxic organic mercury, many people were stricken with a terrible syndrome in the 1950s. Minamata disease, as it came to be known, is characterized by numbness of the extremities and the area around the mouth, constriction of the field of vision, loss of hearing, motor and speech disorders, loss

of muscle coordination, convulsions, and sometimes mental aberrations. People congenitally afflicted with the disease are often mentally retarded.

Bitter Sea incorporates photographs of the plant, the victims, and the protesters, which serve as a painful testimony to the horrors of mercury poisoning. In 1972, more than a decade after the poisoning of Minamata Bay had been recognized, American photographer W. Eugene Smith (1918–1978) went to Japan to document the gruesome story of Minamata Bay. Although he succeeded—his photographs are heartbreaking—goons from the Chisso Corporation beat him so severely that he was partially blinded and never fully recovered his sight.

Between 1962 and 1970, two northwest Ontario communities were warned that fish caught in the English-Wabigoon river system had record-high levels of mercury from a chemical plant up the river. By the mid-1980s, the bands received a compensation package of almost $17 million from the Dryden Chemical Company and the provincial and federal governments. They are still advised not to eat fish from the river. Almost everybody agrees that mercury is bad for your health.

After reading *Diagnosis: Mercury,* I have concluded that there really is no "safe" level of mercury, and I'm going to stop eating tuna. Why would anyone want to continue ingesting such a deadly substance? Let me repeat the symptoms of mercury poisoning: insomnia, nervousness, mild tremor, impaired judgment and coordination, decreased mental efficiency, emotional liability, headache, fatigue, loss of sex drive, depression, severe paresthesia, trouble speaking, trouble walking, tunnel vision, hearing loss, blindness, microcephaly, spasticity, paralysis, and coma. I'm going to tell my children to stop eating tuna, and anybody else who will listen. It's not likely that the Japanese, who buy and consume thousands of tons of bluefin tuna annually will be scared off by mercury warnings, but it is fantastic (as in "fantasy") to think that if enough people stop eating tuna, the fishermen would not be able to sell their catch, they would stop fishing, and the endangered bluefin would be saved from extinction. Maybe this is just a mercury-induced hallucination (until very recently, I'd eaten as much tuna as anybody), but in this epilogue to the revised edition of *Tuna*, I applaud the publication and frightening conclusions of *Diagnosis: Mercury*. I am tempted to quote the entire book here, but the best I can do is recommend that you read it.

Richard Ellis
July 2009

REFERENCES

Adam, M. S., J. Sibert, D. Itano, and K. Holland. 2003. Dynamics of bigeye *(Thunnus obesus)* and yellowfin *(T. albacares)* tuna in Hawaii's pelagic fisheries: Analysis of tagging data with a bulk transfer model incorporating size-specific attrition. *Fishery Bulletin* 101:215–28.

Adler, J. 2002. The great salmon debate. *Newsweek* 140(18):54–56.

Allain, V. 2005. What do tuna eat? A tuna diet study. *SPC Fisheries Newsletter* 112:20–22.

Allen, S., and D. A. Demer. 2003. Detection and characterization of yellowfin and bluefin tuna using passive-acoustical techniques. *Fisheries Research* 63:393–403.

Allport, S. 2006. *The Queen of Fats: Why Omega-3s Were Removed from the Western Diet and What We Can Do to Replace Them.* University of California Press.

Altringham, J. D., and B. A. Block. 1997. Why do tuna maintain elevated slow muscle temperatures? Power output of muscle isolated from endothermic and ectothermic fish. *Journal of Experimental Biology* 200:2617–27.

Altringham, J. D., and R. E. Shadwick. 2001. Swimming and muscle function. In B. A. Block and E. D. Stevens, eds., *Tuna: Physiology, Ecology, and Evolution,* 313–44. Academic Press.

Amoe, J. 2005. Fiji tuna and billfish fisheries. *Fisheries Department, Ministry of Fisheries and Forests, Fiji* FR/WP-12:1–6.

Amorim, A. F., and C. A. Arfelli. 2001. Analysis of Santos (São Paulo), fleet from southern Brazil. *Collective Volume of Scientific Papers. ICCAT* 54(4):263–71.

Anderson, A. 1990. *The Atlantic Bluefin Tuna . . . Yesterday, Today & Tomorrow.* Ocean Sport Fishing.

Anderson, J. L. 2002. Aquaculture and the future: Why fisheries economists should care. *Marine Resources Economics* 17:133–51.

Anderson, J. M., and N. K. Chhabra. 2002. Maneuvering and stability performance in a robotic tuna. *Integrative and Comparative Biology* 42:118–26.

Anon. 1986. Eating fish vital to health. *SAFISH* 10(3): 17–18.

Anon. 2004. Farmed fish vex sushi lovers. *Yomiuri Shimbun,* December 27.

Anon. 2005 *National report: Japan.* Technical Compliance Committee. Commission for the Conservation and Management of Highly Migratory Fish Stocks in the Eastern and Central Pacific Ocean.

Anuska-Pereira, M., A. F. Amorin, and C. A. Arfelli. 2005. Tuna fishing in the south and southeast off Brazil from 1971 to 2001. *Col.Vol.Sci.Pap.ICCAT* 58(5):1715–23.

Apple, R. W. 2002. How to grow a giant tuna. *New York Times,* April 3.

Archer, F., T. Gerrodette, A. Dizon, K. Abella, and S. Southern. 2001. Unobserved kill of nursing dolphin calves in a tuna purse-seine fishery. *Marine Mammal Science* 17(3):540–54.

Associated Press. 2001. Giant tuna sells for record $173,600. http://www.flmnh.ufl.edu/fish/InNews/GiantTuna.htm.

Baker, C. S., and P. J. Clapham. 2004. Modelling the past and future of whales and whaling. *Trends in Ecology and Evolution* 19(7):365–71.

Baker, C. S., G. M. Lento, F. Cipriano, M. L. Dalebout, and S. R. Palumbi. 2000. Scientific whaling: Source of illegal products for market? *Science* 290:1695–96.

Bandini, R. 1939. *Veiled Horizons: Stories of Big Game Fish of the Sea.* Derrydale Press.

Bardach, J. 1968. *Harvest of the Sea.* Harper & Row.

Barrett, P. 2007. Yellowfin revelations. *Marlin,* October: 64–69.

Batalyants, K. Y. 1989. On spawning of skipjack tuna (*Katsuwonus pelamis* L.). *Col.Vol.Sci.Pap.ICCAT* 30(1):20–27.

Baum, J. K., J. M. McPherson, and R. A. Myers. 2005. Farming need not replace fishing if stocks are rebuilt. *Nature* 437:26.

Bayliff, W. H. 1993. A review of the biology and fisheries for northern bluefin tuna, *Thunnus thynnus,* in the Pacific Ocean. In R. S. Shomura, J. Majkowski, and S. Langi, eds., *Interactions of Pacific Tuna Fisheries, vol. 2 of Papers on Biology and Fisheries.* FAO, Rome.

Beardsley, G. L. 1975. A review of the status of the stocks of Atlantic bluefin tuna. *Col.Vol.Sci.Pap.ICCAT* 4:161–72.

Bearzi, G., E. Politi, S. Agazzi, and A. Azzellino. 2006. Prey depletion caused by overfishing and the decline of marine megafauna in the eastern Ionian Sea coastal waters (central Mediterranean). *Biological Conservation* 127:373–582.

Beckett, J. S. 1974. High recovery rates of small bluefin tuna *(Thunnus thynnus)* tagged in the northwest Atlantic. *Col.Vol.Sci.Pap.ICCAT* 2:232–33.

Beebe, W. 1906. *The Bird, Its Form and Function.* Holt.

Bergin, A., and M. Howard. 1994. Southern bluefin tuna fishery: Recent developments in international management. *Marine Policy* 18(3):263–73.

Bernal, D., K. A. Dickson, R. E. Shadwick, and J. B. Graham. 2001. Analysis of the evolutionary convergence for high performance swimming in lamnid sharks and tunas. *Comparative Biochemistry and Physiology* 129:695–726.

Bernal, D., C. Sepulveda, O. Mathieu-Costello, and J. B. Graham. 2003. Comparative studies of high performance swimming in sharks. I. Red muscle morphometrics, vascularization and ultrastructure. *Journal of Experimental Biology* 206:2831–43.

Berners, J. 1486. *The Book of St. Albans.* Westminster. 1966 reprint, Abercrombie & Fitch; University Microfilms.

Bestor, T. C. 2000. How sushi went global. *Foreign Policy* 121:54–63.

———. 2004. *Tsukiji: The Fish Market at the Center of the World.* University of California Press.

Biddle, J. B. 1874. *Materia Medica, for the Use of Students.* Lindsay & Blakiston.

Bigelow, H. B., and W. C. Schroeder. 1953. Fishes of the Gulf of Maine. *U.S. Fish and Wildlife Service Fisheries Bulletin* 74:1–577.

Blake, R. W., K. H. S. Chan, and E. W. Y. Kwok. 2005. Finlets and the steady swimming performance of *Thunnus albacares. Journal of Fish Biology* 67(5):1434–45.

Blanc, M., A. Desurmont, and S. Beverly. 2005. *Onboard Handling of Sashimi-Grade Tuna: A Practical Guide for Crew Members.* Secretariat of the Pacific Community (Fisheries).

Blank, J. M., J. M. Morrisette, P. S. Davie, and B. A. Block. 2002. Effects of temperature, epinephrine, and Ca^{2+} on the hearts of yellowfin tuna. *Journal of Experimental Biology* 205:1881–88.

Blank, J. M., J. M. Morrisette, A. M. Landeira-Fernandez, S. B. Blackwell, T. D. Williams, and B. A. Block. 2004. *In situ* performance of Pacific bluefin tuna hearts in response to acute temperature change. *Journal of Experimental Biology* 207:881–90.

Block, B. A., D. Booth, and F. G. Carey. 1992. Direct measurement of swimming speeds and depth of blue marlin. *Journal of Experimental Biology* 166:267–84.

Block, B. A., A. Boustany, S. Teo, A. Walli, C. J. Farwell, T. Williams, E. D. Prince, M. Stokesbury, H. Dewar, A. Seitz, and K. Wong. 2003. Distribution of western tagged Atlantic bluefin tuna determined from archival and pop-up satellite tags. *Col. Vol. Sci. Pap. ICCAT* 55(3):1127–39.

Block, B. A., and F. G. Carey. 1985. Warm brain and eye temperatures in sharks. *Comparative Biochemistry and Physiology* 156:229–36.

Block, B. A., H. Dewar, S. B. Blackwell, T. D. Williams, E. D. Prince, C. J. Farwell, A. Boustany, S. L. H. Teo, A. Seitz, A. Walli, and D. Fudge. 2001. Migratory movements, depth preferences, and thermal biology of Atlantic bluefin tuna. *Science* 293:1310–14.

Block, B. A., H. Dewar, C. Farwell, and E. D. Prince. 1998. A new satellite technology for tracking the movements of Atlantic bluefin tuna. *Proceedings of the National Academy of Sciences* 95(16):9384–89.

Block, B. A., and J. R. Finnerty. 1994. Endothermy in fishes: A phylogenetic analysis of constraints, predispositions, and selection pressures. *Environmental Biology of Fishes* 40(3):283–302.

Block, B. A., J. R. Finnerty, A. F. R. Stewart, and J. Kidd. 1993. Evolution of endothermy in a fish: Mapping physiological traits on a molecular phylogeny. *Science* 260:210–14.

Block, B. A., J. E. Keen, B. Castillo, H. Dewar, E. V. Freund, D. J. Marcinek, R. W. Brill, and C. Farwell. 1997. Environmental preferences of yellowfin tuna *(Thunnus albacares)* at the northern extent of its range. *Marine Biology* 130(1):119–32.

Block, B. A., and S. Miller. 2007. Unveiling the secret life of an ocean giant. In *2007 World Record Game Fishes,* 84–92. IGFA.

Block, B. A., S. L. H. Teo, A. Walli, A. Boustany, M. J. W. Stokesbury, C. J. Farwell, K. C. Weng, H. Dewar, and T. D. Williams. 2005. Electronic tagging and population structure of Atlantic bluefin tuna. *Nature* 434:1121–27.

Bohnsack, B. L., R. B. Ditton, J. R. Stoll, R. J. Chen, R. Novak, and L. S. Smutko. 2002. The economic effects of the recreational bluefin tuna fishery in North Carolina. *North American Journal of Fisheries Management* 22:165–76.

Borgstrom, G. 1964. *Japan's World Success in Fishing.* Fishing News (Books).

Bransden, M. P., C. G. Carter, and B. F. Nowak. 2001. Alternative methods for nutrition research on the southern bluefin tuna, *Thunnus maccoyii* (Castelnau): Evaluation of Atlantic salmon, *Salmo salar* L., to screen experimental feeds. *Aquaculture Research* 32(1):174–82.

Bregazzi, R. M. 2004. *The Tuna Ranching Intelligence Unit—September 2004.* Advanced Tuna Ranching Technologies. Madrid.

———. 2005. *The Tuna Ranching Intelligence Unit—Special November 2005 ICCAT Sevilla-Spain Meeting Edition.* Advanced Tuna Ranching Technologies. Madrid.

———. 2006. *The Plunder of Bluefin Tuna in the Mediterranean and East Atlantic in 2004 and 2005: Uncovering the Real Story.* World Wide Fund for Nature.

Brill, R. W. 1996. Selective advantages conferred by the high performance physiology of tunas, billfishes, and dolphin fish. *Comparative Biochemistry and Physiology* 113A(1):3–15.

Brill, R. W., K. A. Bigelow, M. K. Musyl, K. R. Fritsches, and E. J. Warrant. 2005. Bigeye tuna *(Thunnus obesus)* behavior and physiology and their relevance to stock assessments and fishery biology. *Col.Vol.Sci.Pap.ICCAT* 57(2):142–61.

Brill, R. W., B. A. Block, C. H. Boggs, K. A. Bigelow, E. V. Freund, and D. J. Marcinek. 1999. Horizontal movements and depth distribution of large adult yellowfin tuna *(Thunnus albacares)* near the Hawaiian Islands, recorded using ultrasonic telemetry: Implications for the physiological ecology of pelagic fishes. *Marine Biology* 133:395–408.

Brill, R. W., and P. G. Bushnell. 2001. The cardiovascular system of tunas. In B. A. Block and E. D. Stevens, eds., *Tuna: Physiology, Ecology, and Evolution,* 79–120. Academic Press.

Brill, R. W., and M. E. Lutcavage. 2001. Understanding environmental influences on movements and depth distributions of tunas and billfishes can significantly improve population assessments. *American Fisheries Society Symposium* 25:179–98.

Brill, R. W., M. Lutcavage, G. Metzger, P. Bushnell, M. Arendt, L. Lucy, C. Watson, and D. Foley. 2001. Horizontal and vertical movements of juvenile bluefin tuna *(Thunnus thynnus),* in relation to oceanographic conditions of the western North Atlantic, determined by ultrasonic telemetry. *Fishery Bulletin* 100:155–67.

Browder, J. A., B. E. Brown, and M. L. Parrack. 1991. The U.S. longline fishery for yellowfin tuna in perspective. *Col.Vol.Sci.Pap.ICCAT* 36:223–40

Browder, J. A., and G. P. Scott. 1992. History of the western Atlantic U.S. yellowfin fishery. *Col.Vol.Sci.Pap.ICCAT* 38:195–202.

Bruni, F. 2004. Sushi at Masa is a Zen thing. *New York Times,* December 29.

Buck, E. H. 1995. *Atlantic Bluefin Tuna: International Management of a Shared Resource.* Congressional Research Service. Washington, D.C.

Bullis, H. R., and F. J. Mather. 1956. Tunas of the genus *Thunnus* of the northern Caribbean. *American Museum Novitates* 1765:1–12.

Bundy, A. 2001. Fishing on ecosystems: The interplay of fishing and predation in Newfoundland-Labrador. *Canadian Journal of Fisheries and Aquatic Sciences* 58(6):1153–67.

Bundy, A., G. R. Lilly, and P. A. Shelton. 2000. A new balance model of the Newfoundland-Labrador Shelf. *Canadian Technical Report of Fisheries and Aquatic Sciences* 2301:1–157.

Burros, M. 2003. Issues of purity and pollution leave farmed salmon looking less rosy. *New York Times,* May 28.

———. 2006. Advisories on fish and pitfalls of good intent. *New York Times,* February 15.

———. 2008. High mercury levels are found in tuna sushi sold in Manhattan. *New York Times,* January 23.

Bushnell, P. G., and D. R. Jones. 1994. Cardiovascular and respiratory physiology of tuna: Adaptations for support of exceptionally high metabolic rates. *Environmental Biology of Fishes* 40(3):303–18.

Butler, M. J. A. 1977. The trap (mackerel) and impoundment (bluefin) fishery in St. Margaret's Bay, Nova Scotia: Its development. *Col.Vol.Sci.Pap.ICCAT* 6(2):237–41.

———. 1978. St. Margaret's Bay Bluefin Research Programme: A progress report. *Col.Vol.Sci.Pap.ICCAT* 7(2):371–74.

———. 1982. Plight of the bluefin tuna. *National Geographic* 162(2):220–39.

Butler, M. J. A., and J. A. Mason. 1978. Behavioural studies on impounded bluefin tuna. *Col.Vol.Sci.Pap.ICCAT* 7(2):379–82.

Butler, M. J. A., and D. Pincock. 1978. Ultrasonic monitoring of bluefin tuna impounded in St. Margaret's Bay. *Col.Vol.Sci.Pap.ICCAT* 7(2):375–78.

Byatt, A., A. Fothergill, and M. Holmes. 2001. *The Blue Planet.* BBC Worldwide Limited.

Campbell, D., D. Brown, and T. Battaglene. 2000. Individual transferable catch quotas: Australian experience in the southern bluefin tuna fishery. *Marine Policy* 24:109–17.

Carey, F. G. 1973. Fish with warm bodies. *Scientific American* 228(2):36–44.

Carey, F. G., J. G. Casey, H. L. Pratt, D. Urquhart, and J. E. McCosker. 1985. Temperature, heat production and heat exchange in lamnid sharks. *Memoirs of the Southern California Academy of Sciences* 9:92–108.

Carey, F. G., and Q. H. Gibson. 1983. Heat and oxygen exchange in the retia mirabile of the bluefin tuna, *Thunnus thynnus. Comparative Biochemistry and Physiology* 74(2):333–42.

Carey, F. G., J. W. Kanwisher, O. Brazier, G. Gabrielson, J. G. Casey, and H. L. Pratt. 1982. Temperature and activities of a white shark, *Carcharodon carcharias. Copeia* 1982(2):254–60.

Carey, F. G., J. W. Kanwisher, and E. D. Stevens. 1984. Bluefin tuna warm their viscera during digestion. *Journal of Experimental Biology* 109:1–20.

Carey, F. G., and K. D. Lawson. 1973. Temperature regulation in free-swimming bluefin tuna. *Comparative Biochemistry and Physiology* 44(2):375–78.

Carey, F. G., and R. J. Olson. 1982. Sonic tracking experiments with tunas. *Col.Vol.Sci.Pap.ICCAT* 17(2):458–66.

Carey, F. G., and J. M. Teal. 1966. Heat conservation in tuna fish muscle. *Proceedings of the National Academy of Sciences* 56(5):1464–69.

Carroll, M. T., J. L. Anderson, and J. Martínez-Garmendia. 2001. Pricing U.S. North Atlantic bluefin tuna and implications for management. *Agribusiness* 17(2):243–54.

Carson, R. L. 1943. Food from the sea: Fish and shellfish of New England. *U.S. Department of the Interior Conservation Bulletin* 33:1–74.

Castro, J. J., J. A. Santiago, and A. T. Santana-Ortega. 2002. A general theory on fish aggregation to floating objects: An alternative to the meeting point hypothesis. *Reviews in Fish Biology and Fisheries* 11:255–77.

Caton, A. E. 1993a. Commercial and recreational components of the southern bluefin tuna *(Thunnus maccoyii)* fishery. In R. S. Shomura, J. Majkowski, and S. Langi, eds., *Interactions of Pacific Tuna Fisheries, vol. 2 of Papers on Biology and Fisheries.* FAO, Rome.

———. 1993b. Review of aspects of southern bluefin tuna biology, population, and fisheries. In R. S. Shomura, J. Majkowski, and S. Langi, eds., *Interactions of Pacific Tuna Fisheries, vol. 2 of Papers on Biology and Fisheries.* FAO, Rome.

———. 2000. The Commonwealth-managed fisheries of Australia: An overview. *Infofish* 6:58–62.

Cayre, P. 1991. Behaviour of yellowfin tuna *(Thunnus albacares)* and skipjack tuna *(Katsuwonus pelamis)* around fish aggregating devices (FADs) in the Comoros Islands as determined by ultrasonic tagging. *Aquatic Living Resources* 4(1):1–12.

Chase, B. C. 2002. Differences in diet of Atlantic bluefin tuna *(Thunnus thynnus)* at five seasonal feeding grounds on the New England continental shelf. *Fisheries Bulletin* 100:168–80.

Chavez, F. P., J. Ryan, S. E. Lluch-Cota, and M. Ñiquen C. 2003. From anchovies to sardines and back: Multidecadal change in the Pacific Ocean. *Science* 299:217–21.

Cheshire, A., and J. Volkman. 2004. Australians net benefits of sustainable fish farming. *Nature* 432:671.

Childers, J., and S. Aalbers. 2006. Summary of the 2005 U.S. North and South Pacific albacore troll fisheries. *NMFS Southwest Fisheries Center Administrative Report* LJ-06-06:1–28.

Chivers, C. J. 1998. Empty waves. *Wildlife Conservation* 101(4):36–44.

Chow, S., H. Okamoto, N. Miyabe, K. Hiramatsu, and N. Barut. 2000. Genetic divergence between Atlantic and Indo-Pacific stocks of bigeye tuna *(Thunnus obesus)* and admixture around South Africa. *Molecular Ecology* 9(2):221–43.

Christensen, V., S. Guénette, J. J. Heymans, C. J. Walters, R. Watson, D. Zeller, and D. Pauly. 2002. Estimating fish abundance of the North Atlantic, 1950 to 1999. In S. Guénette, V. Christensen, and D. Pauly, eds., *Fisheries Impacts on North Atlantic Ecosystems: Models and Analyses. Fisheries Centre Research Reports* 9(4):1–25.

Clarkson, T. W. 1998. Human toxicology of mercury. *Journal of Trace Elements in Experimental Medicine* 11(2-3):303–17.

———. 2002. The three modern faces of mercury. *Environmental Health Perspectives* 110(Suppl. 1):11–23.

Clarkson, T. W., and J. J. Strain. 2003. Nutritional factors may modify the toxic action of methyl mercury in fish-eating populations. *Journal of Nutrition* 133:1539–43.

Clayton, M. 2006. New questions about safety of tuna imports. *Christian Science Monitor,* July 12.

Clean Seas. 2005. *Prospectus.* Clean Seas Tuna Ltd.

———. 2006. *Annual Report.* Clean Seas Tuna Ltd.

Clemens, H. B. 1961. The migration, age, and growth of Pacific albacore *(Thunnus germo)* 1951–1958. *California Fish and Game Fish Bulletin* 115:1–128.

Clover, C. 2008. Bluefin tuna—magnificent fish too valuable to save. *The Telegraph,* November 27.

Clover, C. 2006. *The End of the Line: How Overfishing Is Changing the World and What We Eat.* New Press.

Coan, A. L., G. T. Sakagawa, D. Prescott, P. Williams, and G. Yamasaki. 1999. *The 1998*

U.S. Tropical Tuna Purse Seine Fishery in the Central Western Pacific Ocean. South Pacific Regional Tuna Treaty, March 24–30, 1999. Koror, Republic of Palau.

Coghlan, A. 2002. Extreme mercury levels revealed in whalemeat. *New Scientist* 173:11.

———. 2003. Shops in Japan are selling mercury-ridden dolphin flesh as whalemeat. *New Scientist* 178:7.

———. 2006. Diabetes spotlight falls on fish. *New Scientist* 197:18.

Collette, B. B., and C. E. Nauen. 1983. *Scombrids of the World: An Annotated and Illustrated Catalogue of Tunas, Mackerels, Bonitos, and Related Species Known to Date.* FAO Fisheries Synopsis 125(2), Rome.

Collette, B. B., C. Reeb, and B. A. Block. 2001. Systematics of the tunas and mackerels (scombridae). In B. A. Block and E. D. Stevens, eds., *Tuna: Physiology, Ecology, and Evolution,* 5–33. Academic Press.

Cone, M. 2006. Warning on tuna cans is rejected. *Los Angeles Times,* May 13.

Corriero, A., S. Karakulak, N. Santamaria, M. Deflorio, D. Spedicato, P. Addis, S. Desantis, F. Cirillo, A. Fenech-Farrugia, R. Vassallo-Agius, J. M. de la Serna, Y. Oray, A. Cau, P. Megalofonou, and G. Metrio. 2005. Size and age at sexual maturity of female bluefin tuna (*Thunnus thynnus* L. 1758) from the Mediterranean Sea. *Journal of Applied Ichthyology* 21(6):483–86.

Corson, T. 2007. *The Zen of Fish: The Story of Sushi from Samurai to Supermarket.* HarperCollins.

Cort, J. L. 1991. Age and growth of the bluefin tuna *Thunnus thynnus* (L.) of the northeast Atlantic. *Col.Vol.Sci.Pap.ICCAT* 35(2):213–30.

Cort, J. L., and F. X. Bard. 1980. Description of the bluefin tuna fishery in the Bay of Biscay. *Col.Vol.Sci.Pap.ICCAT* 11:168–73.

Cousteau, J.-Y., and F. Dumas. 1953. *The Silent World.* Pocket Books.

Cury, P., O. Anneville, F. X. Bard, A. Fonteneau, and C. Roy. 1998. Obstinate North Atlantic bluefin tuna *(Thunnus thynnus):* An evolutionary perspective to consider spawning migration. *Col.Vol.Sci.Pap.ICCAT* 50(1):239–47.

Cushing, D. H. 1988. *The Provident Sea.* Cambridge University Press.

Dagorn, L., P. Bach, and E. Josse. 2004. Movement patterns of large bigeye tuna *(Thunnus obesus)* in the open ocean, determined using ultrasonic telemetry. *Marine Biology* 136(2):361–71.

Dagorn, L., and P. Fréon. 1999. Tropical tuna associated with floating objects: A simulation study of the meeting point hypothesis. *Canadian Journal of Fisheries and Aquatic Sciences.* 56(6):984–93.

Dagorn, L., E. Josse, and P. Bach. 2000. Individual differences in horizontal movements of yellowfin tuna *(Thunnus albacares)* in nearshore areas in French Polynesia, determined using ultrasonic telemetry. *Aquatic Living Resources* 13(4):193–202.

Dai, X., L. Zhao, and L. Xu. 2003. A review of available bluefin tuna information for China: 1994–2001. *Col.Vol.Sci.Pap.ICCAT* 55(3):1233–41.

Dalton, R. 2002. US pushes fish farming into deep water. *Nature* 420:451.

———. 2004a. Plan to cull aquarium tuna dead in the water. *Nature* 431:233.

———. 2004b. Fishing for trouble. *Nature* 431:502–04.

———. 2004c. Plans to track tuna canned amid claims of cash shortfall. *Nature* 432:539.

———. 2005a. Satellite tags give fresh angle on tuna quota. *Nature* 434:1056–57.

———. 2005b. Bill on deep-sea fish farms brings wave of disapproval. *Nature* 435:1014.

———. 2006. Fish futures. *Conservation in Practice* 7(3):22–27.

Darby, A. 2006. Revealed: how Japan caught and hid $2 billion worth of rare tuna. *Sydney Morning Herald,* August 12.

Davidson, P. W., G. J. Meyers, C. Cox, C. Axtell, C. Shamlaye, J. Sloan-Reeves, E. Cernichiari, L. Needham, A. Choi, Y. Wang, M. Berlin, and T. W. Clarkson. 1998. Effects of prenatal and postnatal methylmercury exposure from fish consumption on neurodevelopment: Outcomes at 66 months of age in the Seychelles Development Study. *Journal of the American Medical Association* 280(8):701–07.

Davidson, S. 2008. Looks good for tuna fingerlings in 2009. *Port Lincoln Times,* November 20.

Davis, J. E. 2005. Not a fish story. *San Francisco Chronicle,* August 3.

Davis, T. L. O., and J. H. Farley. 2001. Size distribution of southern bluefin tuna *(Thunnus maccoyii)* by depth on their spawning ground. *Fishery Bulletin* 99:381–86.

Davis, T. L. O., and C. A. Stanley. 2002. Vertical and horizontal movements of southern bluefin tuna *(Thunnus maccoyii)* in the Great Australian Bight observed with ultrasonic telemetry. *Fishery Bulletin* 100:448–65.

Dean, C. 2006a. Fish farms also harbor deadly lice. *New York Times,* October 3.

———. 2006b. Study sees "global collapse" of fish species. *New York Times,* November 3.

Debelle, P. 2006. Fishy behavior doesn't worry the millionaires of Port Lincoln. *The Age,* August 19.

Dekker, A., and K. Truelove. 1994. Australia plays a leading role in SBT Convention. *Australian Fisheries,* July 1994:6–7.

de Leiva, J. I., and J. Majkowski, 2005. *Tuna resources.* FAO, Rome.

De Metrio, G., G. P. Arnold, J. M. de la Serna, B. A. Block, P. Megalofonou, M. Lutcavage, I. Oray, and M. DeFlorio. 2005. Movements of bluefin tuna (*Thunnus thynnus* L.) tagged in the Mediterranean Sea with pop-up satellite tags. *Col.Vol.Sci.Pap.ICCAT* 58(4):1337–40.

De Metrio, G., I. Oray, G. P. Arnold, M. Lutcavage, M. DeFlorio, J. L. Cort, S. Karakulak, N. Anbar, and M. Ultanur. 2004. Joint Turkish-Italian research in the eastern Mediterranean: Bluefin tuna tagging with pop-up satellite tags. *Col.Vol.Sci.Pap.ICCAT* 56(3):1163–67.

Delgado, C. L., N. Wada, M. W. Rosegrant, S. Meijer, and M. Ahmed. 2003. *The Future of Fish: Issues and Trends to 2020.* International Food Policy Research Institute.

Dewar, H., and J. B. Graham. 1994a. Studies of tropical tuna swimming in a large water tunnel. I. Energetics. *Journal of Experimental Biology* 192:13–31.

———. 1994b. Studies of tropical tuna swimming in a large water tunnel. III. Kinematics. *Journal of Experimental Biology* 192:45–59.

Dewar, H., J. B. Graham, and R. W. Brill. 1994. Studies of tropical tuna swimming in a large water tunnel. II. Thermoregulation. *Journal of Experimental Biology* 192:33–44.

Dickson, K. A. 1995. Unique adaptations of the metabolic chemistry of tunas and billfishes for life in the pelagic environment. *Environmental Biology of Fishes* 42(1):65–97.

Dickson, K. A., and J. B. Graham. 2004. Evolution and consequences of endothermy in fishes. *Physiological and Biochemical Zoology* 77(6):998–1018.

Dizon, A. E., and R. W. Brill. 1979. Thermoregulation in tunas. *American Zoologist* 19(1):249–65.

Donley, J. M., C. A. Sepulveda, P. Konstantinitis, S. Gemballa, and R. E. Shadwick. 2004. Convergent evolution in mechanical design of lamnid sharks and tunas. *Nature* 429:31–33.

Donoghue, M., R. R. Reeves, and G. S. Stone, eds. 2003. *Report of the Workshop on Interaction Between Cetaceans and Longline Fisheries.* New England Aquarium Press.

Earle, S. A. 1991. Sharks, squids, and horseshoe crabs—the significance of marine biodiversity. *BioScience* 41(7):506–09.

———. 1995. *Sea Change: A Message of the Oceans.* Fawcett Columbine.

Ellis, R. 1982. *Dolphins and Porpoises.* Knopf.

———. 1987. Australia's southern seas. *National Geographic* 171(3):286–319.

———. 2003a. *The Empty Ocean.* Island Press.

———. 2003b. End of the line: Eating the tuna out of existence. *The Ecologist* 33(8):43–55.

———. 2005a. *Tiger Bone & Rhino Horn: The Destruction of Wildlife for Traditional Chinese Medicine.* Island Press.

———. 2005b. *Singing Whales and Flying Squid: The Discovery of Marine Life.* Lyons Press.

Essington, T. E., D. E. Schindler, R. J. Olson, J. F. Kitchell, C. Boggs, and R. Hilborn. 2002. Alternative fisheries and the predation rate of yellowfin tuna in the eastern Pacific Ocean. *Ecological Applications* 12(3):724–34.

Fabricant, F. 2001. Sashimi that isn't: A star is born. *New York Times,* September 12.

Fackler, M. 2007. Waiter, there's deer in my sushi. *New York Times,* June 25.

Faiella, G. 2003. *Fishing in Bermuda.* Macmillan Caribbean.

Farley, J. H., and T. L. O. Davis. 1998. Reproductive dynamics of southern bluefin tuna, *Thunnus maccoyii. Fishery Bulletin* 96(2):223–36.

Farrington, S. K. 1937. *Atlantic Game Fishing.* Kennedy Brothers.

———. 1942. *Pacific Game Fishing.* Coward-McCann.

———. 1949. *Fishing the Atlantic, Offshore and On.* Coward-McCann.

———. 1953. *Fishing the Pacific, Offshore and On.* Coward-McCann.

———. 1971. *Fishing with Hemingway and Glassell.* McKay.

———. 1974. *The Trail of the Sharp Cup.* Dodd, Mead.

Farwell, C. J. 2001. Tunas in captivity. In B. A. Block and E. D. Stevens, eds., *Tuna: Physiology, Ecology, and Evolution,* 391–412. Academic Press.

Fierstine, H. L., and V. Walters. 1968. Studies in locomotion and anatomy of scombroid fishes. *Memoirs of the Southern California Academy of Sciences* 6:1–31.

Fisheries Agency of Japan. 2005. Report on the progress in the implementation of the measures to eliminate illegal unreported and unregulated large scale tuna longline fishing vessels. *Col.Vol.Sci.Pap.ICCAT* 58(5):1776–80.

Fisheries Data Analysis Division, Southeast Fisheries Center (NMFS). 1980. A description of the fishery for Atlantic bluefin tuna by United States fishermen. *Col.Vol.Sci.Pap.ICCAT* 11:186–92.

Fitch, J. E., and R. J. Lavenberg. 1971. *Marine Food and Game Fishes of California.* University of California Press.

Fonteneau, A., J. Ariz, D. Gaertner, V. Nordstrom, and P. Pillares. 2000. Observed changes in the species composition of tuna schools in the Gulf of Guinea between

1981 and 1999, in relation with the Fish Aggregating Device fishery. *Aquatic Living Resources* 13:253–57.

Fréon, P., and L. Dagorn. 2004. Review of fish associative behaviour: Toward a generalization of the meeting point hypothesis. *Reviews in Fish Biology and Fisheries* 10(2):183–207.

Fritsches, K. A., and E. Warrant. 2004. Do tuna and billfish see colors? *PFRP Newsletter* 9(1):1–4.

Fromentin, J.-M. 2003. The East Atlantic and Mediterranean bluefin tuna stock management: Uncertainties and alternatives. *Scientia Marina* 67(Suppl. 1):51–62.

———. 2006. Size limits regulation for tuna: Should we also consider the protection of large fish? *Col.Vol.Sci.Pap.ICCAT* 59(2):590–92.

Fromentin, J.-M., H. Farrugio, M. DeFlorio, and G. De Metrio. 2003. Preliminary results of aerial surveys of bluefin tuna in the western Mediterranean Sea. *Col.Vol.Sci.Pap.ICCAT* 55(3):1019–27.

Fromentin, J.-M., and A. Fonteneau. 2001. Fishing effects and life history traits: A case study comparing tropical versus temperate tunas. *Fisheries Research* 53(2):133–50.

Fromentin, J.-M., A. Fonteneau, and V. Restrepo. 2006. Ecosystem approach to fisheries: A brief overview and some considerations for its application to ICCAT. *Col.Vol.Sci.Pap.ICCAT* 59(2):682–89.

Fromentin, J.-M., and J. E. Powers. 2005. Atlantic bluefin tuna: Population dynamics, ecology, fisheries and management. *Fish and Fisheries* 6(4):281–98.

Fromentin, J.-M., and C. Ravier. 2005. The East Atlantic and Mediterranean bluefin tuna stock: Looking for sustainability in a context of large uncertainties and strong political pressures. *Bulletin of Marine Science* 76(2):353–61.

Galaz, T., and A. De Maddalena. 2004. On a great white shark, *Carcharodon carcharias* (Linnaeus, 1758), trapped in a tuna cage off Libya, Mediterranean Sea. *Annales, Series historia naturalis* 14(2):159–164.

Garcia, A., J. M. de la Serna, J. L. López-Jurado, F. Alemany, and E. Rodriguez-Marín. 2002. Bluefin tuna egg and larval survey in the Balearic Sea June 2001 (TUNIBAL 06/01). *Col.Vol.Sci.Pap.ICCAT* 54(2):425–31.

Garcia, S. M., and R. J. R. Grainger. 2005. Gloom and doom? The future of marine capture fisheries. *Philosophical Transactions of the Royal Society B.* 360:21–46.

Garrell, M. H. 1993. The pursuit of bluefins meets reality. *Sea Frontiers* 31(6):18–21.

Gerrodette, T. 2002. Tuna-dolphin issue. In W. F. Perrin, B. Würsig, and J. G. M. Thewissen, eds., *Encyclopedia of Marine Mammals,* 1269–73. Academic Press.

Gerrodette, T., and J. Forcada. 2005. Non-recovery of two spotted and spinner dolphin populations in the eastern tropical Pacific Ocean. *Marine Ecology Progress Series* 291:1–21.

Goadby, P. 1975. *Big Fish and Blue Water: Gamefishing in the Pacific.* Angus & Robertson. Sydney.

Godsil, H. C. 1938. The high seas tuna fishery of California. *California Fish and Game Fish Bulletin* 51:1–44.

Goldburg, R., and R. Naylor. 2005. Future seascapes, fishing, and fish farming. *Frontiers in Ecology and Environment* 3(1):21–28.

Goldwater, L. J. 1972. *Mercury: A History of Quicksilver.* York Press.

Gorton, S. 2005. Fishery shut, 19 dolphins killed. *Port Lincoln Times,* August 30.

Graham, J. B., H. Dewar, N. C. Lai, W. R. Lowell, and S. M. Arce. 1990. Aspects of shark swimming performance determined using a large water tunnel. *Journal of Experimental Biology* 151:175–92.

Graham, J. B., and K. A. Dickson. 2001. Anatomical and physiological specializations for endothermy. In B. A. Block and E. D. Stevens, eds., *Tuna: Physiology, Ecology, and Evolution,* 121–65. Academic Press.

———. 2004. Tuna comparative physiology. *Journal of Experimental Biology* 207:4015–24.

Graham-Rowe, D. 2006. China and its friends push tiger farm idea. *New Scientist* 92:16–17.

Gray, J. 1933. Studies in animal locomotion. I. The movements of fish with special reference to the eel. *Journal of Experimental Biology* 10:88–104.

Greenberg, P. 2006a. Green to the gills. *New York Times Magazine,* June 18:32–39.

———. 2006b. Other fish to fry. *New York Times,* September 8.

Grey, L. 1982. *Angler's Eldorado: Zane Grey in New Zealand.* Walter J. Black.

Grey, R. C. 1930. *Adventures of a Deep-Sea Angler.* Harper & Brothers. 2002 edition, Derrydale Press.

Grey, Z. 1919. *Tales of Fishes.* Harper & Brothers. 1990 edition, Derrydale Press.

———. 1925. *Tales of Fishing Virgin Seas.* Harper & Brothers.

———. 1927. *Tales of Swordfish and Tuna.* Harper & Brothers.

———. 1931. *Tales of Tahitian Waters.* Harper & Brothers. 1990 edition, Derrydale Press.

Gruber, S. H., D. I. Hamasaki, and B. L. Davis. 1975. Window to the epiphysis in sharks. *Copeia* 2:378–80.

Guerra, A. S. 1980. Description of the bluefin tuna fishery *(Thunnus thynnus)* in the Canary Islands. *Col.Vol.Sci.Pap.ICCAT* 11:174–77.

Gunn, J., and B. A. Block. 2001. Advances in acoustic, archival and satellite tagging of tunas. In B. A. Block and E. D. Stevens, eds., *Tuna: Physiology, Ecology, and Evolution,* 167–224. Academic Press.

Günther, A.C.L.G. 1880. *An Introduction to the Study of Fishes.* Adam and Charles Black.

Hampton, J., A. Lewis, and P. Williams. 2002. *The Western and Central Pacific Tuna Fishery: 2000 Overview and Status of Stocks.* Tuna Fisheries Assessment Report/Secretariat of the Pacific Community. Oceanic Fisheries Programme Report No. 3.

Hampton, J., J. R. Sibert, P. Kleiber, M. N. Maunder, and S. J. Harley. 2005. Decline of Pacific tuna exaggerated? *Nature* 434:E1–2.

Hanrahan, B., and F. Juanes. 2001. Estimating the number of fish in Atlantic bluefin tuna *(Thunnus thynnus)* schools using models derived from captive school observations. *Fishery Bulletin* 99:420–31.

Harada, Y. 1999. A flag-waving squabble. *Sumudra* 24:30–31.

———. 2002. Tuning tuna. *Sumudra* 33:32–34.

Harder, B. 2003. Whale meat in Japan is loaded with mercury. *Science News* 163(23):365.

Hardin, G. 1968. The tragedy of the commons. *Science* 162:1243–48.

Hayden, T. 2003. Empty oceans: Why the world's seafood supply is disappearing. *U.S. News and World Report* 134(20):38–45.

Hayes, E. A. 1997. *A Review of the Southern Bluefin Tuna Fishery: Implications for Ecologically Sustainable Management.* TRAFFIC Oceania.

Hazin, F. H. V., M. K. Broadhurst, and H. G. Hazin. 2000. Preliminary analysis of the feasibility of transferring new longline technology to small artisanal vessels off northeastern Brazil. *Marine Fisheries Review* 62(1):27–34.

Hazin, F. H. V., J. R. Zagaglia, M. K. Broadhurst, P. E. P. Travassos, and T. R. Q. Bezerra. 1998. Review of small-scale pelagic longline fishery off northeastern Brazil. *Marine Fisheries Review* 60(3):1–8.

Hearn, W. S., and T. Polacheck. 2003. Estimating long-term growth-rate changes of southern bluefin tuna *(Thunnus maccoyii)* from two periods of tag-return data. *Fishery Bulletin* 101:58–74.

Heilner, V. C. 1953. *Salt Water Fishing.* Knopf.

Hempel, G., and D. Pauly. 2002. Fisheries and fisheries science in their search for sustainability. In J. G. Field, G. Hempel, and C. P. Summerhayes, eds., *Oceans 2020: Science, Trends, and the Challenge of Sustainability,* 109–35. Island Press.

Hespe, M. 2004. Bluefin Tuna: Australian Port Lincoln's Crop. *Japan Times,* September 17.

Hester, F. 2002. Atlantic bluefin tuna: Some considerations on mixing on the feeding grounds. *Col.Vol.Sci.Pap.ICCAT* 54(2):400–06.

Hibbeln, J. R. 1998. Fish consumption and major depression. *The Lancet* 351:1213.

Hibbeln, J. R., J. M. Davis, C. Steer, P. Emmett, I. Rogers, C. Williams, and J. Golding. 2007. Maternal seafood consumption in pregnancy and neurodevelopmental outcomes in childhood (ALSPAC study): An observational cohort study. *The Lancet* 369(9561):578–85.

Hightower, J. M. 2008. *Diagnosis: Mercury: Money, Politics & Poison.* Island Press.

Hilborn, R., T. A. Branch, B. Ernst, A. Magnusson, C. V. Minte-Vera, M. D. Scheuerell, and J. L. Valero. 2003. State of the world's fisheries. *Annual Review of Environment and Resources* 28:359–99.

Hisada, K., C. Shingu, and T. Yonemori. 1989. Recent status of the southern bluefin tuna stock. *Col.Vol.Sci.Pap.ICCAT* 8(2):454–60.

Hites, R. A., J. A. Foran, D. O. Carpenter, M. C. Hamilton, B. A. Knuth, and S. J. Schwager. 2004. Global assessment of organic contaminants in farmed salmon. *Science* 303:226–29.

Hobsbawn, P. I., J. D. Findlay, S. Rowcliffe, and A. Bodsworth. 2005. Australia's annual review of the southern bluefin tuna fishery. Working Paper CCSBT-EC/051/SBT Fisheries Australia, presented at the Second Meeting of the Extended Commission for the Conservation of Southern Bluefin Tuna. Attachment 8-1:1–13.

Holland, K. N., R. W. Brill, and R. K. C. Chang. 1990. Horizontal and vertical movements of yellowfin and bigeye tuna associated with fish aggregating devices. *Fishery Bulletin* 88(3):493–507.

Holland, K. N., R. W. Brill, R. K. C. Chang, J. R. Sibert, and D. A. Fournier. 1992. Physiological and behavioral thermoregulation in bigeye tuna *(Thunnus obesus). Nature* 358:410–12.

Holland, K. N., P. Kleiber, and S. M. Kajiura. 1999. Different residence times of yellowfin tuna, *Thunnus albacares,* and bigeye tuna, *T. obesus,* found in mixed aggregations over a seamount. *Fishery Bulletin* 97:392–95.

Holland, K. N., and J. R. Sibert. 2005. Physiological thermoregulation in bigeye tuna, *Thunnus obesus. Environmental Biology of Fishes* 40(3):317–27.

Holloway, M. 2002. Blue revolution (fish farming). *Discover* 23(9):57–63.

Holt, S. J. 1969. The food resources of the ocean. *Scientific American* 221(2):178–94.

Honma, M. 1973. Overall fishing intensity and catch by length class of yellowfin tuna in Japanese Atlantic longline fishery, 1956–1970. *Col.Vol.Sci.Pap.ICCAT* 1:59–77.

———. 1976. Catch statistics of Japanese Atlantic purse seine fishery, 1974 and 1975. *Col.Vol.Sci.Pap.ICCAT* 5(1):23–25.

Honma, M., and Z. Suzuki. 1974a. Catch statistics and sample length composition in Japanese Atlantic tuna purse seine fishery, 1967–1969. *Col.Vol.Sci.Pap.ICCAT* 2:10–14.

———. 1974b. Catch statistics and sample length composition in Japanese Atlantic tuna purse seine fishery, 1971 and 1972, with a brief review of the fishery since 1964. *Col.Vol.Sci.Pap.ICCAT* 2:15–24.

Hori, T. 1996. *Tuna and the Japanese.* Japan External Trade Organization.

Hosaka, E. Y. 1944. *Sport Fishing in Hawaii.* Bond's.

Hsieh, C.-H., C. S. Reiss, J. R. Hunter, J. R. Beddington, R. M. May, and G. Sugihara. 2006. Fishing elevates variability in the abundance of exploited species. *Nature* 443:859–62.

Humston, R., J. S. Ault, M. E. Lutcavage, and D. B. Olson. 2000. Schooling and migration of large pelagic fishes relative to environmental data. *Fisheries Oceanography* 9:136–46.

Hurley, P. C. F., and T. D. Iles. 1980. A brief description of Canadian fisheries for Atlantic bluefin tuna. *Col.Vol.Sci.Pap.ICCAT* 11:93–97.

International Game Fish Association. 2006. *World Record Game Fishes.* IGFA. Dania Beach, Florida.

Issenberg, S. 2007. *The Sushi Economy.* Gotham.

Itano, D. G. 2000. *The Reproductive Biology of Yellowfin Tuna (Thunnus albacares) in Hawaiian Waters and the Western Tropical Pacific Ocean: Project Summary.* Joint Institute for Marine and Atmospheric Research Contribution 00-328.

Itoh, T. 2006. Sizes of adult bluefin tuna *Thunnus orientalis* in different areas of the western Pacific Ocean. *Fisheries Science* 72:53–62.

Itoh, T., S. Tsuji, and A. Nitta. 2003a. Migration patterns of young Pacific bluefin tuna *(Thunnus orientalis)* determined with archival tags. *Fishery Bulletin* 101(3):514–34.

———. 2003b. Swimming depth, ambient water temperature preference, and feeding frequency of young Pacific bluefin tuna *(Thunnus orientalis)* determined with archival tags. *Fishery Bulletin* 101(3):535–44.

Itzkoff, D. 2008. Piven leaves show amid concerns for his health. *New York Times,* December 19.

IUCN. 2006. *2006 IUCN Red List of Threatened Species.* www.iucnredlist.org. Downloaded on October 19, 2006.

Jackson, J. B. C., and K. G. Johnson. 2001. Measuring past biodiversity. *Science* 293:2401–03.

Jackson, J. B. C., M. X. Kirby, W. H. Berger, K. A. Bjorndal, L. W. Botsford, B. J. Bourque, R. H. Bradbury, R. Cooke, J. Erlandson, J. A. Estes, T. P. Hughes, S. Kidwell, C. B. Lange, H. S. Lenihan, J. M. Pandolfi, C. H. Peterson, R. S. Steneck, M. J. Tegner, and R. R. Warner. 2001. Historical overfishing and the recent collapse of coastal ecosystems. *Science* 293:629–38.

Japan's tuna crisis. 2007. *New York Times,* June 26.

Jordan, D. S., and B. W. Evermann. 1902. *American Food and Game Fishes.* Doubleday, Page.

———. 1926. A review of the giant mackerel-like fishes, tunnies, spearfishes, and swordfishes. *Occasional Papers of the California Academy of Sciences* 12:1–113.

Joseph, J. 1994. The tuna-dolphin controversy in the eastern Pacific Ocean: Biological, economic and political impacts. *Ocean Development and International Law* 25(1):1–30.

———. 1998a. A brief history of tuna research. *Col.Vol.Sci.Pap.ICCAT* 50(1): 183–93.

———. 1998b. On scientific advice and tuna management. *Col.Vol.Sci.Pap.ICCAT* 5(2):855–61.

Joseph, J., and J. W. Greenough. 1979. *International Management of Tuna, Porpoise, and Billfish.* University of Washington Press.

Joseph, J., W. Klawe, and P. Murphy. 1988. *Tuna and Billfish: Fish Without a Country.* Inter-American Tropical Tuna Commission.

Joseph, J., and D. Squires. 2003. Dolphin mortality in the eastern Pacific tuna fishery: a case study of bycatch issues in marine fisheries. Unpublished manuscript.

Josse, E., A. Bertrand, and L. Dagorn. 1999. An acoustic approach to study tuna aggregated around fish aggregating devices in French Polynesia: Methods and validation. *Aquatic Living Resources* 12(5):303–13.

Josse, E., L. Dagorn, and A. Bertrand. 2000. Typology and behaviour of tuna aggregations around fish aggregating devices from acoustic surveys in French Polynesia. *Aquatic Living Resources* 13:183–92.

Josupeit, H., and C. Catarci. 2004. The world tuna industry—an analysis of imports, prices, and of their combined impact on tuna catches and fishing capacity. FAO, Rome.

Kaji, T., M. Tanaka, Y. Takahishi, M. Oka, and N. Ishibashi. 1993. Preliminary observations on development of Pacific bluefin tuna *Thunnus thynnus thynnus* (Scombridae) raised in the laboratory, with special reference to the digestive system. *Marine and Freshwater Research* 47(2):261–69.

Karakulak, S., I. Oray, A. Corriero, A. Aprea, D. Spedicato, D. Zubani, N. Santamaria, and G. De Metrio. 2004. First information on the reproductive biology of the bluefin tuna *(Thunnus thynnus)* in the eastern Mediterranean. *Col.Vol.Sci.Pap .ICCAT* 56(3):1158–62.

Kataviæ, I., V. Tièina, and V. Franièviæ. 202. A preliminary study of the growth rate of bluefin tuna from Adriatic when reared in the floating cages. *Col.Vol.Sci.Pap.ICCAT* 54(2):472–76.

Katz, S. L., D. A. Syme, and R. E. Shadwick. 2001. Enhanced power in yellowfin tuna. *Nature* 410:770–71.

Kawamura, G., S. Masuma, N. Tezuka, M. Koiso, T. Jimbo, and K. Namba. 2003. Morphogenesis of sense organs in the bluefin tuna *Thunnus orientalis.* In H. I. Browman and A. B. Skiftesvik, eds., *The Big Fish Bang: Proceedings of the 26th Annual Larval Fish Conference, Os, Norway, 22–26 July 2002,* 123–35. Institute of Marine Research, Bergen.

Keith-Reid, R. 2002. Yellow fin tuna under threat. *Pacific,* February: 22–28.

Kessler, A. 1976. The hunt for tuna. *Oceans* 9(4):50–57.

Kitagawa, T., S. Kimura, H. Nakata, and H. Yamada. 2004. Overview of research on tuna thermo-physiology using electric tags. *Memoirs of National Polar Research* (Special Issue) 58:69–79.

Klawe, W., I. Barrett, and B. M. H. Klawe. 1963. Hæmoglobin content of the blood of six species of scombroid fishes. *Nature* 198:96.

Kleiner, K. 2002. All fished out. *New Scientist* 173(2331):11.

Klimley, A. P., and C. F. Holloway. 1999. School fidelity and homing synchronicity of yellowfin tuna, *Thunnus albacares. Marine Biology* 133:307–17.

Klimley, A. P., S. J. Jorgensen, A. Muhlia-Melo, and S. C. Beavers. 2003. The occurrence of yellowfin tuna *(Thunnus albacares)* at Espiritu Santo Seamount in the Gulf of California. *Fishery Bulletin* 101:684–92.

Kluger, J. 2006. Mercury rising. *Time* 168(10):52–55.

Kobayashi, N., R. J. Barnard, S. M. Henning, D. Elashoff, S. T. Reddy, P. Cohen, P. Leung, J. Hong-Gonzalez, S. J. Freedland, J. Said, D. Gui, N. P. Seeram, L. M. Popoviciu, D. Bagga, D. Heber, J. A. Glaspy, and W. J. Aronson. 2006. Effect of altering dietary omega-6/omega-3 fatty acid ratios on prostate cancer membrane composition, Cyclooxygenase-2, and Prostaglandin E_2. *Clinical Cancer Research* 12:4662–70.

Kondo, I. 2001. *Rise and Fall of the Japanese Coast Whaling.* Sanyosha.

Korsmeyer, K. E., and H. Dewar. 2001. Tuna metabolism and energetics. In B. A. Block and E. D. Stevens, eds., *Tuna: Physiology, Ecology, and Evolution,* 36–78. Academic Press.

Krkošek, M., M. A. Lewis, A. Morton, L. N. Frazer, and J. P. Volpe. 2006. Epizootics of wild fish induced by farm fish. *Proceedings of the National Academy of Sciences* 103:15506–15510.

Krkošek, M., A. Morton, and J. P. Volpe. 2005. Nonlethal assessment of juvenile pink and chum salmon for parasitic sea lice infections and fish health. *Transactions of the American Fisheries Society* 134:711–16.

Kummer, C. 2007. The rise of the sardine. *The Atlantic* 300(1):154–58.

Kwei, E. A. 1991. Pricing of yellowfin and other tunas. *Col.Vol.Sci.Pap.ICCAT* 36:506–14.

La Mesa, M., M. Sinopoli, and F. Andaloro. 2005. Age and growth rate of juvenile bluefin tina *(Thunnus thynnus)* from the Mediterranean Sea (Sicily, Italy). *Scientia Marina* 69(2):241–49.

La Monte, F. 1945. *North American Game Fishes.* Doubleday.

———. 1952. *Marine Game Fishes of the World.* Doubleday.

———. 1965. *Giant Fishes of the Open Sea.* Holt, Rinehart and Winston.

Lato, D. 2003. $6m tuna lost as cage collapses. *GROWfish.* www.growfish.com .au/grow/pages/News/2003/mar2003/56603.htm

Lee, M., ed. 2000. *Seafood Lover's Almanac.* National Audubon Society.

Leech, M. 2007. Bluefin tuna: A failure of management. *Marlin,* March:34–36.

Linthicum, D. S., and F. G. Carey. 1972. Regulation of brain and eye temperatures by the bluefin tuna. *Comparative Biochemistry and Physiology* 43(2):425–30.

Lipton, J. 1968. *An Exaltation of Larks.* Viking.

Lo, N. C. H., and T. D. Smith. 1986. Incidental mortality of dolphins in the Eastern tropical Pacific, 1959–72. *Fishery Bulletin* 84(1):27–34.

Loew, E. R., W. N. McFarland, and D. Margulies. 2002. Developmental changes in the visual pigments of the yellowfin tuna, *Thunnus albacares. Marine and Freshwater Behaviour and Physiology* 35:235–46.

Lowe, T. E., R. W. Brill, and K. L. Cousins. 2000. Blood-binding characteristics of bigeye tuna *(Thunnus obesus),* a high-energy demand teleost that is tolerant of low ambient oxygen. *Marine Biology* 136(6):1087–98.

Lund, E., D. Engeset, E. Alsaker, G. Skeie, A. Hjårtaker, A.-K. Lundebye, and E. Niebor. 2004. Cancer risk and salmon intake. *Science* 305:477.

Lutcavage, M. E. 2001. Bluefin spawning in Central North Atlantic? *Pelagic Fisheries Research Program Newsletter* 6(2):1–3.

Lutcavage, M. E., R. W. Brill, J. Porter, P. Howey, E. Murray, A. Mendillo, W. Chaprales, M. Genovese, and T. Rollins. 2001. Summary of pop-up satellite tagging of giant bluefin tuna in the joint US-Canadian program, Gulf of Maine and Canadian Atlantic. *Col.Vol.Sci.Pap.ICCAT* 52(1):759–70.

Lutcavage, M. E., R. W. Brill, G. B. Skomal, B. C. Chase, J. L. Goldstein, and J. Tutein. 2000. Tracking adult North American bluefin tuna *(Thunnus thynnus)* in the northwestern Atlantic using ultrasonic telemetry. *Marine Biology* 137(2):347–58.

Lutcavage, M. E., and S. Kraus. 1995. The feasibility of direct photographic assessment of giant bluefin tuna, *Thunnus thynnus,* in New England waters. *Fishery Bulletin* 93:495–503.

MacKenzie, B. R., and R. A. Myers. 2007. The development of the northern European fishery for North Atlantic bluefin tuna *Thunnus thynnus* during 1900–1950. *Fisheries Research* 77:229–239.

MacLeish, W. H. 1989. *The Gulf Stream: Encounters with the Blue God.* Houghton Mifflin.

Maggio, T. 2001. *Mattanza: The Ancient Sicilian Ritual of Bluefin Tuna Fishing.* Penguin.

Magnuson, J. J., C. Safina, and M. P. Sissenwine. 2001. Whose fish are they anyway? *Science* 293:1267–68.

Mahan, S. and J. Savitz. 2007. Cleaning up: taking mercury-free chlorine production to the bank. *Oceana,* July 2007.

Margulies, D., V. Scholey, S. Hunt, and J. Wexler. 2005. Achotines Lab studies diets for larval, juvenile yellowfin tuna. *Global Aquaculture Advocate* 8(2):87.

Margulies, D., V. P. Scholey, J. B. Wexler, R. J. Olson, A. Nakazawa, and J. M. Suter. 1997. Captive spawning of the yellowfin tuna and the development of their eggs and larvae. *Tuna Newsletter* (NMFS Southwest Fisheries Center) 126:4–5.

Marra, J. 2005. When will we tame the oceans? *Nature* 436:175–76.

Martínez-Garmendia, J., and J. L. Anderson. 2005. Conservation, markets, and fisheries policy: The North Atlantic bluefin tuna and the Japanese sashimi market. *Agribusiness* 21(1):17–36.

Martínez-Garmendia, J., J. L. Anderson, and M. T. Carroll. 2000. Effect of harvesting alternatives on the quality of U.S. North American bluefin tuna. *North American Journal of Fisheries Management* 20:908–22.

Mason, J. H. 1975. U.S. Atlantic bluefin tuna tagging, October 1971 through October 1974. *Col.Vol.Sci.Pap.ICCAT* 4:133–40.

Mather, F. J. 1954. Northerly occurrences of warmwater fishes in the Western Atlantic. *Copeia* 1954(4):292–93.

———. 1960. Recaptures of tuna, marlin and sailfish tagged in the western North Atlantic. *Copeia* 1960(2):149–51.

———. 1980a. Note on the relationship between recently acquired mark-recapture data and existing age estimates for Atlantic bluefin tuna. *Col.Vol.Sci.Pap.ICCAT* 9(2):470–77.

———. 1980b. A preliminary note on migratory tendencies and distributional patterns of Atlantic bluefin tuna based on recently acquired and cumulative tagging results. *Col.Vol.Sci.Pap.ICCAT* 9(2):478–90.

Mather, F. J., W. R. Bartlett, and J. S. Beckett. 1967. Transatlantic migrations of young bluefin tuna. *Journal of the Fisheries Research Board of Canada* 24(9):1991–97.

Mather, F. J., and C. G. Day. 1954. Observations of the pelagic fishes of the tropical Atlantic. *Copeia* 1954(3):179–88.

Mather, F. J., and R. H. Gibbs. 1957. Distributional records of fishes from the waters off New England and the Middle Atlantic states. *Copeia* 1957(3):242–44.

———. 1958. Distribution of the Atlantic bigeye tuna, *Thunnus obesus,* in the western North Atlantic and the Caribbean Sea. *Copeia* 1958(3):237–39.

Mather, F. J., and J. M. Mason. 1973. Recent information on tagging and tag returns for tunas and billfishes in the Atlantic Ocean. *Col.Vol.Sci.Pap.ICCAT* 1:501–31.

———. 1976. Results of United States cooperative tagging of Atlantic bluefin tuna: October 1974 through October 1975. *Col.Vol.Sci.Pap.ICCAT* 11:180–83.

Mather, F. J., J. H. Mason, and A. C. Jones. 1974. Distribution, fisheries and life history relevant to the identification of Atlantic bluefin tuna stocks. *Col.Vol.Sci.Pap .ICCAT* 2:234–58.

Mather, F. J., B. J. Rothschild, and G. J. Paulik. 1973. Preliminary analysis of bluefin tagging data. *Col.Vol.Sci.Pap.ICCAT* 1:413–44.

Mather, F. J., and H. A. Schuck. 1951. A blackfin tuna *(Parathunnus atlanticus)* from North Carolina waters. *Copeia* 1951(3):248.

———. 1952. Additional notes on the distribution of the blackfin tuna *(Parathunnus atlanticus). Copeia* 1952(4):267.

Matsen, B. 1990. *Deep Sea Fishing: The Lure of Big Game Fish.* Thunder Bay Press.

Matsuda, Y. 1998. History of the Japanese tuna fisheries and a Japanese perspective on Atlantic bluefin tuna. *Col.Vol.Sci.Pap.ICCAT* 50(2):733–51.

Matthiessen, P. 1959. *Wildlife in America.* Viking.

McClane, A. J. 1965. *McClane's Standard Fishing Encyclopedia.* Holt, Rinehart and Winston.

McDowell, N. 2002. Stream of escaped farm fish raises fear for wild salmon. *Nature* 416:571.

McGarry, A. 2008. Tuna firm breeds "holy grail." *The Australian,* March 5.

McGhee, K. 2004. Taking stock: Southern bluefin on the line. *ECOS* (*CSIRO Newsletter*) 119:24–27.

McPherson, G. R. 1991. Reproductive biology of yellowfin tuna in the eastern Australian Fishing Zone, with special reference to the north-western Coral Sea. *Marine and Freshwater Research* 42(5):465–77.

Migdalski, E. C. 1958. *Angler's Guide to the Salt Water Game Fishes, Atlantic and Pacific.* Ronald Press.

Migdalski, E. C., and G. S. Fichter. 1976. *The Fresh & Salt Water Fishes of the World.* Knopf.

Milius, S. 2000. Carnivorous fish nibble at farming gain. *Science News* 158(1):7.

Minasian, S. 1977. Dolphins and/or tuna. *Oceans* 10(3):60–63.

Mishima, A. 1992. *Bitter Sea: The Human Cost of Minimata Disease.* Kosei.

Miyabe, N. 2003. Description of the Japanese longline fishery and its fishery statistics in the Mediterranean Sea during recent years. *Col.Vol.Sci.Pap.ICCAT* 55(1):131–37.

Miyabe, N., M. Ogura, T. Matsumoto, and Y. Nishikawa. 2004. *National Tuna Fisheries Report of Japan as of 2004.* SCTB17 Working Paper NFR-11. National Research Institute of Far Seas Fisheries, Shizuoka.

Miyabe, N., and H. Okamoto. 2005. Documentation of data provision and processing for the Japanese tuna fisheries in the eastern Pacific Ocean. *IATTC Document* DC-1-02a:1-6.

Miyake, P. M. 1987. Historical review of Atlantic tuna catches and some thoughts on tuna management. *Col.Vol.Sci.Pap.ICCAT* 26(2):433–55.

———. 2006. Why do we have to control the global fishing capacity? *OPRT Newsletter* 10:1–4.

Miyake, P. M., J. M. de la Serna, A. Di Natale, A. Farrugia, I. Katavic, N. Miyabe, and V. Ticina. 2003. General review of bluefin tuna farming in the Mediterranean area. *Col.Vol.Sci.Pap.ICCAT* 55(1):114–24.

Miyake, P. M., N. Miyabe, and H. Nakano, 2004. *Historical Trends in Tuna Catches in the World.* FAO Fisheries Technical Paper No. 467. FAO, Rome.

Miyashita, S., Y. Sawada, N. Hattori, H. Nakatsukasa, T. Okada, O. Murata, and H. Kumai. 2000. Mortality of northern bluefin tuna *Thunnus thynnus* due to trauma caused by collision during growout culture. *Journal of the World Aquaculture Society* 31(4):632–39.

Montaigne, F. 2007. Still waters. *National Geographic* 211(4):42–51.

Morgan, R. 1955. *World Sea Fisheries.* Pitman.

Mori, M., T. Katsukawa, and H. Matsuda. 2001. Recovery plan for an exploited species, southern bluefin tuna. *Population Ecology* 43:125–32.

Mowat, F. 1984. *Sea of Slaughter.* Atlantic Monthly Press.

Moyle, P. B., and J. J. Cech. 2004. *Fishes: An Introduction to Ichthyology.* Prentice-Hall.

Munro, M. 2007. Tale of tagged tuna: One swims to Cuba, the other to Gibraltar. *Edmonton Journal Digital,* August 6. digital.edomontonjournal.com

———. 2007b. Tuna fishery like the last days of the buffalo. *Edmonton Journal Digital,* October 7. digital.edmontonjournal.com

Musyl, M. K., R. W. Brill, C. H. Boggs, D. S. Curran, T. K. Kazama, and M. P. Seki. 2003. Vertical movements of bigeye tuna *(Thunnus obesus)* associated with islands, buoys, and seamounts near the main Hawaiian islands from archival tagging data. *Fisheries Oceanography* 12(3):152–68.

Myers, R. A., K. G. Bowen, and N. J. Barrowman. 1999. Maximum reproductive rate of fish at low population sizes. *Canadian Journal of Fisheries and Aquatic Science* 56(12):2404–19.

Myers, R. A., and C. A. Ottensmeyer. 2005. Extinction risk in marine species. In E. Norse and L. B. Crowder, eds., *Marine Conservation Biology,* 58–79. Island Press.

Myers, R. A., and B. Worm. 2003. Rapid worldwide depletion of predatory fish communities. *Nature* 423:280–83.

———. 2005. Extinction, survival or recovery of large predatory fishes. *Philosophical Transactions of the Royal Society B.* 360:13–20.

Mylonas, C. 2005. Research breakthrough—saving the bluefin tuna. Hellenic Centre for Marine Research. www.hcmr.gr/english_site/news/archive/show_news2.php3?id=46

Nagai, T., S. Hayasi, and T. Yonemori. 1987. A proposal to management of bluefin tuna stock in the western Atlantic based on information obtained by September, 1986. *Col.Vol.Sci.Pap.ICCAT* 26(2):276–82.

National Research Council. 1992. *Dolphins and the Tuna Industry.* National Academy Press.

———. 1994. *An Assessment of Atlantic Bluefin Tuna.* National Academy Press.

Naylor, R. L. 2006a. Environmental safeguards for open-ocean aquaculture. *Issues in Science and Technology,* spring 2006:53–58.

———. 2006b. Offshore aquaculture legislation. *Science* 313:1363.

Naylor, R. L., and M. Burke. 2005. Aquaculture and ocean resources: Raising tigers of the sea. *Annual Review of Environment and Resources* 30:185–218.

Naylor, R. L., R. J. Goldburg, H. Mooney, M. C. M. Beveridge, J. Clay, C. Folke, N. Kautsky, J. Lubchenko, J. Primavera, and M. Williams. 1998. Nature's subsidies to shrimp and salmon farming. *Science* 282:883–84.

Naylor, R. L., R. J. Goldburg, J. H. Primavera, N. Kautsky, M. C. M. Beveridge, J. Clay, C. Folke, J. Lubchenko, H. Mooney, and M. Troell. 2000. Effect of aquaculture on world fish supplies. *Nature* 405:1017–24.

Neill, W. H., R. K. C. Chang, and A. E. Dizon. 1976. Magnitude and ecological implications of thermal inertia in skipjack tuna, *Katsuwonus pelamis* (Linnaeus). *Environmental Biology of Fishes* 1(1):61–80.

Nestle, M. 2006. *What to Eat.* North Point.

Newlands, N. K., M. E. Lutcavage, and T. J. Pitcher. 2004. Analysis of foraging movements of the Atlantic bluefin tuna *(Thunnus thynnus):* Individuals will switch between two modes of search behaviour. *Population Ecology* 46:39–53.

Nicoll, R. 1993. Port Lincoln carves out a bright future. *Australian Fisheries Newsletter,* November:14–20.

Nishida, T., S. Chow, and P. Grewe. 1998. Review and research plan on the stock structure of yellowfin tuna *(Thunnus albacares)* and bigeye tuna *(Thunnus obesus)* in the Indian Ocean. *Indian Ocean Tuna Commission Proceedings* 1:230–36.

Nootmorn, P. 2004. Reproductive biology of bigeye tuna in the eastern Indian Ocean. *Indian Ocean Tuna Commission Proceedings* 7:1–6.

Norman, J. R., and F. C. Fraser. 1938. *Giant Fishes, Whales and Dolphins.* Norton.

Norman, J. R., and P. H. Greenwood. 1963. *A History of Fishes.* Hill and Wang.

Northridge, S. P. 1991. *Driftnet Fisheries and their Impact on Non-Target Species: A Worldwide Review.* FAO Fisheries Technical Paper 320:1–115.

Olson, R. J., and V. P. Scholey. 1990. Captive tunas in a tropical marine research laboratory: Growth of late-larval and early-juvenile black skipjack *Euthynnus lineatus. Fishery Bulletin* 88:821–28.

Onishi, N. 2006. Farming bluefin tuna, through thick stocks and thin. *New York Times,* September 26:A4.

Oppian. c. A.D. 180. *Halieutica.* 1928 translation by A.W. Mair. Loeb Classical Library, Harvard University Press.

Oray, I. K., and F. S. Karakulak. 1997. Some remarks on the bluefin tuna (*Thunnus thynnus* L. 1758) fishery in Turkish waters in 1993, 1994, 1995. *Col.Vol.Sci.Pap .ICCAT* 46(2):357–62.

Parfit, M. 1995. Diminishing returns: Exploiting the ocean's bounty. *National Geographic* 188(5):2–37.

Parmentier, E., O. Colleye, M. L. Fine, B. Frédérich, P. Vandewalle, and A. Herrel. 2007. Sound production in the clownfish *(Amphiprion clarkii). Science* 316:1006.

Parrack, M. L., S. L. Brunenmeister, and S. Nichols. 1979. An analysis of Atlantic bluefin catches, 1960–1976. *Col.Vol.Sci.Pap.ICCAT* 8(2):391–420.

Parrack, M. L., and P. L. Phares. 1979. Aspects of the growth of the Atlantic bluefin tuna determined from mark-recapture data. *Col.Vol.Sci.Pap.ICCAT* 8(2):356–66.

Paul, L. 1997. *Marine Fishes of New Zealand.* Reed.

Paul, L. M. B. 1994. *High Seas Driftnetting: A Plunder of the Global Common.* Earthtrust.

Pauly, D. 1995. Anecdotes and the shifting baseline syndrome of fisheries. *Trends in Ecology and Evolution* 10(10):430.

———. 2007. Ransom Aldrich Myers (1952–2007). *Nature* 447:160.

Pauly, D., J. Alder, E. Bennett, V. Christensen, P. Tyedmers, and R. Watson. 2003. The future for fisheries. *Science* 302:1359–61.

Pauly, D., V. Christensen, J. Dalsgaard, R. Froese, and F. Torres. 1998. Fishing down marine food webs. *Science* 279:860–63.

Pauly, D., V. Christensen, R. Froese, and M.-L. Palomares. 2000. Fishing down aquatic food webs. *American Scientist* 88(1):46–51.

Pauly, D., and J. Maclean. 2002. *In a Perfect Ocean: The State of Fisheries and Ecosystems in the North Atlantic Ocean.* Island Press.

Pauly, D., and M.-L. Palomares. 2005. Fishing down marine food webs: It is far more pervasive than we thought. *Bulletin of Marine Science* 76(2):197–212.

Pauly, D., and R. Watson. 2003. Counting the last fish. *Scientific American* 289(1):42–47.

Perrin, W. F. 1968. The porpoise and the tuna. *Sea Frontiers* 14(3):166–74.

———. 1969. Using porpoise to catch tuna. *World Fishing* 18(6):42–45.

Peterson, C. L., W. L. Klawe, and G. D. Sharp. 1973. Mercury in tunas: A review. *Fishery Bulletin* 71(3):603–13.

Pierce, W. G. 1989. *Going Fishing: The Story of the Deep-Sea Fishermen of New England.* International Marine.

Pliny. First century A.D. *Natural History.* 10 vols. Loeb Classical Library, Harvard University Press.

Polovina, J. J. 1996. Decadal variation in the trans-Pacific migration of the northern bluefin tuna *(Thunnus thynnus)* coherent with climate-induced change in prey abundance. *Fisheries Oceanography* 5(2):114–19.

Porch, C. E. 2005. The sustainability of Western Atlantic bluefin tuna: A warm-blooded fish in a hot-blooded fishery. *Bulletin of Marine Science* 76(2):363–84.

Porch, C. E., and S. C. Turner. 1999. Virtual population analyses of Atlantic bluefin tuna with alternative models of transatlantic migration. *Col.Vol.Sci.Pap.ICCAT* 49(1):291–305.

Porch, C. E., S. C. Turner, and R. D. Methot. 1994. Estimates of the abundance and mortality of west Atlantic bluefin tuna using the stock synthesis model. *Col.Vol.Sci.Pap.ICCAT* 42(1):229–39.

Porch, C. E., S. C. Turner, and J. E. Powers. 2001. Virtual population analyses of Atlantic bluefin tuna with alternative models of transatlantic migration: 1970–1997. *Col.Vol.Sci.Pap.ICCAT* 52(1):1022–45.

Powell, K. 2003. Eat your veg. *Nature* 426:378–79.

Powers, J. E., and J. Cramer. 1996. An exploration of the nature of Atlantic tuna mixing. *Col.Vol.Sci.Pap.ICCAT* 45(2):173–81.

Preikshot, D., and D. Pauly. 2005. Global fisheries and marine conservation: Is coexistence possible? In E. Norse and L. B. Crowder, eds., *Marine Conservation Biology,* 185–97. Island Press.

Preston, G. L., L. B. Chapman, and P. G. Watt. 1998. *Vertical Longlining and Other Meth-*

ods of Fishing around Fish Aggregating Devices (FADs): A Manual for Fishermen. Secretariat of the Pacific Commission.

Radcliffe, W. 1921. *Fishing from the Earliest Times.* John Murray. 1974 reprint, Ares Publishers, Chicago.

Raloff, J. 2005. Empty nets. *Science News* 167(23):360–62.

Ravier, C., and J.-M. Fromentin. 2001a. Long-term fluctuations in the eastern Atlantic and Mediterranean bluefin tuna population. *ICES Journal of Marine Science* 58(6):1299–1317.

———. 2001b. Trends in NE Atlantic and Mediterranean bluefin abundance. *PFRP Newsletter* 6(4):1–6.

———. 2004. Are the long-term fluctuations in Atlantic bluefin tuna *(Thunnus thynnus)* population related to environmental changes? *Fisheries Oceanography* 13(3):145–56.

Ray, J. C., K. Redford, R. Steneck, and J. Berger, eds. 2005. *Large Carnivores and the Conservation of Biodiversity.* Island Press.

Reid, T. R. 1995. Tsukiji: The great Tokyo fish market. *National Geographic* 188(5):38–55.

Reiger, G. 1973. *Profiles in Saltwater Angling.* Prentice-Hall.

———. 1976. The bluefin tuna. *Motorboat,* November:47.

Reischauer, E. O. 1974. *Japan: The Story of a Nation.* Tuttle.

Relini, O., L. F. Garibaldi, C. Cima, and G. Palandri. 1995. Feeding of the swordfish, the bluefin and other pelagic nekton in the western Ligurian Sea. *Col.Vol.Sci.Pap.ICCAT* 44(1):283–86.

Rembold, C. M. 2004. The health benefits of eating salmon. *Science* 305:475.

Revkin, A. C. 2003. Commercial fishing is cited in decline of oceans' big fish. *New York Times,* May 14.

———. 2005. Tracking the imperiled bluefin from ocean to sushi platter. *New York Times,* May 3.

———. 2006. Farms' output grows closer to matching fishing harvests. *New York Times,* September 4.

Richards, W. J. 1976. Spawning of bluefin tuna *(Thunnus thynnus)* in the Atlantic Ocean and adjacent seas. *Col.Vol.Sci.Pap.ICCAT* 5(2):267–78.

———. 1977. A further note on Atlantic bluefin tuna spawning. *Col.Vol.Sci.Pap .ICCAT* 6(2):335–36.

Richardson, S. 1994. Warm blood for cold water. *Discover* 15(1):42–43.

Ringle, K. 1986. Seafood summit: Tokyo's Tsukiji fishmarket. *Oceans* 19(5):12–17.

Rivas, L. R. 1953. The pineal apparatus of tunas and related scombrid fishes as a possible light receptor controlling phototactic movements. *Bulletin of Marine Science of the Gulf and Caribbean* 3:168–80.

———. 1977. Age composition anomalies as evidence for transoceanic migrations by intermediate age groups of the North Atlantic bluefin tuna *(Thunnus thynnus). Col.Vol.Sci.Pap.ICCAT* 5(2):290–96.

———. 1979. Proposed terminology for size groups of the North Atlantic bluefin tuna *(Thunnus thynnus). Col.Vol.Sci.Pap.ICCAT* 8(2):441–46.

Rivas, L. R., and F. J. Mather. 1976. A comparison of Eastern and Western Atlantic bluefin tuna *(Thunnus thynnus)* with reference to stock differences. *Col.Vol.Sci.Pap.ICCAT* 5(2):290–96.

Rivkin, M. 2005. *Big Game Fishing Headquarters: A History of the IGFA.* IGFA Press.
———. 2007. *Angling and War: The Collision of Big-Game Fishing and World War II.* (In press.)
Robbins, M. W. 2006. The catch. *Mother Jones* 31(2):49–53.
Roberts, C. M. 1997. Ecological advice for the global fisheries crisis. *Trends in Ecology and Evolution* 12(1):35–38.
———. 2007. *The Unnatural History of the Sea.* Island Press.
Robins, C. M., and A. E. Caton. 1998. Review of Australian tuna fisheries in the Indian Ocean. *Proceedings of the Indian Ocean Tuna Consultation* 1:76–78.
Rodriguez-Marín, E., J. M. de la Serna, A. Garcia, C. Pla, and A. Gracia. 2004. Spanish bluefin tuna research activities. *Col.Vol.Sci.Pap.ICCAT* 56(3):1175–81.
Rodriguez-Roda, J. 1980. Description of the Spanish bluefin *(Thunnus thynnus)* trap fishery. *Col.Vol.Sci.Pap.ICCAT* 11:180–83.
Roe, S., and M. Hawthorne. 2005. How safe is tuna? *Chicago Tribune,* December 13.
Romanov, E. V., and V. V. Zamorov. 2002. First record of a yellowfin tuna *(Thunnus albacares)* in the stomach of a longnose lancetfish *(Alepisaurus ferox). Fishery Bulletin* 100(2):386–89.
Rosenthal, E. 2006a. Fishing depletes Mediterranean tuna, conservationists say. *New York Times,* July 16.
———. 2006b. In Europe it's fish oil after heart attacks, but not in U.S. *New York Times,* October 3.
Roughley, T. C. 1951. *Fish and Fisheries of Australia.* Angus and Robertson.
Russell, D., and B. Keating. 1984. Sushi today, gone tomorrow: The plight of the bluefin tuna. *Amicus Journal,* winter:38–46.
Ryther, J. H. 1981. Mariculture, ocean ranching, and other culture-based fisheries. *BioScience* 31(3):223–30.
Safina, C. 1993. Bluefin tuna in the West Atlantic: Negligent management and the making of an endangered species. *Conservation Biology* 7(2):229–33.
———. 1995. The world's imperiled fish. *Scientific American* 273(5):46–53.
———. 1997. *Song for the Blue Ocean.* Henry Holt.
———. 1998a. Scorched-earth fishing. *Issues in Science and Technology* 14(3):33–36.
———. 1998b. Fish market mutiny. *New York Times,* April 14.
———. 2001. Tuna conservation. In B. A. Block and E. D. Stevens, eds., *Tuna: Physiology, Ecology, and Evolution,* 413–59. Academic Press.
Safina, C., A. Rosenberg, R. A. Myers, T. J. Quinn, and J. S. Collie. 2005. U.S. ocean fish recovery: Staying the course. *Science* 309:707–08.
Sakagawa, G. T., and A. L. Coan. 1974. A review of some aspects of the bluefin tuna *(Thunnus thunnus thynnus)* fisheries of the Atlantic Ocean. *Col.Vol.Sci.Pap.ICCAT* 2:259–313.
Samson, J. 1973. *Line Down! The Special World of Big-Game Fishing.* Winchester.
Sara, R. 1980. Bluefin tuna trap fishing in the Mediterranean. *Col.Vol.Sci.Pap.ICCAT* 11:129–44.
———. 1998. A route three-thousand years long: Genetic grouping, social manifestations, body language of bluefin tuna, through the under-water filming of the Favignana trap fishery (video). *Col.Vol.Sci.Pap.ICCAT* 50(2):493.
Schaefer, K. M. 1984. Swimming performance, body temperature and gastric evacuation times of the black skipjack, *Euthynnus lineatus. Copeia* 4:1000–05.

———. 2000. Assessment of skipjack tuna *(Katsuwonus pelamis)* spawning activity in the eastern Pacific Ocean. *Fishery Bulletin* 99:343–50.

———. 2001. Reproductive biology of tunas. In B. A. Block and E. D. Stevens, eds, *Tuna: Physiology, Ecology, and Evolution,* 225–70. Academic Press.

Schick, R. S., J. Goldstein, and M. E. Lutcavage. 2004. Bluefin tuna *(Thunnus thynnus)* in relation to sea surface temperature fronts in the Gulf of Maine (1994–96). *Fisheries Oceanography* 13(4):225–38.

Schiermeier, Q. 2002. How many more fish in the sea? *Nature* 419:662–65.

———. 2003. Fish farms' threat to salmon stocks exposed. *Nature* 425:753.

Schindler, D. E., T. E. Essington, J. F. Kitchell, C. Boggs, and R. Hilborn. 2002. Sharks and tunas: Fisheries impacts on predators with contrasting life styles. *Ecological Applications* 12(3):735–48.

Scholey, V., D. Margulies, R. J. Olson, J. B. Wexler, J. M. Suter, and S. Hunt. 2001. Lab culture and reproduction of yellowfin tuna in Panama. *Global Aquaculture Advocate* 4(2):17–18.

Scholey, V., D. Margulies, J. Wexler, and S. Hunt. 2003. Panamanian lab hosts research on tuna, other marine species. *Global Aquaculture Advocate* 6(1):75–76.

———. 2004. Larval tuna research mimics ocean conditions in lab. *Global Aquaculture Advocate* 7(1):38.

Scott, T. D., C. J. M. Glover, and R. V. Southcott. 1974. *The Marine and Freshwater Fishes of South Australia.* Government Printer, South Australia.

Scully, M. 2002. *Dominion.* St. Martin's.

Seabrook, J. 1994. Death of a giant: Stalking the disappearing bluefin tuna. *Harper's* 288(1729):48–56.

Secor, D. H. 2002. Is Atlantic bluefin tuna a metapopulation? *Col.Vol.Sci.Pap.ICCAT* 54(2):390–99.

Sepulveda, C. A., K. A. Dickson, and J. B. Graham. 2003. Swimming performance studies on the eastern Pacific bonito, *Sarda chiliensis,* a close relative of the tunas (family Scombridae). *Journal of Experimental Biology* 206:2739–48.

Shadwick, R. E. 2005. How tunas and lamnid sharks swim: An evolutionary convergence. *American Scientist* 93(6):524–31.

Sharp, G. D. 2001. Tuna oceanography—an applied science. In B. A. Block and E. D. Stevens, eds, *Tuna: Physiology, Ecology, and Evolution,* 345–89. Academic Press.

Shelly, K. C. 2003. *Lynn Bogue Hunt: A Sporting Life.* Derrydale.

Shingu, C., and K. Hisada. 1977. A review of the Japanese Atlantic longline fishery for bluefin tuna and the consideration of the present status of the stock. *Col.Vol.Sci.Pap.ICCAT* 6(2):366–84.

———. 1978. Recent status of the medium and large bluefin tuna population in the Atlantic Ocean. *Col.Vol.Sci.Pap.ICCAT* 7(2):266–75.

———. 1979. Analysis of the Atlantic bluefin tuna stock caught by longline fishery, based on the data up to 1978. *Col.Vol.Sci.Pap.ICCAT* 8(2):421–29.

———. 1980. Analysis of the Atlantic bluefin tuna stock. *Col.Vol.Sci.Pap.ICCAT* 9(2):595–600.

Shingu, C., K. Hisada, S. Kume, and M. Honma. 1975. Biological information on Atlantic bluefin tuna caught by longline fishery and some views on the management of the resource. *Col.Vol.Sci.Pap.ICCAT* 4:145–60.

Sibert, J. R., J. Hampton, P. Kleiber, and M. Maunder. 2006. Biomass, size, and trophic status of top predators in the Pacific. *Oceanographic Science* 314:1773–76.

Sibert, J. R., M. K. Musyl, and R. W. Brill. 2003. Horizontal movements of bigeye tuna *(Thunnus obesus)* near Hawaii determined by Kalman filter analysis of archival tagging data. *Fisheries Oceanography* 12(32):141–51.

Silvani, L., M. Gazo, and A. Aguilar. 1999. Spanish driftnet fishing and incidental catches in the Mediterranean. *Biological Conservation* 90(1):79–85.

Sissenwine, M. P., P. M. Mace, J. E. Powers, and G. P. Scott. 1998. A commentary on Western Atlantic bluefin tuna assessments. *Transactions of the American Fisheries Society* 127:838–55.

Sloan, S. 1993. The "Connie Jean" incident. *The Edge Big Game Fishing Report,* fall 1993:22–23.

———. 2003. *Ocean Bankruptcy: World Fisheries on the Brink of Disaster.* Lyons.

Small, M. 2002. The happy fat. *New Scientist* 175(2357):34–37.

Smith, P. 1973. The bluefin tuna: The fat man who won't dance. *Sports Afield,* June:54–55, 108–12.

Smith, P. J., L. Griggs, and S. Chow. 2001. DNA identification of Pacific bluefin tuna *(Thunnus orientalis)* in the New Zealand fishery. *New Zealand Journal of Freshwater and Marine Research* 35:843–50.

Smith, T. D., ed. 1979. *Report of the Status of Porpoise Stocks Workshop* (August 27–31, 1979, La Jolla, California). NMFS Southwest Fisheries Center Administrative Report LJ-79-41.

———. 2001. Examining cetacean ecology using historical fishery data. In P. Holm, T. D. Smith, and D. J. Starkey, eds., *Research in Maritime History,* no. 21:207–14. International Maritime History Association/Census of Marine Life.

Steele, J. H., and M. Schumacher. 1999. On the history of marine fisheries: Report of the Woods Hole workshop. *Oceanography* 12(3):28–29.

———. 2000. Ecosystem structure before fishing. *Fisheries Research* 44:201–05.

Steneck, R. S. 1998. Human influences on coastal ecosystems: Does overfishing create trophic cascades? *Trends in Ecology and Evolution* 13(11):429–30.

Stevens, E. D., and F. G. Carey. 1981. One why of the warmth of warm-bodied fish. *American Journal of Regulatory, Integrative and Comparative Physiology* 240(3):151–55.

Stevens, E. D., F. G. Carey, and J. Kanwisher. 1978. Changes in visceral temperature of Atlantic bluefin tuna. *Col.Vol.Sci.Pap.ICCAT* 7(2):383–88.

Stokesbury, M. J. W., R. Cosgrove, A. Boustany, D. Browne, S. L. H. Teo, R. K. O'Dor, and B. A. Block. 2007. Results of satellite tagging of Atlantic bluefin tuna, *Thunnus thynnus,* off the coast of Ireland. *Hydrobiologia* 582:91–97.

Stokesbury, M. J. W., S. L. H. Teo, A. Seitz, R. K. O'Dor, and B. A. Block. 2004. Movement of Atlantic bluefin tuna *(Thunnus thynnus)* as determined by satellite tagging initiated off New England. *Canadian Journal of Fisheries and Aquatic Sciences* 61(10):1976–87.

Sugano, M., and F. Hirahara. 2000. Polyunsaturated fatty acids in the food chain in Japan. *American Journal of Clinical Nutrition* 71(1):189–96.

Sugimoto, M. 2002. Morphological color changes in fish: Regulation of pigment cell density and morphology. *Microscopic Research Techniques* 58(6):496–503.

Summers, A. P. 2004. Fast fish. *Nature* 429:31–33.

Sun, C.-L., W.-R. Wang, and S. Yeh. 2005. *Reproductive Biology of Yellowfin Tuna in the Central and Western Pacific Ocean.* Western and Central Pacific Fisheries Commission, WCPFC-SCI/B1/WP-1.

Susca, V., A. Corriero, M. DeFlorio, C. R. Bridges, and G. DeMetrio. 2001. New results on the reproductive biology of the bluefin tuna *(Thunnus thynnus)* in the Mediterranean. *Col.Vol.Sci.Pap.ICCAT* 52(2):745–51.

Suzuki, Z. 1980. Bluefin fisheries and stocks in the Atlantic, 1970–81. *Col.Vol.Sci.Pap.ICCAT* 20(2):399–416.

———. 1992. Critical review of the stock assessment of bluefin tuna in the western Atlantic. *Col.Vol.Sci.Pap.ICCAT* 39(3):710–16.

———. 1993. A review of the biology and fisheries for yellowfin tuna *(Thunnus albacares)* in the western and central Pacific Ocean. In R. S. Shomura, J. Majkowski, and S. Langi, eds., *Interactions of Pacific Tuna Fisheries, vol. 2 of Papers on Biology and Fisheries.* FAO, Rome.

Syme, D. A., and R. E. Shadwick. 2002. Effects of longitudinal body position and swimming speed on mechanical power of deep red muscle from skipjack tuna *(Katsuwonus pelamis). Journal of Experimental Biology* 205:189–200.

Takagi, M., T. Okamura, S. Chow, and N. Taniguchi. 2001. Preliminary study of albacore *(Thunnus alalunga)* stock differentiation inferred from microsatellite DNA analysis. *Fishery Bulletin* 99:697–701.

Taylor, L. R. 1970. Untaxing taxonomy. *Oceans* 3(6):104–06.

Teal, J., and M. Teal. 1975. *The Sargasso Sea.* Atlantic–Little, Brown.

Terry, P., P. Lichtenstein, M. Feychting, A. Ahlbom, and A. Wolk. 2001. Fatty fish consumption and risk of prostate cancer. *The Lancet* 357:1764–66.

Thiele, C. 1969. *Blue Fin.* Harper & Row.

Ticina, V., L. Grubisic, I. Katavic, I. Jeftimijades, and V. Franicevic. 2003. Tagging of small bluefin tuna in the growth-out floating cage—report of the research activities on tuna farming in the Adriatic Sea during 2002. *Col.Vol.Sci.Pap.ICCAT* 55(3):1278–81.

Tinsley, J. B. 1984. *The Sailfish: Swashbuckler of the Open Seas.* University of Florida Press.

Tosches, N. 2007. If you knew sushi. *Vanity Fair,* June:120–40.

Triantafyllou, M. S., A. H. Techet, Q. Zhu, D. N. Beal, F. S. Hover, and D. K. P. Yue. 2002. Vorticity control in fish-like propulsion and maneuvering. *Integrative and Comparative Biology* 42:1026–31.

Triantafyllou, M. S., and G. S. Triantafyllou. 1995. An efficient swimming machine. *Scientific American* 272(3):64–72.

Trivedi, B. 2006. The good, the fad and the unhealthy. *New Scientist* 191(2570):42–49.

Tudela, S. 2002a. *Tuna Farming in the Mediterranean: The "Coup de Grâce" to a Dwindling Population?* World Wildlife Fund Mediterranean Program Office, Rome.

———. 2002b. Grab, cage, fatten, sell. *Samudra,* July:9–17.

Tudela, S., A. K. Kai, F. Maynou, M. El Andalossi, and P. Guglielmi. 2005. Driftnet fishing and biodiversity conservation: The case study of the large-scale Moroccan driftnet fleet operating in the Alboran Sea (SW Mediterranean). *Biological Conservation* 121:65–78.

Tuomisto, J. T., J. Tuomisto, M. Tainio, M. Niittynen, P. Verkasalo, Vartiainen, H. Kiviranta, and J. Pekkanen. 2004. Risk-benefit analysis of eating farmed salmon. *Science* 305:476.

Uozumi, Y. 2006. Crisis for tuna resources triggered by biased scientific paper—an argument over Myers paper. *OPRT Newsletter* 9:1–3.

Vaughan, H. 1982. *The Australian Fisherman's Companion.* Landsdowne Press.

Volpe, J. P. 2005. Dollars without sense: The bait for big-money tuna ranching around the world. *BioScience* 55(4):301–02.

von Brandt, A. 1972a. *Fish Catching Methods of the World.* Fishing News (Books).

———. 1972b. *Revised and Enlarged Fish Catching Methods of the World.* Fishing News (Books).

Waldman, P. 2005. Mercury and tuna: U.S. advice leaves lots of questions. *Wall Street Journal,* August 1:A1–6.

Walters, V., and H. Fierstine. 1964. Measurements of swimming speeds of yellowfin tuna and wahoo. *Nature* 202:208–09.

Ward, P., and R. A. Myers. 2005. Shifts in open-ocean fish communities coinciding with the commencement of commercial fishing. *Ecology* 86(4):835–47.

Ward, R. D., N. G. Elliott, and P. M. Grewe. 1995. Allozyme and Mitochondrial DNA separation of the Pacific northern bluefin tuna, *Thunnus thunnus orientalis* (Temminck and Schlegel), from southern bluefin tuna, *Thunnus maccoyii* (Castelnau). *Marine and Freshwater Research* 46:921–30.

Watson, R., and D. Pauly. 2001. Systematic distortions in world fisheries catch trends. *Nature* 414:534–36.

Weaver, D. E. 2004. Contaminant levels in farmed salmon. *Science* 305:478.

Webber, H. H. 1968. Mariculture. *BioScience* 18(10):940–45.

Weber, P. 1993. *Abandoned Seas: Reversing the Decline of the Oceans.* Worldwatch.

———. 1994. *Net Loss: Fish, Jobs, and the Environment.* Worldwatch.

Weng, K. C., and B. A. Block. 2004. Diel vertical migration of the bigeye thresher shark *(Alopias superciliosus),* a species possessing orbital retia mirabilia. *Fishery Bulletin* 102(1):221–29.

Western and Central Pacific Fisheries Commission. 2007. Estimates of annual catches in the WCPFC statistical area. *Scientific Committee Second Regular Session, 13–24 August 2007.*

Westneat, M. W., and S. A. Wainwright. 2001. Mechanical design for swimming: Muscle, tendon and bone. In B. A. Block and E. D. Stevens, eds., *Tuna: Physiology, Ecology, and Evolution,* 271–311. Academic Press.

Wexler, J. B., D. Margulies, S. Masuma, N. Tezuka, K. Teruya, M. Oka, M. Kanematsu, and H. Nikaido. 2001. Age validation and growth of yellowfin tuna, *Thunnus albacares,* larvae reared in the laboratory. *IATTC Bulletin* 22(1):52–71.

Wexler, J. B., V. P. Scholey, R. J. Olson, D. Margulies, A. Nakazawa, and J. M. Suter. 2003. Tank culture of yellowfin tuna, *Thunnus albacares:* Developing a spawning population for research purposes. *Aquaculture* 220:327–53.

Whitehead, S. S. 1931. Fishing methods for the bluefin tuna *(Thunnus thynnus)* and an analysis of the catches. *California Fish and Game Fish Bulletin* 33:1–34.

Whynott, D. 1995. *Giant Bluefin.* North Point Press.

———. 1999. The most expensive fish in the sea. *Discover* 20(4):80–85.

———. 2000. Something's fishy about this robot. *Smithsonian* 31(5):54–60.

Wild, A. 1993. A review of the biology and fisheries for yellowfin tuna, *Thunnus albacares,* in the eastern Pacific Ocean. In R. S. Shomura, J. Majkowski, and S. Langi, eds., *Interactions of Pacific Tuna Fisheries, vol. 2 of Papers on Biology and Fisheries.* FAO, Rome.

Williams, J. G., F. P. Chavez, J. Ryan, S. E. Lluch-Cota, and M. Ñiquen C. 2003. Sardine fishing in the early 20th century. *Science* 300:2032–33.

Wilson, S. G., M. E. Lutcavage, R. W. Brill, M. P. Genovese, A. B. Cooper, and A. W. Everly. 2005. Movements of bluefin tuna *(Thunnus thynnus)* in the northwestern Atlantic Ocean recorded by pop-up satellite archival tags. *Marine Biology* 146:409–23.

Worm, B., E. B. Barbier, N. Beaumont, J. E. Duffy, C. Folke, B. S. Halpern, J. B. C. Jackson, H. K. Lotze, F. Micheli, S. R. Palumbi, E. Sala, K. A. Selkoe, J. J. Stachowicz, and R. Watson. 2006. Impacts of biodiversity loss on ocean ecosystem services. *Science* 314:787–90.

Worm, B., M. Sandow, A. Oschlies, H. K. Lotze, and R. S. Myers. 2005. Global patterns of predator diversity in the open ocean. *Science* 309:1365–69.

Wright, K. 2005. Our preferred poison (mercury). *Discover* 26(3):58–65.

WWF Mediterranean. 2008. End of the line for tuna commission, time for trade measures. *Bluefin Tuna Bulletin* 65, November 24.

Wylie, P. 1990. *Crunch and Des: Classic Stories of Saltwater Fishing.* Lyons.

Yang, P. T. 1980. National report of China (Taiwan), 1978. *Col.Vol.Sci.Pap.ICCAT* 9(3):675–76.

Yao, M. 1988. A note on Japanese longline fisheries in the Atlantic Ocean. *Col.Vol.Sci.Pap.ICCAT* 27:222–29.

Young, J. W., R. W. Bradford, T. D. Lamb, and V. D. Lyne. 1996. Biomass of zooplankton and micronekton in the southern bluefin tuna fishery grounds off eastern Tasmania, Australia. *Marine Environmental Progress Series* 138:1–14.

Young, J. W., T. D. Lamb, D. Le, R. W. Bradford, and A. W. Whitlaw. 1997. Feeding ecology and interannual variations in diet of southern bluefin tuna, *Thunnus maccoyii,* in relation to coastal and oceanic waters of eastern Tasmania, Australia. *Environmental Biology of Fishes* 50:275–91.

Yuen, H. S. H. 1966. Swimming speeds of yellowfin and skipjack tuna. *Transactions of the American Fisheries Society* 95:203–09.

———. 1970. Behavior of skipjack tuna, *Katsuwonus pelamis,* as determined by tracking with ultrasonic telemetry. *Journal of the Fisheries Research Board of Canada* 27:2071–79.

Zeidberg, L. D., and B. H. Robison. 2007. Invasive range expansion by the Humboldt squid, *Dosidicus gigas,* in the eastern North Pacific. *Proceedings of the National Academy of Sciences of the USA* 104(31):12948–50.

Zharov, V. L., N. F. Paliy, V. I. Sauskan, and V. G. Yurov. 1973. Results of Soviet fisheries investigations on Atlantic tuna. *Col.Vol.Sci.Pap.ICCAT* 1:549–55.

INDEX

Note: Page numbers in *italics* refer to illustrations.

PHOTOGRAPHIC CREDITS

All illustrations not listed here are courtesy of the author.

Peter Barrett: 47
Roberto Mielgo Bregazzi, ATRT, Madrid: 164
Alex Buttigeig: 28
CIA: 162
International Game Fish Association: 51, 85, 88, 91, 93, 96, 99, 102
Terry Maas, Freedive.net: 193
Monterey Bay Aquarium: 69
NOAA: 49, 110, 114, 115, 199
Sam Ogden, Photo Researchers, Inc.: 40
Collection Paul Ricard, Bandol, France: 113
Rob Staunton, Stehr Group Australian Fisheries: 274
Scott Taylor: 37
Tuna Hunting Fishing Charters, Gloucester, Mass.: 108
Hoyt Watson, Woods Hole Oceanographic Institution: 33
Courtesy Kiwi White: 13, 123, 169, 237
Neil Williams: 17